AF806942

Advances in Industrial Control

More information about this series at http://www.springer.com/series/1412

Antonella Ferrara · Simona Sacone
Silvia Siri

Freeway Traffic Modelling and Control

Springer

Antonella Ferrara
Department of Electrical, Computer and Biomedical Engineering
University of Pavia
Pavia
Italy

Simona Sacone
Department of Informatics, Bioengineering, Robotics and Systems Engineering
University of Genoa
Genoa
Italy

Silvia Siri
Department of Informatics, Bioengineering, Robotics and Systems Engineering
University of Genoa
Genoa
Italy

ISSN 1430-9491 ISSN 2193-1577 (electronic)
Advances in Industrial Control
ISBN 978-3-030-09364-8 ISBN 978-3-319-75961-6 (eBook)
https://doi.org/10.1007/978-3-319-75961-6

Softcover re-print of the Hardcover 1st edition 2018

Printed on acid-free paper

This Springer imprint is published by Springer Nature
The registered company is Springer International Publishing AG
The registered company address is: Gewerbestrasse 11, 6330 Cham, Switzerland

To our wonderful families who have constantly supported and endured us during the time we have dedicated to the writing of this book

Series Editors' Foreword

The series *Advances in Industrial Control* aims to report and encourage technology transfer in control engineering. The rapid development of control technology has an impact on all areas of the control discipline. New theory, new controllers, actuators, sensors, new industrial processes, computer methods, new applications, new philosophies…, new challenges. Much of this development work resides in industrial reports, feasibility study papers and the reports of advanced collaborative projects. The series offers an opportunity for researchers to present an extended exposition of such new work in all aspects of industrial control for wider and rapid dissemination.

The field of automotive vehicles and the infrastructure that they use is undergoing unparalleled and extensive technological change. There is a dramatic push towards electric motive power either as the sole energy source or in combination with petrol or diesel combustion in vehicles. Information technology is revolutionizing both the art of driving and the transport infrastructure itself. With human drivers being replaced by automated AI driving systems, totally autonomous vehicles cannot be far away. Consequently, it is very timely to be introducing this monograph *Freeway Traffic Modelling and Control* by authors: Antonella Ferrara, Simona Sacone and Silvia Siri to the *Advances in Industrial Control* monograph series.

The monograph is divided into three parts:

- Part I: After an opening introductory chapter that lays out the main terms and issues to be considered by the authors, the text moves on to present more detailed fundamental traffic system definitions and characteristics in Chap. 2.
- Part II: The modelling of freeway traffic systems is presented in five thorough chapters. Part II closes with a chapter on state estimation for freeway traffic.
- Part III: The monograph then concludes with a four-chapter presentation of freeway traffic control. The contents here are a mix of state-of-the-art freeway traffic control theory with introductions to the relevant control techniques, a review of technology and global applications, and new research results original to the authors. An exciting part of this part of the book is the work of the authors

on event-triggered control for this field. Although a much-talked-about and discussed control method, it is of considerable interest to have an application described and demonstrated.

The monograph *Freeway Traffic Modelling and Control* is well organized and the writing is of an admirable clarity and precision. The authors consider this monograph to be a reference text for the field of freeway traffic modelling, identification, estimation and control. The authors also include material on actual freeway traffic technology and real-world applications. With this interesting mix of state-of-the-art material and the new control directions, this is a perfect entry for the *Advances in Industrial Control* series.

Other monographs in the *Advances in Industrial Control* series that are concerned with automotive vehicles and/or the related traffic infrastructure include the following extremely wide-ranging examples:

- *Hybrid Predictive Control for Dynamic Transport Problems* by Alfredo Núñez, Doris A. Sáez and Cristián E. Cortés (ISBN 978-1-4471-4350-5, 2013);
- *Robust Control Design for Active Driver Assistance Systems* by Péter Gáspár, Zoltán Szabó, Józef Bokor and Balázs Németh (ISBN 978-3-319-46124-3, 2017);
- *Hybrid Systems, Optimal Control and Hybrid Vehicles* by Thomas J. Böhme and Benjamin Frank (ISBN 978-3-319-51315-7, 2017);

and soon to be published:

- *Feedback Control Theory for Dynamic Traffic Assignment* (second edition) by Pushkin Kachroo and Kaan Özbay (ISBN 978-3-31969-229-6, 2018).

About the Authors

Antonella Ferrara is a Full Professor at the Department of Electrical, Computer and Biomedical Engineering, University of Pavia, Italy.

Dr. Simona Sacone is an Associate Professor at the Department of Informatics, Bioengineering, Robotics and Systems Engineering, University of Genoa, Italy.

Dr. Silvia Siri is an Assistant Professor at the Department of Informatics, Bioengineering, Robotics and Systems Engineering, University of Genoa, Italy.

The authorial team has published widely on the subject of freeway traffic modelling and control and is currently active in control conferences and societies.

Glasgow, Scotland, UK

Michael J. Grimble
Michael A. Johnson
Industrial Control Centre

Preface

Researchers have been studying new methods and solutions to monitor and control freeway traffic systems for some decades. Yet, present technological developments make the transfer of research results from the speculative level to the applicative one more and more effective and reliable. This is the reason why, recently, the research in this field has significantly revitalised, with the aim of solving already treated issues and open problems according to a modern and more efficient perspective. Yet, in spite of the renewed attention to freeway traffic modelling and control, a modern reference book on the subject was missing.

This book has the aim of filling this gap. It is conceived to provide the reader with an overview of the main modelling and control approaches for freeway traffic systems developed in the last decades, starting from the earliest methods until the most recent scientific results suitable for field implementations. Even if the book is focused on a specific application, it has a scientific setting and treats all the theoretical elements associated with freeway traffic control in a rigorous mathematical way. It addresses the basis and the developments of the most widely used traffic models, as well as the fundamentals of the control techniques developed for freeway systems, making reference to systems theory, control theory, optimisation and state estimation issues. To complement the theoretical part, the book includes some explicit references to technological elements and implementation issues for freeway traffic systems.

Inspired by the present-day developments in Green Information and Communications Technology, the book also focuses on sustainable and environment-friendly freeway traffic control systems, in which suitable polluting emission models are involved in the decision-making. In order to properly define emission models for freeway traffic systems, it is useful to distinguish different classes of vehicles, characterised by different fuel types, occupations and priorities. For these reasons, many of the traffic models and the control frameworks presented in the book are of the multi-class type.

These innovative models and control algorithms can be beneficial in contributing to the creation of a novel perspective of 'green traffic systems' for freeways, to be combined, as soon as possible, with the corresponding concept already developed in urban traffic control. This in order to suit the requirements of a sustainable and dependable integrated transportation paradigm, which has to include any type of traffic and transportation mode to attain a more efficient, smoother, greener and less expensive system, really inaugurating the era of 'smart mobility'.

In the last part of the book, a discussion on the evolution trends of freeway traffic control, in view of the new technologies which have revolutionised data collection, data processing, communications and computing, is also provided. The emerging freeway traffic control paradigms are illustrated making reference to innovative concepts like those of intelligent and cooperative vehicles, smart infrastructures, traffic-oriented big data technologies, cloud-based data mining techniques, cyber-security and privacy-preserving methodologies.

The book is specifically addressed to people working in the context of traffic monitoring and control, as well as to researchers, students, technicians and practitioners in transportation and control engineering. In addition, this book could be useful also to freeway traffic managers, freeway stakeholders and public authorities in the field of traffic and transportation. It is also intended as a valuable textbook for courses related to transport engineering, traffic management and control.

The present book is subdivided into three parts and includes eleven chapters.

Part I is an introduction to the book and provides the reader with some basic insights on freeway traffic systems. After a brief description on the main critical issues in freeway traffic and the related possible solutions, some basic elements and some common traffic phenomena are described, as well as a classification of traffic models is provided.

Part II focuses on models representing the dynamic behaviour of freeway traffic. First-order and second-order macroscopic traffic flow models are illustrated in detail, considering also the extensions to multi-class frameworks. The most wide-spread microscopic and mesoscopic traffic models are also described, together with traffic emission models that can be of interest in the design of environment-friendly traffic control systems. This part of the book ends with a description of state estimation techniques for freeway traffic systems.

Part III is devoted to freeway traffic control methods. The most important control methodologies adopted in freeway networks are outlined, considering different control measures, such as ramp metering (applied to regulate the flow entering the mainstream from the on-ramps), mainstream control (devoted to control vehicles in the mainstream), route guidance (applied to properly route vehicles in freeway networks) and their possible combination. Emerging control strategies for freeway systems, based on the innovative concepts mentioned previously in this preface, are discussed. The book is concluded with a look at the future that is rapidly becoming the present of freeway traffic management and control systems.

To conclude this preface, we would like to deeply thank Cecilia Pasquale for her contribution to the research activity which the book is based on, for the careful reading of all the chapters, as well as for the many interesting and clever corrections and suggestions. We are also very grateful to Paola Goatin who contributed in revising the book sections on continuous models and provided us with valuable comments and hints. We thank Giulia Piacentini as well for helping us in revising the manuscript. Also, the valuable scientific collaborations with Bart De Schutter have surely inspired and contributed to the book writing. We are grateful to Carlos Canudas de Wit for the motivating and enlightening collaboration on traffic control. Last but not least, we wish to thank Markos Papageorgiou for the extremely instructive and inspiring discussions and for the research activity carried out together in the last years.

Savona and Pavia, Italy
December 2017

Antonella Ferrara
Simona Sacone
Silvia Siri

Contents

Part III Freeway Traffic Control

Abbreviations

ACC	Adaptive Cruise Control
ACTM	Asymmetric Cell Transmission Model
ADAS	Advanced Driver Assistance Systems
AHS	Automated Highway Systems
ARZ	Aw–Rascle–Zhang
ATMS	Advanced Traffic Management Systems
CA	Cellular Automata
CACC	Cooperative Adaptive Cruise Control
CAV	Connected and Automated Vehicles
CTM	Cell Transmission Model
FCD	Floating Car Data
FHOCP	Finite-Horizon Optimal Control Problem
GHR	Gazis–Herman–Rothery
GPS	Global Positioning System
HDV	Heavy-Duty Vehicles
ICT	Information and Communication Technology
IoT	Internet of Things
IoV	Internet of Vehicles
ITS	Intelligent Transportation Systems
IVHS	Intelligent Vehicle Highway Systems
LMI	Linear Matrix Inequality
LN-CTM	Link-Node Cell Transmission Model
LTM	Link Transmission Model
LWR	Lighthill–Whitham–Richards
ME	Mainstream Emissions
METANET	Modèle d'Écoulement de Trafic sur Autoroute NETworks
MFD	Macroscopic Fundamental Diagram
MLD	Mixed Logical Dynamical
MPC	Model Predictive Control
MS	Mean Speed

NCS	Networked Control System
PCE	Passenger Car Equivalents
PW	Payne–Whitham
RE	Ramp Emissions
RGIS	Route Guidance and Information System
RSU	Road Side Unit
SAE	Society of Automotive Engineers
SIoT	Social Internet of Things
SIoV	Social Internet of Vehicles
SMM	Switching-Mode Model
TCP/IP	Transmission Control Protocol/Internet Protocol
TE	Total Emissions
TTD	Total Travel Distance
TTS	Total Time Spent
TTT	Total Travel Time
TWT	Total Waiting Time
V2I	Vehicle-to-Infrastructure
V2V	Vehicle-to-Vehicle
VACS	Vehicle Automation and Communication Systems
VANET	Vehicular Ad Hoc Network
VLM	Variable-Length Cell Transmission Model
VMS	Variable Message Sign
XFCD	eXtended Floating Car Data

Part I
Introduction

Chapter 1
Freeway Traffic Systems

1.1 Sustainable Mobility

Transportation systems have always played a relevant role in the social and economical development of all the countries, but only in recent years the need for traffic mobility systems has grown considerably. Although this increase surely provides positive effects on the human progress, at the same time, the increase in traffic mobility is the source of several negative externalities such as pollutant emissions, congestion, safety reduction and environmental deterioration.

These phenomena have been observed for both the advanced economies and the developing countries, where the significant growth in the number of vehicles has generated a negative impact on the quality of people's life. In particular, the increased number of vehicles has caused both an increment of pollution and an intensification of congestion phenomena, due to the inefficiency of the existing infrastructures that are often unable to handle the growth of traffic demand. Indeed, it is not always possible, both for physical and economic constraints, to modify the existing infrastructure to meet the current traffic demand.

In order to address all these problems, it is required to implement management and control tools which allow to improve the system performance and the quality of drivers' life, without requiring significant infrastructural changes to the present traffic network. In addition, the present mobility systems must be designed to be sustainable, i.e. they must pursue the economic development, the social welfare, and the environmental safeguard, guaranteeing the needs of the current society and the future one.

The concept of *sustainable development* has been recently introduced in the international legislation, in order to overcome the limited logic of simply pursuing economic benefits in the short term without considering the consequences of political actions in the long period. The case of road transport systems is particularly interesting, since the road represents the most widespread option to move passengers and to supply goods (see Fig. 1.1), though being one of the most critical choices, since it produces several negative effects with implications for the entire society.

A. Ferrara et al., *Freeway Traffic Modelling and Control*, Advances in Industrial Control, https://doi.org/10.1007/978-3-319-75961-6_1

Fig. 1.1 A road stretch in A1 freeway, close to Rome, Italy (courtesy of Autostrade per l'Italia SpA, photo from Archivio Videofotografico Autostrade per l'Italia)

The importance of these issues emerges also from the actions taken by the European Commission which has promoted, in recent years, several studies [1–3] and the White Paper on Transport [4]. Analogously, in the U.S., following the Clean Air Act, different regulations have been introduced since the 60s in order to limit the emissions of pollutants. A recent comparative study focusing on the main differences between the European and the U.S. legislations about emissions in the automotive sector is reported in [5].

All the countries all over the world, according to their legislation, have developed guidelines and regulations in the area of sustainable development, by highlighting the possible actions to be implemented in order to reduce emissions. In particular, the main common priorities are referred to the reduction of energy consumptions, emissions of greenhouse gases and pollutants, the containment of noise and network congestion, the compliance with minimum standards of safety and minimum standards of functionality.

1.2 Criticalities of Freeway Traffic Systems

Among the negative impacts associated with the growth of freeway traffic systems, a major issue is surely related to recurrent and non-recurrent congestion phenomena (see Fig. 1.2) which, in turns, cause an increase of the time spent by travellers in

Fig. 1.2 A freeway stretch in I-405, city of Los Angeles, U.S. (courtesy of Michael Ballard)

the network, of fuel consumption, of environmental impact, as well as a higher probability of accidents. The main critical factors connected with freeway traffic systems are analysed in detail in the following subsections.

1.2.1 Congestion Phenomena

Traffic congestion is a major criticality in modern freeway systems, causing serious infrastructure degradation in and around metropolitan areas. Despite the significant advances in the area of *Information and Communication Technology* (ICT), it seems that the full exploitation of such innovative technologies to mitigate motorway traffic congestion has not been completely achieved yet. Urban and interurban freeways were originally conceived so as to provide virtually unlimited mobility to road users. However, the dramatic expansion of car ownership has led to daily recurrent and non-recurrent freeway congestions of thousands of kilometres in length around the world. Such congestions substantially reduce the available infrastructure capacity at the rush hours, i.e. at the time in which this capacity is most urgently needed, causing delays, increased environmental pollution and reduced traffic safety. Similar effects are observed in the frequent case of non-recurrent congestions caused by incidents, road works (see Fig. 1.3), and so on.

Traffic jams occur whenever a high number of vehicles attempt to use a common roadway with limited capacity (Fig. 1.4). Such events may have various effects on the

Fig. 1.3 Road works for deep renovation of asphalt pavement on A1 freeway, Italy (courtesy of Autostrade per l'Italia SpA, photo from Archivio Videofotografico Autostrade per l'Italia)

Fig. 1.4 Congestion forming in I-5 North at State Route 55 in Santa Ana, U.S. (courtesy of Michael Ballard)

performance and the quality of the system, causing the formation of queues and the consequent increase of travel times. In some cases, the consequences of traffic jams are even worse, leading to deadlock states characterised by excessive delays in travel times, high reductions of the safety level and a consequent strong increase of fuel consumptions. Moreover, in the literature several studies recognise that, in addition to the economic disadvantage caused by the loss of time and the increase of fuel consumption, it is possible to observe a further social damage due to the increasing level of stress of the drivers, which suffer the frustration produced by the frequent exposure to congestion phenomena. Indeed, users lose confidence on the reliability of the system considering the higher time spent to reach their destination as a *wasted time*, which could have been used for other activities [6, 7].

Congestion phenomena are mainly classified as recurring and non-recurring congestions, the main difference being related to the predictability of the occurrence of the congestion [8]. *Recurrent congestion* is predictable since it is a direct consequence of the traffic daily routine. This is generally caused by a traffic demand close or greater than the one for which the saturation of the infrastructure occurs, and in freeway contexts it is often associated with the movement of commuters during rush hours. It is worth noting that drivers acquire some experience about recurrent congestion phenomena and they plan their travel choices accordingly.

Non-recurrent congestion is instead connected with random events which are not predictable by users. For such reason, non-recurrent congestion leads to a greater frustration of road users compared with recurrent congestion. Non-recurrent congestion is normally generated by the following causes:

- *traffic accidents*, which interrupt the normal flow of traffic and block one or more lanes. Besides vehicle crashes, traffic incidents include all the events that cause traffic disturbances such as vehicle malfunctions on roadways. In this case, the congestion happens, primarily, for the capacity reduction caused by the interruption of one or more lanes and, secondarily, by the slowdowns caused by drivers that decrease their speed to observe the accident or the rescue operations. The severity of the congestion depends both on the number of lanes which have been obstructed and on the duration of the event;
- *adverse weather conditions*, which have a negative influence on the driving behaviour. Events such as rain or snow reduce the pavement adherence, as well as fog or intense rain phenomena produce a considerable reduction of visibility. These aspects generate a strong decrease of the mean speed and a high increase in the frequency of traffic incidents;
- *demand fluctuations*, i.e. the variability of traffic flows due to demand peaks that happen daily, weekly and seasonally, with particular reference to the holiday periods and the emergency evacuations;
- *work zones*, implying a reduction of the road capacity due to activities of construction and maintenance. The impact of the works on the reliability of the infrastructure depends on their extension and duration; especially the short term activities have a greater effect on non-recurrent congestion, since the drivers are not able to reschedule their choices taking into consideration the suffered delay;

- *special events*, which normally generate a high traffic demand in a limited time period, with consequent creation of non-recurrent congestion in the vicinity of stadiums, sports centres, shopping centres and others.

In addition to delays, another important aspect that must be taken into account as a negative effect of congestion is related to the *reliability* of the roadway [9]. The reliability of a roadway is related to the variance of the travel times experienced by the drivers: reliable roadways have travel times with low variability, whereas high variabilities in travel times make a road unreliable. The reduction of reliability of a freeway is normally due to non-recurrent congestion phenomena, while delays derived from recurrent congestion do not affect reliability as much, because they are rather constant and predictable. Only recently, the concept of reliability has become an important measure for roadway performance. One of the main reasons for the importance of measuring and managing freeway reliability is that drivers have less tolerance for unexpected delays than for expected ones. Many drivers prefer to choose a path with reliable congestion than a path with unreliable travel times, even if the reliable path is characterised by longer expected travel times.

1.2.2 Pollutant Emissions

Despite the significant technological progress, the levels of pollutant emissions generated by road transport are surely a major cause of risks for the human health and for the environment, especially in urban areas (Fig. 1.5). This is specifically due to the increase of traffic flows and the almost exclusive use of fossil fuels [4, 10].

In particular, the process of air degradation may be caused by three different sources:

- *chemical pollution*, generated by the adoption of heat engines and strictly related to the process of combustion in which several toxic substances are produced, such as carbon monoxide, hydrocarbons, nitrogen oxides, sulphur oxides and particulate matter;
- *thermal pollution*, caused by the exposure of the ecosystem to heat sources and greenhouse gases, such as nitrous oxide, methane, sulphur hexafluoride and especially carbon dioxide;
- *noise pollution*, connected with excessive noise, with effects on the psychological state and physical health of the exposed population.

The previously cited pollutants, besides providing significant environmental damages, involve serious repercussions on human health [11]. *Carbon monoxide* is produced by an inefficient fuel combustion basically caused by a lack of oxygen. The risk derived from the presence of carbon monoxide in the air is due to its high affinity with the haemoglobin in the human blood, which provokes the formation of carboxyhemoglobin causing inadequate oxygenation of the blood cells. The presence of *hydrocarbons* in the air is instead due to the presence of phenomena leading to the

Fig. 1.5 A freeway stretch in Villeneuve d'Ascq, close to Lille, France

incomplete combustion of hydrocarbons. These hydrocarbons, when introduced into the environment, cause chemical reactions in the air, contributing to the formation of photochemical smog and greenhouse effect. Among the substances produced in the regular combustion process, there are also *nitrogen oxides*. Most of them are constituted by nitrogen monoxide, while the formation of nitrogen dioxide occurs in a secondary reaction and is not strictly linked with the phase of combustion. Finally, the *particulate matter* is constituted by microscopic solid particles suspended in the air and produced mainly in the phase of combustion in diesel engines. The harmfulness of these powders is related to their small dimension, since this particulate matter can penetrate deeply in the pulmonary alveoli causing a severe damage.

Particular attention must also be paid to the production of *carbon dioxide*. Although it is not toxic nor harmful, its accumulation in the atmosphere is one of the major causes of the *greenhouse effect* and the resulting *global warming*. Carbon dioxide is a thermal pollutant resulting from the complete combustion of the carbon present in the fuel. The transport sector represents a significant source in the production of carbon dioxide, which has highly increased in the last decades [12]. For these reasons, stricter standards have been progressively adopted to limit vehicle emissions in many countries. Nonetheless, nowadays the level of carbon dioxide detected in some areas (especially in urban districts) is still far from the normative limits and, therefore, further actions are required to improve air quality.

1.2.3 Safety Issues

If traffic congestion has a negative impact on the economy and the quality of citizens' life, its effects on traffic safety are less evident. For this reason, some studies have been carried out to analyse the relation between road safety and congestion in freeways, often leading to different outcomes. According to the classification reported in [13], the studies focused on the relationship between traffic conditions and safety (in terms of crash severity, frequency and type) can be divided in two categories. The former group compares congestion and safety levels at different locations or different time periods, on the basis of data obtained during long observation periods. The studies of the latter group use instead short observation periods in order to analyse which traffic conditions lead to crashes. Hence, they are more suitable to capture short-term variations in traffic flows, which normally correspond to the formation or recovering phase of congestion.

Some studies in the literature aim to find quantitative relations between crashes and variables specifically referred to traffic conditions, such as traffic flow or traffic density [14, 15]. A recent research on the relation between crashes and traffic density has been conducted on the freeways of some American countries [15], by distinguishing accidents into two categories according to their severity. The data collected in this research show a U-shaped relationship between crash rate and traffic density. As a matter of fact, at low traffic densities, *single-vehicle crashes* are more likely to happen, because the interactions among vehicles are rare and drivers are inattentive and travel at very high speeds. On the other hand, at high traffic densities, the interactions among vehicles highly increase so that *rear-end crashes* and *multiple-vehicle crashes* due to lane changing behaviours become more frequent.

1.2.4 Freight Transport Issues

All the negative effects associated with traffic and described before are much more critical and relevant when considering a particular typology of traffic, that is the transport of goods by road. Although European policies and similar rules in other countries have encouraged the *modal shift* towards more sustainable means of transport (such as the rail mode), the use of road still remains the preferred choice for short and medium-range freight transportation. This is due to many reasons, such as the higher flexibility of road transport, which is the most suitable mode to meet the requirements of the fragmentation of industrial production.

As aforementioned, freight transport by road, that is normally realised with trucks, has negative impacts both from the social point of view and from the point of view of safety and environmental safeguard (Fig. 1.6). Analogously to passenger traffic, greenhouse gas emissions and congestion phenomena are considered the most serious environmental and sustainability issues related to freight transport and logistics.

Fig. 1.6 High percentage of trucks in A10 freeway, in Savona, Italy

In this context, it is possible to state that sustainability of freight transport seems more difficult to be achieved than for passenger transport. This is due to a variety of factors, including the long time horizons necessary to implement major technological changes in heavy vehicles, the need for significant price changes to induce modal shift and the lack of innovation in freight transport modes. The main guidelines of the developed countries towards sustainability in freight transport and logistics are related to promote intermodal transportation, to improve efficiency and environmental performance of the existing modes, to develop alternative fuels technologies, and to exploit opportunities provided by ICT in order to find innovative and sustainable solutions [16, 17].

1.3 Actions to Improve Freeway Traffic Systems

Different possible actions and interventions in freeway traffic systems have been studied and implemented worldwide in order to improve traffic circulation and safety, and to mitigate the environmental impact. These solutions can be directly related to the design of the infrastructure or the development, in the current automotive sector, of new technologies for producing safer and more compatible vehicles. In addition, another possibility is to exploit the concepts of information technology and traffic engineering (electronic surveillance, vehicle communications systems, traffic analysis and control theory) to increase the efficient use of the present transport infrastructure.

Any type of action realised in a freeway traffic system should be developed taking into account that the mobility of people and goods is a complicated problem, since it affects a large number of actors with interests and objectives that are often incompatible and conflicting. The developed measures should address the needs of the main stakeholders, that are:

- government or local *authorities*, which plan to improve the quality of life in terms of environment safeguard, accessibility (both to areas and services) and decongestion of traffic, without compromising the socio-economic vitality of the country;
- *road users*, who are involved in the twofold role of drivers and citizens and require high standards of services and quality of life;
- *managers of road service* and *automotive industry*, whose main aim is the maximisation of profits and the minimisation of costs.

1.3.1 Infrastructure Design

A first possibility to tackle the problems deriving from recurrent and non-recurrent congestion phenomena is related to changes of the freeway infrastructure. In particular, for events of recurrent congestion, some design solutions entail the extension of the existing infrastructure adding traffic lanes, introducing alternative routes and modifying the road geometry where bottlenecks occur. To address non-recurrent congestion, it is possible to define specific actions. For each cause of non-recurrent congestion, several possible solutions have been deeply examined in recent studies [9].

The main non-recurrent design treatments act on the *geometry* of the roadway, though not implying a massive intervention on the infrastructure. These strategies adopt technical solutions (such as insertion of emergency lanes, emergency crossovers, crash investigation sites, alternating shoulders, ramp widening, and so on), which improve the freeway accessibility both for road users and for road operators who are in charge of carrying out works and rescue operations in a quick and safe manner. Some of these interventions can be applied not only to deal with non-recurrent congestion but also to regulate the traffic flow in recurrent congestion situations. For instance, the *movable traffic barriers* (which are concrete barriers that can be shifted from one side of a lane to another one, to change the designated purpose or direction of travel flow for that lane) have potential benefits in case of non-recurrent congestion situations, such as work zones and major incidents, but their most common applications are to alleviate recurrent congestion due to an unbalanced flow during peak periods.

This wide range of geometric design treatments can actually reduce delays and improve travel time reliability, but they also imply negative consequences. First of all, the realisation of civil works necessarily provokes a loss of soil which could have instead an agricultural or ecosystem use and, also, it generates a territorial fragmentation interrupting the natural habitat. The construction of new infrastructures also produces a modification of the landscape and can have a strong hydrogeological impact, in terms of contamination of surface and ground water.

Finally, it is important to point out that all these measures for improving the freeway infrastructure have some indirect impacts [18]. For instance, the increased capacity of the infrastructure may actually improve the traffic performance in the short term, but may also attract a higher level of traffic demand, further worsening the current level of congestion and environmental pollution.

1.3.2 Technological Solutions on Vehicles

In addition to design actions on the infrastructure, the performance of the freeway traffic system can also be improved by exploiting new technological solutions in the design of vehicles. The two main directions of the technological development in the automotive industry are related to the reduction of fuel consumption and pollutant emissions, on the one hand, and to the increase of safety and comfort for passengers, on the other hand.

Considering fuel consumptions and pollutant emissions, different aspects have been developed, both related to vehicle technologies and devoted to the adopted fuels [19]. The main aim is to achieve a good compromise between energy efficient use and production costs, also considering aspects such as safety and reliability. In this context, the technologies already available or under development offer high potential for reducing pollutants in the long term, but many of them require a partial or total redesign of the vehicle, yielding high production costs that make these technologies difficult to be applied.

The introduction of vehicles with alternative propulsion is in the direction towards decarbonisation and replacement of fossil fuels [20, 21]. A first interesting technology is the *electric* propulsion system, mainly used in the urban context (Fig. 1.7). Although electric vehicles can recover energy during the braking phase, the battery performance severely limits their spread in the market on a larger scale at present, but the strong development in storage technologies allows to think that a larger diffusion of electric vehicles will happen in the near future. The *hybrid electric* propulsion system, i.e. an electric propulsion system combined with a conventional internal combustion engine, allows to overcome the limits of electric vehicles. A third possibility is given by the application of *fuel cells* for automotive field. According to this latter technology, the electricity for the electric motor is produced by an electrochemical device, where hydrogen is used as fuel and the oxygen present in the air is used as combustive agent.

As aforementioned, a further step towards green mobility is the progressive development of *alternative fuels* that in general allow a considerable abatement of carbon dioxide emissions. Specifically, the most promising alternatives are natural gas, liquefied petroleum gas, hydrogen, mixtures of hydrogen and natural gas and biofuels.

Fig. 1.7 Electric vehicle and charging station in Savona University Campus, Italy

It is worth noting that the urgency to limit the use of fossil fuels is not only due to the need of environmental protection, but is also encouraged by the increasing price of oil. Another contribution to alleviate fuel consumption is given by the enhancement of vehicles, due to the application of technological devices which allow the achievement of a better performance, for example the reduction of vehicle mass thanks to innovative materials, the increase of the efficiency of the transmission system, the improvement of vehicle aerodynamics and others.

Considering safety, security and comfort aspects, the most common systems developed to automate and enhance safety and driving conditions on vehicles are called *Advanced Driver Assistance Systems* (ADAS). For instance, in order to avoid collisions and accidents or to attenuate their effects, modern technologies are devoted to alert the driver about potential problems or to take over control of the vehicle. Other features are adaptive light control, adaptive cruise control, automatic parking, collision avoidance, intelligent speed adaptation, platooning systems, cooperative merging, and so on. These ADAS are either built into cars or available as add-on packages or aftermarket solutions. All these systems are based on multiple data sources, including automotive imaging, radar, image processing, computer vision, and in-car networking. These systems aiming at assisting, improving and easing the driving tasks are often called also *Vehicle Automation and Communication Systems* (VACS). The interested reader can find a very broad classification of VACS in [22], with specific attention to the freeway traffic management perspective.

In the most advanced solutions, additional inputs can come from sources which are external from the vehicle, such as other vehicles or the infrastructure, known respectively as *Vehicle-to-Vehicle* (V2V) or *Vehicle-to-Infrastructure* (V2I) systems. Vehicles equipped with V2V technology can wirelessly broadcast information and receive messages from other vehicles in the proximity, for instance about their position and speed. The communication among vehicles and with the infrastructure raises important issues about the cooperation among drivers, in order to follow objectives that can refer to a system perspective more that an individual logic. These next-generation vehicles including a high level of wireless connectivity and automated driving capability are sometimes known as *Connected and Automated Vehicles* (CAV).

A final issue that has become more and more relevant in the last years for ADAS is related to vehicle *cybersecurity*. Analogously to computers, modern vehicles must be protected from hacking, malicious cyber-attacks, and any other unauthorised access to retrieve driver data or to manipulate vehicle functionality. Vulnerabilities may exist for example within the wireless communication functions of a vehicle, within a mobile device connected to the vehicle via USB, Bluetooth, or Wi-Fi, or within a third-party device connected through a vehicle diagnostic port, so that a hacker could remotely exploit these vulnerabilities and gain access to the controller or to possible data stored in the vehicle.

1.3.3 Application of ICT

Even though infrastructure interventions and technological advances on vehicles can allow to improve the performance of road traffic systems, they cannot provide a complete solution to traffic problems. In particular, the delay in the adoption of appropriate infrastructures can be the result of many factors such as the high investment costs, the excessive duration of design and construction phases, the environmental incompatibility and the lack of space. At the same time, technological solutions on vehicles, such as VACS, can ameliorate comfort and traffic conditions for drivers, but the maximum exploitation of such technologies can be obtained by introducing management and control tools, which act according to a system perspective.

Indeed, the effective utilisation and exploitation of the road infrastructure is possible only if suitable management systems exploiting ICT are applied. The ICT-based applications are a valid opportunity to enhance the system efficiency at operational, economical and environmental level, by exploiting the present infrastructure and the available technologies in order to improve the performance of the whole traffic system.

The advanced applications which aim to provide innovative services for transport and traffic management, by enabling the users to be better informed, safer and more coordinated, are often known with the name of *Intelligent Transportation Systems* (ITS). The main applications of ITS are related to public and private transport management through optimisation and control tools, information to travellers, improvement and control of vehicle safety, emergency management, promotion of environmentally efficient use of the road network. The development of ITS represents a real opportunity to effectively address the forecasted growth in traffic demand and the inability to meet the mobility needs only through infrastructure investments. Thanks to these applications, road users can benefit from the information and directions provided by such systems and, meanwhile, the managers of road service and local authorities can take advantage of the increased capabilities of collecting, monitoring, and disseminating data.

ITS tools for managing and controlling freeway traffic systems (see Sect. 1.4) are first of all devoted to the reduction of traffic congestion. Recurrent congestion may be managed by smoothing peak demands through techniques such as ramp metering, mainstream traffic control, driver information and guidance systems that inform motorists about congestion situations ahead or about alternative routes. The management of non-recurring congestion is more difficult, because of its unpredictable nature; however, the control techniques adopted to manage recurrent congestion may also be beneficial in reducing the effects of non-recurrent congestion.

More sophisticated ITS applications encourage the use of the road network in an environmentally sustainable way. This can be achieved by adopting appropriate tools to regulate the traffic flow and, at the same time, to obtain a lower level of fuel consumption and pollutant emissions. Some strategies are based on the communication of the optimal speed to minimise the energy effort, the indication of alternative paths which are more efficient from the energetic point of view, the introduction of dedicated lanes for particular categories of traffic (heavy vehicles, public transport, and so on), or the implementation of strategies such as vehicle platooning to reduce fuel consumption.

1.4 Management and Control of Freeway Traffic

In order to manage, operate, and maintain freeway facilities in an efficient way, surveillance and control methods are often integrated with ICT tools into suitable freeway traffic management programs (Fig. 1.8). These systems are often referred to as *Advanced Traffic Management Systems* (ATMS). ATMS can be classified in different ways but, for the purposes of this book, it seems useful to distinguish them between road-based and vehicle-based traffic control systems.

Road-based traffic control systems are nowadays the most commonly utilised: such systems allow to regulate the traffic flow in a freeway system by controlling all the vehicles together, i.e. by acting at a macroscopic level. These systems for dynamic traffic control intervene in traffic in order to improve the performance of the traffic networks, i.e. to increase safety, to improve traffic flows, to reduce travel times, to make travel times more reliable, or to reduce emissions and noise production [23]. The control measures that are normally employed in freeway networks are *ramp management* (in particular ramp metering, applied with traffic lights at the on-ramps), *mainstream control* (including variable speed limits, lane control, congestion warning, keep-lane instructions, and so on), and *route guidance* (normally displaying specific indications at intersections) [24].

On the other hand, *vehicle-based* traffic control systems are the most modern alternative which will become more and more relevant in the near future, but the real

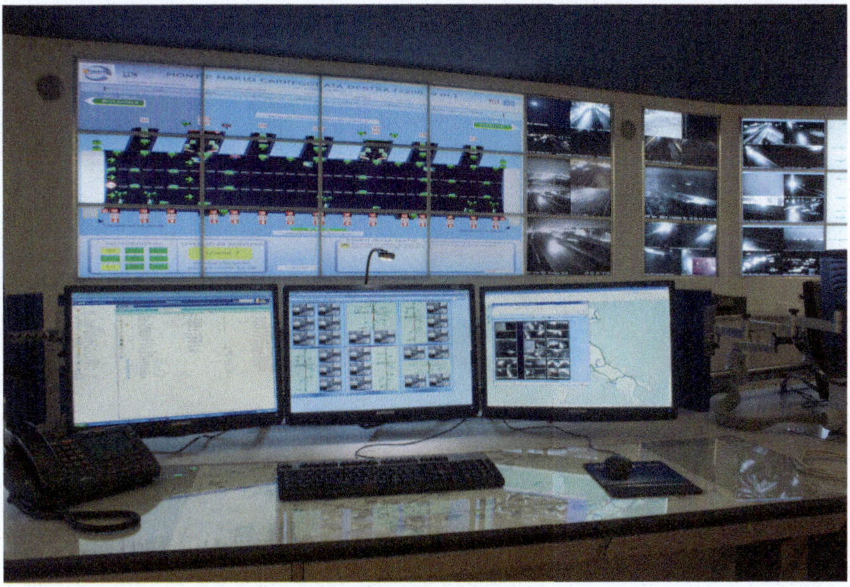

Fig. 1.8 The radio room located in the Bologna regional freeway management centre, Italy (courtesy of Autostrade per l'Italia SpA, photo from Archivio Videofotografico Autostrade per l'Italia)

application of which still is only at prototype levels. Vehicle-based control systems will shift the macroscopic control to specific control actions imposed to each vehicle. This type of control is based on the development of *Intelligent Vehicles* (IV), which are equipped with sensors to make measurements and try to achieve more efficient vehicle operation, either by assisting the driver or by taking partial or complete control of the vehicle [25].

This section presents the most widely used road-based freeway control actions i.e. ramp management, mainstream control, and route guidance, and an overview of vehicle-based traffic control systems.

1.4.1 Ramp Management

Ramp management can be defined as the application of control devices, such as traffic signals, signs and gates, to regulate the number of vehicles entering or leaving the freeway. Ramp management strategies may be used to control the access at the on-ramps or to regulate the rate of vehicles entering the freeway, but they can also be applied to off-ramps.

The most widespread strategy belonging to this category is *ramp metering* (Fig. 1.9), which has been used in the U.S. since the early 1960s. Ramp metering is realised by placing traffic signals at on-ramps to control the rate at which vehi-

Fig. 1.9 A ramp metering installation in downtown San Diego to State Highway 94, U.S. (courtesy of Michael Ballard)

cles enter the freeway. The ramp metering controller, through suitable algorithms, computes the metering rate to be applied; such metering rate is implemented by appropriately setting the phase lengths of the traffic signals present at the on-ramps. According to the adopted ramp metering policy, it is possible to distinguish different types of ramp metering, e.g. single-lane with one vehicle per green, single-lane with multiple vehicles per green, and dual-lane.

Ramp metering can be implemented with different purposes. A first use is related to regulate the merging process of the on-ramp traffic by breaking the platoons and by spreading the on-ramp traffic demand over time, in order to mitigate shock waves. Another important use of ramp metering strategies is to prevent breakdowns. When traffic density is high, it is possible to prevent traffic breakdowns on the freeway via ramp metering by properly adjusting the metering rate in order to maintain the mainstream density below a critical value. More in general, ramp metering systems can be used to increase the throughput of vehicles, to increase the average speed along the freeways, to reduce the total number of crashes, to reduce vehicle emissions and fuel consumption. Some of the ramp metering benefits, in terms of safety, mobility and productivity, as well as in terms of environmental effects, are reported in [26] for some cities in the U.S., where very high improvements have been revealed.

Besides all these benefits, the application of ramp metering strategies can have some disadvantages. Among them, it can be cited the fact that drivers may use parallel facilities to avoid ramp meters, sometimes corresponding to longer trips. Moreover, ramp metering strategies may result in unfair policies, in which the users of specific on-ramps are highly penalised compared with others, or can lead to the shift of traffic congestion from one location to another. Also, it is highly relevant to consider that the on-ramp queues have a storage upper bound, due to practical space limitations. If the queue storage capacity is low, the potential of ramp metering can be strongly limited.

1.4.2 Mainstream Control

While ramp metering acts on the freeway ramps in order to regulate traffic, it is possible to control traffic conditions also through strategies acting on the mainstream. These strategies have been applied in many European countries, such as in Germany, in the Netherlands, and in the United Kingdom, and in some U.S. states, with the common objective of homogenising the traffic flow in order to reduce congestion phenomena and to fully exploit the freeway capacity.

Fig. 1.10 Variable speed limits at the junction Kleinpolderplein of the A13 and A20 freeways, the Netherlands (courtesy of Rijkswaterstaat, Photo: Essencia Communication/Rob de Voogd)

Such strategies can be of different types, the most common one being the definition of *variable speed limits* to be displayed on Variable Message Signs (VMSs) (Fig. 1.10). A first obvious scope of speed limits is related to the increase of safety, since it is well known that speed reduction leads to improved safety conditions on freeways, and also because it has been shown that speed limits reduce the frequency of lane changes. With this scope, speed limits are applied in potentially dangerous situations, such as upstream the congested areas or during adverse weather conditions. Variable speed limits can also be applied to homogenise the traffic behaviour, i.e. to reduce the speed differences. Another important objective of variable speed limits is the prevention of traffic breakdowns, by avoiding high densities in the mainstream.

Besides variable speed limits, mainstream control includes also other strategies, such as *mainline metering* adopting traffic lights along the mainline in order to reduce breakdown phenomena. In addition, *lane control* can be applied to prevent the use of lanes upstream critical areas, such as accident locations or highly congested on-ramps. Another possibility is to give "keep your lane" indications to the drivers, so that they are not allowed to change lanes, leading to lower disturbances in the freeway traffic flow. The use of *peak lanes* is also common in many cases: the hard shoulder lane of a freeway (which is normally used only by vehicles in emergency) is opened to traffic during peak hours. In this way, the capacity of the road is increased, but safety may be reduced. In some situations, the shoulder lane is opened only to dedicated vehicles, such as public transport, freight transport, or high occupancy vehicles. Another popular strategy is the use of *reversible lanes* (also called tidal flow lanes): a freeway lane can be used in both directions and the current direction is determined dynamically on the basis of the highest traffic demand.

1.4.3 Route Guidance

Another possibility to control freeway traffic is to efficiently distribute the traffic demand over the network, by properly routing traffic flows on alternative paths. A traffic network can include many origins and destinations with multiple paths connecting each origin-destination pair. During peak hours, the travel time on many paths increases and alternative routes (which imply longer times in absence of congestion) may become competitive.

Even though the past experience can be helpful for regular drivers (who are familiar with the traffic conditions in the network) to take routing decisions, daily varying demands, changing environmental conditions, exceptional events and accidents make the traffic conditions very difficult to be predicted and, consequently, the routing decisions very difficult to be taken. This results in situations in which some road links are very congested and, at the same time, other links on alternative paths are rather under-utilised. The use of VMSs to provide en-route information to motorists or to explicitly give them route recommendations can improve the overall network efficiency (Fig. 1.11). The information systems which disseminate to drivers messages

Fig. 1.11 Routing indications in A20 freeway, close to Rotterdam, the Netherlands (courtesy of Rijkswaterstaat, Photo: Essencia Communication/Rob de Voogd)

with information and recommendations to assist them in their route choice decisions are called *Route Guidance and Information Systems* (RGIS).

In many cases, freeway managers prefer to provide real-time information rather than to give explicit route indications. RGIS typically display traffic information such as congestion length, delay on the alternative routes, or travel time to the next common point on the alternative routes. Simply displaying real-time information has the advantage that drivers can make their own routing decisions and do not feel compelled by the controller, but there are many disadvantages. First of all, given the real-time information, it is not easy for drivers to take a routing decision in few seconds, and this decision is much harder for those drivers who are not familiar with the network. Besides, the VMSs have limited space to display the traffic information that must be strongly summarised and can become ineffective in many cases.

1.4.4 Vehicle-Based Traffic Control

In recent years, the fast development of technology in the automotive industry has led to the diffusion of many VACS, as previously introduced in Sect. 1.3.3. These systems are expected to change the features and capabilities of individual vehicles in the next decades, so that a scenario with self-driving vehicles moving in a completely connected road-vehicle infrastructure does not seem completely unrealistic any more. Nevertheless, the most likely scenario for the next future is the one in which freeway systems will be characterised by a *mixed traffic* flow, in which both traditional cars and vehicles provided with VACS of different advancement levels will share the same infrastructure. The mixed-traffic case is not only the most probable framework of the next future but also the most interesting and difficult challenge for researchers and scientists.

At present, many vehicles are equipped with *human-machine interfaces* through which the drivers can receive advice or warnings (e.g. blind spot warning, parking assistance, and so on). There are also *semi-autonomous* systems which can take partial control of vehicle manoeuvres (e.g. avoidance systems which initially warn the driver via seat vibrations and, in case of no reactions by the driver, start to brake). Finally, *fully autonomous* systems can take complete control of vehicle operations (e.g. fully automated adaptive cruise control and anti-lock braking systems).

The vehicles of today and surely those of the future are provided with many sensors, so that they are able to collect lots of data. These types of vehicles are often called *probe vehicles* or *floating cars*, since they are capable to make measurements of the traffic state along their trajectories by adopting suitable technologies, such as Global Positioning System (GPS) devices, radar systems, cameras, and so on.

Hence, the vehicles of the future will easily include sensors and control devices, since they will be able to collect and transmit information, as well as they will easily actuate specific control actions and, consequently, actively interfere with traffic. This will surely change the architecture of traffic management systems, that will require an adaptation of the present traffic management actions and strategies in order to be able to exploit the potential of VACS to further improve traffic conditions in road networks [27].

The possible future impacts of VACS on freeway traffic, as well as the effects of autonomous vehicles provided with high connectivity potentials are analysed in the two recent papers [22, 28], which can provide the reader with an interesting overview of current and future trends for freeway traffic management systems.

References

1. Maibach M, Schreyer C, Sutter D, Van Essen HP, Boon BH, Smokers R, Schroten A, Doll C, Pawlowska B, Bak M (2008) Handbook on estimation of external costs in the transport sector. Report for the European Commission, CE Delft
2. Martino A, Maffii S, Sitran A, Giglio M (2009) The calculation of external costs in the transport sector. Report for the European Parliament's Committee on Transport and Tourism, Brussels
3. Korzhenevych A, Dehnen N, Bröcker J, Holtkamp M, Meier H, Gibson G, Varma A, Cox V (2014) Update of the handbook on external costs on transport. DG MOVE Ricardo-AEA/R/ED57769, Report for the European Commission
4. (2011) White paper: roadmap to a single European transport area - towards a competitive and resource efficient transport system. COM, Brussels
5. Nesbit M, Fergusson M, Colsa A, Ohlendorf J, Hayes C, Paquel K, Schweitzer J-P (2016) Comparative study on the differences between the EU and US legislation on emissions in the automotive sector. European Union

6. Hennessy DA, Wiesenthal DL (1999) Traffic congestion, driver stress and driver aggression. Aggress Behav 25:409–423
7. Lajunen T, Parker D, Summala H (1999) Does traffic congestion increase driver aggression? Trans Res Part F 2:225–236
8. Potts IB, Harwood DW, Hutton JM, Fees CA, Bauer KM, Lucas LM, Kinzel CS, Frazier RJ (2014) Identification and evaluation of the cost-effectiveness of highway design features to reduce nonrecurrent congestion. Transportation Research Board, Washington
9. Potts IB, Harwood DW, Hutton JM, Fees CA, Bauer KM, Lucas LM, Kinzel C, Frazier RJ (2014) Design guide for addressing nonrecurrent congestion. Transportation Research Board, Strategic Highway Research Program 2, Washington
10. Maddison D, Pearce D, Johansson O, Calthrop E, Litman T, Verhoef E (2014) Blueprint 5: true costs of road transport. Routledge
11. Hoek G, Brunekreef B, Goldbohm S, Fischer P, Van Den Brandt PA (2002) Association between mortality and indicators of traffic-related air pollution in the Netherlands: a Cohort study. The Lancet 360:1203–1209
12. IEA (2015) CO_2 emissions from fuel combustion. International Energy Agency, Paris
13. Marchesini P, Weijermars W (2010) The relationship between road safety and congestion on motorways. SWOV Institute for Road Safety Research, Leidschendam
14. Golob TF, Recker WW, Alvarez VM (2004) Freeway safety as a function of traffic flow. Accid Anal Prev 36:933–946
15. Potts IB, Harwood DW, Fees CA, Bauer KM, Kinzel CS (2015) Further development of the safety and congestion relationship for urban freeways. Transportation Research Board, Strategic Highway Research Program 2, Washington
16. The future of sustainable freight transport and logistics. (2010) European Parliament, Brussels
17. Systems sustainable freight transport: opportunities for developing countries. (2015) United Nations, Trade and Development Board, Geneva
18. Noland RB (2001) Relationships between highway capacity and induced vehicle travel. Trans Res Part A 35:47–72
19. Romm J (2006) The car and fuel of the future. Energy Policy 34:2609–2614
20. Chan CC (2007) The state of the art of electric, hybrid, and fuel cell vehicles. Proceedings of the IEEE 95:704–718
21. Ehsani M, Gao Y, Emadi A (2009) Modern electric, hybrid electric, and fuel cell vehicles: fundamentals, theory, and design. CRC Press, Taylor and Francis Group, Boca Raton
22. Diakaki C, Papageorgiou M, Papamichail I, Nikolos I (2015) Overview and analysis of vehicle automation and communication systems from a motorway traffic management perspective. Trans Res Part A 75:147–165
23. Hegyi A, Bellemans T, De Schutter B (2009) Freeway traffic management and control, In: Meyers RA (ed) Encyclopedia of complexity and systems science. Springer, Berlin, pp 3943–3964
24. Papageorgiou M, Diakaki C, Dinopoulou V, Kotsialos A, Wang Y (2003) Review of road traffic control strategies. Proc. IEEE 91:2043–2067
25. Baskar LD, De Schutter B, Hellendoorn J, Papp Z (2011) Traffic control and intelligent vehicle highway systems: a survey. IET Intell Trans Syst 5:38–52
26. Jacobson L, Stribiak J, Nelson L, Sallman D (2006) Ramp management and control handbook. U.S. Department of Transportation, Federal Highway Administration, Washington
27. Papageorgiou M, Diakaki C, Nikolos I, Ntousakis I, Papamichail I, Roncoli C (2015) Freeway traffic management in presence of vehicle automation and communication systems (VACS). In: Meyer G, Beikerin S (eds) Road vehicle automation, vol 2. Springer, Berlin, pp 205–214
28. Mahmassani HS (2016) Autonomous vehicles and connected vehicle systems: flow and operations considerations. Transp Sci 50:1140–1162

Chapter 2
Fundamentals of Traffic Dynamics

2.1 Basic Concepts of Traffic Flow Theory

In order to represent the dynamics and the performance of the traffic stream, some variables, such as flow, speed and density, need to be introduced and discussed, since they represent very important concepts in traffic flow theory. In order to better understand results and analyses of traffic dynamics, it is also useful to become familiar with traffic-related diagrams, such as space–time diagrams, time series or plots representing the relations between flow and density or between speed and density.

2.1.1 Flow, Speed, Density and Related Variables

Flow, speed and density are the most important quantities and measures used to describe the behaviour of traffic flow, seen at an aggregate level. Many other important variables adopted in traffic flow theory derive or are related to these three main quantities. Table 2.1 reports the main variables useful for representing the traffic stream dynamics. For each variable, its meaning is briefly described and the related unit is reported. Note that most of these variables could be measured with different units; for instance, speed can be measured in [km/h] but also in [m/s] or in [miles/h], and so on. The units reported in Table 2.1 are the most common according to the contents of this book.

Speed It is measured in space unit per time unit, for instance, in [km/h], and can have both a microscopic meaning (speed of a vehicle) and a macroscopic interpretation if referred to the movement of the traffic stream. In this latter case, it is necessary to compute an average speed, which can be obtained in two ways, corresponding to two different interpretations of speed:

A. Ferrara et al., *Freeway Traffic Modelling and Control*, Advances in Industrial Control, https://doi.org/10.1007/978-3-319-75961-6_2

Table 2.1 Basic traffic variables

Variable	Description	Unit
Speed	Distance travelled per time unit	[km/h]
Density	Number of vehicles per space unit	[veh/km]
Space headway	Distance between vehicles	[m/veh]
Occupancy	Time percentage in which a location is occupied by vehicles	[%]
Flow	Number of vehicles passing a given location per time unit	[veh/h]
Time headway	Time lag between vehicles	[s/veh]
Demand	Number of vehicles requiring to enter the freeway per time unit	[veh/h]

- *time-mean speed* or *arithmetic mean speed*: instantaneous speeds of different vehicles are measured at a particular location for a given time interval and the mean speed is obtained as average of all the speeds in the sample;
- *space-mean speed* or *harmonic mean speed*: travel times of different vehicles between two given locations are measured at a fixed time instant and the mean speed is computed as the ratio between the distance of the two locations and the average observed travel times.

Density Sometimes called also *concentration*, it is the number of vehicles on a road segment at a fixed time instant and is expressed in vehicles per space unit, referred to the whole freeway (e.g. [veh/km]) or specified per lane (e.g. [veh/km/lane]).

Space headway indicates the distance between two vehicles, given by the distance between the front and the rear bumpers of the vehicles, plus the length of the leading vehicle. It is measured in space unit per vehicle, e.g. [m/veh]. Space headway is a microscopic variable, referred to one single vehicle. Note that the average space headway, which has a macroscopic meaning, is the inverse of density.

Since it is difficult to directly measure density with the available technology, density is often estimated on the basis of a related variable, easier to measure, that is called *occupancy*. Occupancy is defined as the time in which a given location (i.e. the detection zone of a sensor) is occupied by vehicles, and is usually expressed as a percentage.

Traffic Flow Often called also *traffic volume* or *flux*, it is the rate at which vehicles pass through a given location in a road and is expressed in number of vehicles per time unit, normally in [veh/h]. Traffic flow can be referred to the entire freeway or can be specifically associated with a single lane, in this latter case it is given in number of vehicles per time unit per lane, e.g. [veh/h/lane]. Flow is related with mean speed and density with the following well-known relation: *flow is the product of harmonic mean speed and density*.

An important variable associated with traffic flow is *time headway*. It is expressed in time unit per vehicle, e.g. [s/veh]. Time headway has a microscopic meaning, since it indicates the headway in time between the front bumpers of two vehicles following each other. The average time headway, having a macroscopic meaning, can be computed as the inverse of traffic flow.

The same unit of traffic flow is often adopted to represent the *demand*, which is the flow of vehicles that want to enter the freeway, and is again expressed in number of vehicles per time unit. The demand can be associated with any access to the freeway, i.e. it can be referred to the mainstream, to the on-ramps or to other roads merging with the considered freeway.

2.1.2 Traffic Diagrams

In order to represent the evolution of the traffic conditions of a given road portion, some traffic diagrams are used. The most widespread types of diagrams, described hereinafter, are space–time diagrams, time series, flow–density and speed–density diagrams.

Space–Time Diagrams They are used to plot the trajectories of vehicles travelling in a given road segment for a certain time interval, reporting time on the x-axis and space on the y-axis. Analysing a space–time diagram, it is possible to obtain some important information:

- the slope of the trajectory represents the local speed of the vehicle in a given location at a given time instant; if the slope is equal to zero, i.e. the trajectory is horizontal, this means that the corresponding vehicle is stationary;
- two crossing trajectories correspond to an overtaking between the two vehicles;
- the horizontal distance between two trajectories indicates the time headway, whereas the vertical distance between trajectories represents the distance headway;
- the number of trajectories crossing a horizontal segment associated with a given location and a certain time interval corresponds to the traffic flow;
- the number of trajectories crossing a vertical segment associated with a given time instant and a certain space interval corresponds to the traffic density.

An example of this type of diagrams is shown in Fig. 2.1, where the trajectories of five vehicles are depicted. Vehicle A and Vehicle B travel at a rather constant speed, with Vehicle B following Vehicle A and moving at a lower speed. The trajectory of Vehicle C shows that it starts travelling at an almost constant speed, then it slows down and, after, it accelerates to reach a constant speed that is lower than the initial one. Vehicle D, initially moving at a constant speed, slows down until stopping completely, and then it leaves until reaching again a constant speed. Finally, Vehicle E travels at a rather constant speed and, after about 25 s from its departure, it overtakes Vehicle D.

Time Series They are generally used to plot the profiles of aggregated quantities, referred to a given location (often coming from a sensor positioned at that location),

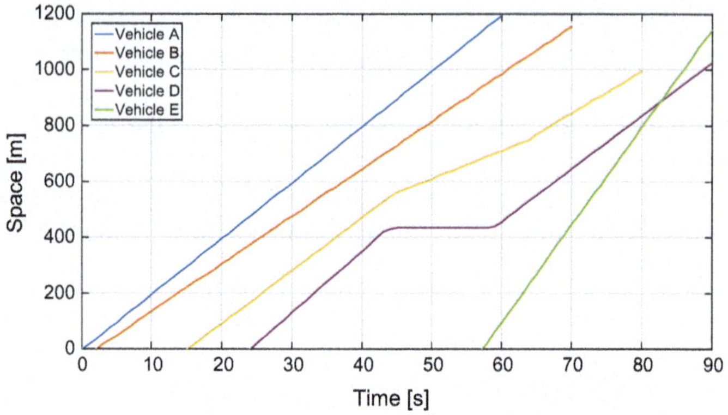

Fig. 2.1 Example of space–time diagram: trajectories of five vehicles

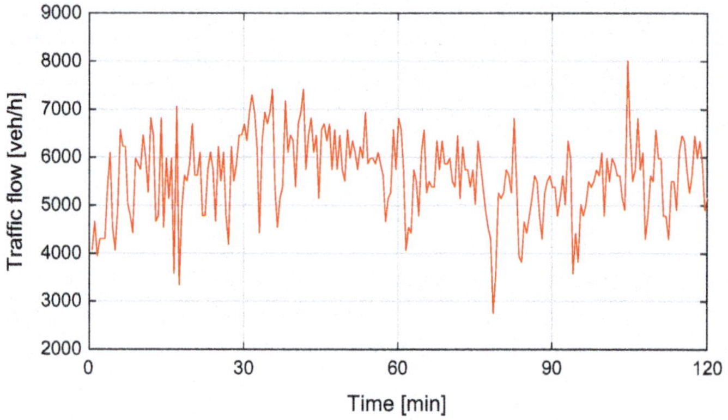

Fig. 2.2 Example of time series: traffic flow in a 2-h period

over a given time period. These graphs are useful to understand the evolution in time of the plotted variables, from which the dynamic behaviour of the traffic system can be inferred.

Figure 2.2 reports an example of the traffic flow profile during a period of two hours, with a sample time of 30 s. Figure 2.3 shows an example of a typical daily time series of the mainstream demand, with sample time of 10 s, while a daily time series of the speed, with a sample time of 5 min, is reported in Fig. 2.4.

Flow–Density and Speed–Density Diagrams Other two common diagrams, often adopted in traffic flow theory, are flow–density and speed–density diagrams. The *flow–density diagram* shows the measured flow over density, for a specific time interval and a specific road portion. Analogously, *speed–density diagrams* plot the measured speed over the traffic density, again for a given time period and a given road segment. These data are generally referred to heterogeneous traffic conditions (different types of vehicles and drivers, e.g. different percentages of trucks) and

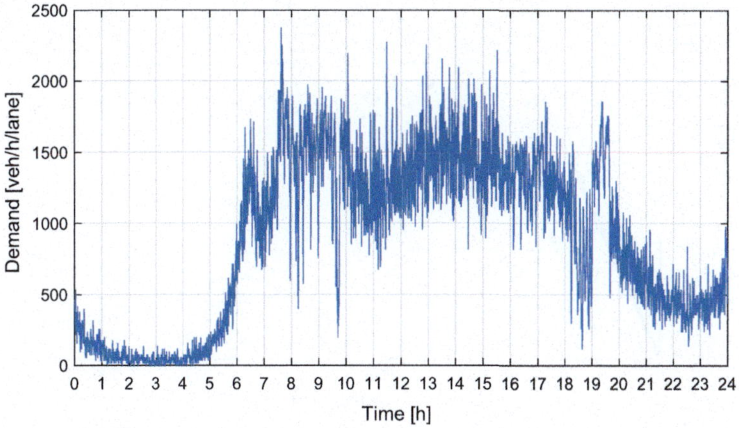

Fig. 2.3 Example of time series: mainstream demand during a day

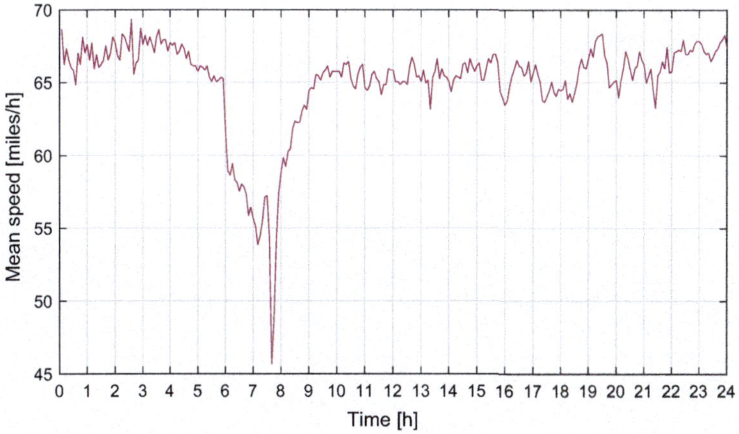

Fig. 2.4 Example of time series: speed during a day

different external circumstances (e.g. variable weather situations or unexpected events). Examples of flow–density and speed–density diagrams, obtained from measurements of a three-lane road portion, are shown, respectively, in Figs. 2.5 and 2.6.

2.1.3 The Fundamental Diagram

A very relevant concept in traffic theory is given by the so-called *Fundamental Diagram*, which represents the theoretical relation between flow and density in a given road or road section, in case of homogeneous and stationary traffic conditions. It is

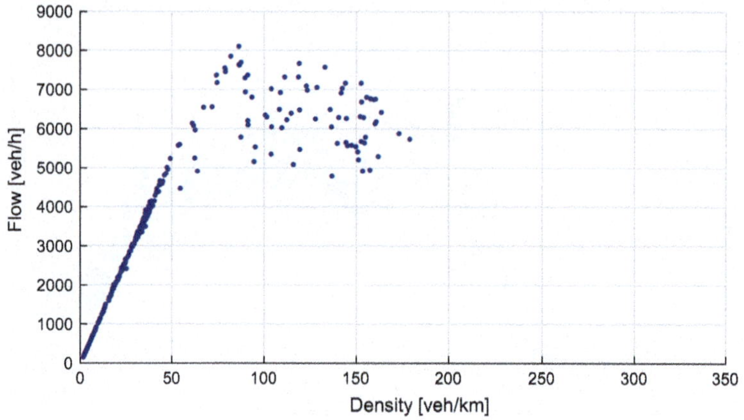

Fig. 2.5 Example of flow–density diagram

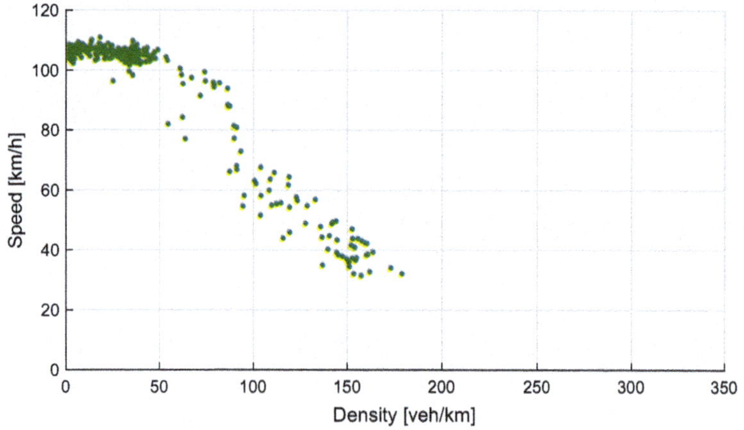

Fig. 2.6 Example of speed–density diagram

important to distinguish the Fundamental Diagram from the flow–density diagram obtained from measured data, since the latter describes non-stationary traffic conditions of different types of vehicles and, also, there can be systematic errors in the measurement process.

An example of the typical form of the Fundamental Diagram is shown in Fig. 2.7. The classical features of this diagram are the following:

- when the traffic density is zero, the flow is zero;
- the slope of the tangent of the curve in zero represents the *free-flow speed*, i.e. the speed of vehicles in absence of traffic or with a very low flow; the free-flow speed is a design feature of a road facility, since it indicates the standard driver behaviour when there are no constraints given by the traffic conditions;

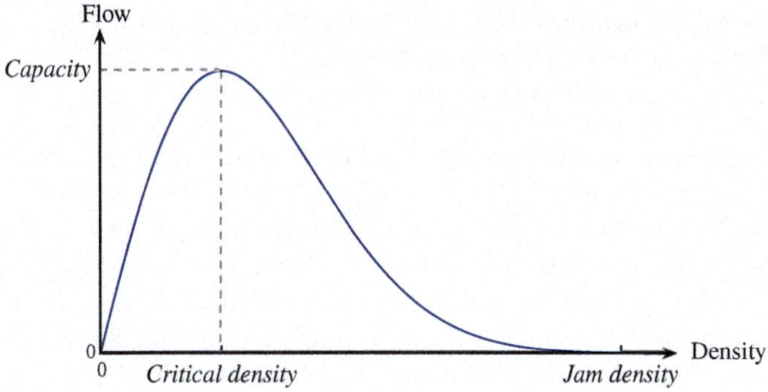

Fig. 2.7 Example of fundamental diagram

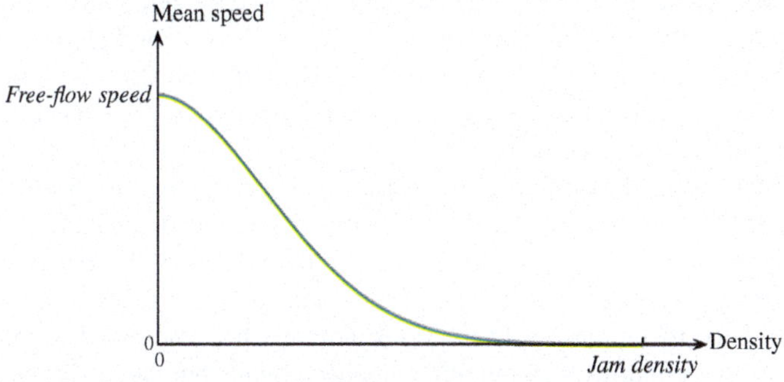

Fig. 2.8 Example of steady-state speed–density relation

- as the density increases, the flow initially increases, with a rather linear behaviour (due to the limited interactions among vehicles), that is maintained to a certain value of density, called *critical density*;
- when the density is equal to the critical value, the flow reaches its maximum point, called *capacity*; the capacity is in general a measure of the maximum throughput of a road;
- beyond the critical value, a further increase in the density corresponds to a decrease in the flow; this is due to the increment of the interactions among vehicles and a consequent reduction of the average speed;
- the traffic flow becomes zero when the density reaches its maximum value, called *jam density*; the jam density can be estimated on the basis of the minimum spacing between vehicles (given by the average minimum gap between vehicles plus the average vehicle length).

Analogously, a steady-state relation between speed and density can be defined, the typical form of which is shown in Fig. 2.8. The steady-state speed–density relation has the following features:

- when the density is zero, vehicles travel at the free-flow speed;
- as the density increases, their speed decreases;
- the speed becomes zero when the jam density is reached.

The idea that there exists one steady-state flow–density curve and one steady-state speed–density curve comes from the so-called Lighthill–Whitham–Richards theory (see Sect. 3.2). The functional forms of these relations have been normally obtained through data-fitting techniques or on the basis of car-following concepts or, also, by considering the analogy of traffic flow dynamics with the behaviour of fluids. Many well-known functional forms have been developed in the literature, starting from the early and simple Greenshields' linear model proposed in 1935 [1]. Since then, many other studies have tried to improve this simplified model in order to obtain a functional form with parameters to be calibrated on the basis of real data. Some famous and consolidated steady-state relationships are the Greenberg's logarithmic model [2], or the exponential models by Underwood, Newell and Drake [3–5]. More recent works on estimating the shape of Fundamental Diagrams are, for instance, [6–8]. The interested reader can find a detailed review of different Fundamental Diagrams and their properties in [9].

As aforementioned, according to the classical theory, a given traffic system is characterised by one steady-state flow–density curve and one steady-state speed–density curve, which are represented as smooth functions (e.g. those reported in Figs. 2.7 and 2.8). However, these uniqueness and smoothness assumptions are not always validated by empirical observations, not only because real data refer to non-stationary traffic conditions of inhomogeneous vehicles but also because of the so-called traffic hysteresis effects and capacity drop phenomena.

The *hysteresis* effects are associated with the observation that measured traffic flows (or speed) do not only depend on the measured density but also on the evolution of the traffic system; hence, the flow–density and speed–density relations seem not to be univocal, especially in case of congested conditions in which real data are more sparse (see Figs. 2.5 and 2.6). The first theory on hysteresis was developed by Newell in 1962, who assumed the presence of two congested branches in the Fundamental Diagram, the deceleration branch and the acceleration branch, with the former being above the latter [10]. Since then, different theories have been developed to correctly capture the hysteresis phenomenon. In [11], a mathematical theory for modelling hysteresis in traffic flow is reported by separating three different phases in traffic flow, i.e. equilibrium, acceleration and deceleration phases. This theory is then complemented in [12], where the existence of two branches in the Fundamental Diagram is confirmed, but it is concluded that hysteresis is better explained in terms of aggressive and timid driver behaviours rather than acceleration and deceleration phases.

An alternative theory is the *three-phase traffic theory* developed by Kerner [13], according to which congested traffic is not a unique phase, as in standard traffic theory, but it consists of two phases, synchronised flow and wide moving jam: the phase of synchronised flow is not described by a line but by an area in the flow–density plane. In other words, in synchronised flow, a given speed is related to infinite possible values of traffic density and a given density is related to infinite possible values of speed, implying that there is no Fundamental Diagram for the case of synchronised flow [14]. This theory provoked some critiques (see, e.g. [15]). This controversy was then discussed in [16], where the authors compared the three-phase traffic theory with two-phase models with Fundamental Diagrams and partly explained such a controversy in terms of different definitions of traffic phases given by the researchers.

The *capacity drop* phenomenon, instead, is normally associated with a bottleneck in the roadway and corresponds to a decrease of the maximum supported capacity: the flow exiting the bottleneck is lower than the capacity, even though the demand upstream the bottleneck is higher than the capacity. This phenomenon has been empirically observed in many freeways [17, 18], where the measured flow exiting a queue is lower than the flow measured before the queue formation and can be also observed in flow–density diagrams where a drop in the capacity is evidenced. The phenomenon of capacity drop is discussed more in detail in Sect. 2.2.4.

Macroscopic Fundamental Diagram Recently, the concept of a steady-state relation among flow, density and speed has been extended to a larger geographical scale, in particular to entire urban areas. The existence of a Macroscopic Fundamental Diagram (MFD) for large urban areas was suggested by empirical experiments, e.g. the field experiments in Yokohama, Japan [19] and those in the Twin Cities Metropolitan Area in Minnesota, U.S. [20]. Traffic hysteresis phenomena have been observed also in Macroscopic Fundamental Diagrams. For instance, in [21] it is shown that, for a fixed average network density, the observed network flows are higher in the onset of congestion and lower in the offset of congestion, while in [22] clockwise hysteresis loops on a day in medium size cities are investigated and discussed.

2.2 Traffic Flow Phenomena

Freeway traffic is a very complex dynamical system, the study of which has involved scientists for many decades and still presents challenging problems for researchers. Traffic dynamics can be different in different countries, depending on many external factors and exogenous conditions, but there are some common aspects and specific traffic phenomena that have been observed worldwide during the years. These traffic flow phenomena will be briefly discussed in this section.

2.2.1 Capacity, Bottlenecks and Breakdown

In the description of the Fundamental Diagram in Sect. 2.1.3, the *capacity* has been defined as the maximum throughput of a road portion. Hence, capacity is a measure of the ability of a transportation facility to carry traffic. For this reason, the concept of capacity plays an important role in various aspects of transportation analysis, both when a new transport facility is designed (to evaluate if it is able to handle the estimated traffic demand) and when management and control systems are applied.

The precise definition of capacity provided by the Highway Capacity Manual has evolved over time (refer to [23] for a detailed description of this historical evolution). In the last version of the Highway Capacity Manual [24], the capacity of a facility is defined as 'the maximum sustainable flow rate at which vehicles or persons reasonably can be expected to traverse a point or uniform segment of a lane or roadway during a specified time period under given roadway, geometric, traffic, environmental, and control conditions'. When real traffic data are observed, it is clear that this definition is not completely adequate to measure capacity. As a matter of fact, real traffic observations show that the maximum flow measured at a given location varies from day to day, resulting in the fact that there is not a unique value of capacity for a facility [25].

Moreover, in many cases, the concept of capacity is confused with the breakdown flow rate, that is, the flow rate measured in the moment in which *breakdown* starts, i.e. when there is a transition from the uncongested to the congested state and a queue appears. Many real data demonstrate instead that the maximum throughput generally occurs much earlier than traffic breakdown.

Even if the capacity and the breakdown flow rate generally have different values and occur in different time instants, it is evident that the concept of capacity is strictly related with the phenomenon of traffic breakdown, because breakdown happens when the traffic demand exceeds the capacity. Hence, it is also necessary to give a precise definition of breakdown. To do that, first of all, the location where congestion starts need to be identified and this is the so-called *bottleneck* location. If instead in a given location there is congestion due to a downstream bottleneck, this is not a breakdown event but simply a queue spillback. Breakdown is usually associated with an amount of sudden drop in traffic flow speed when traffic demand exceeds capacity. Many research studies try to quantify the amount of speed drop and its duration necessary to cause a breakdown (see, e.g. [26, 27]).

Different types of bottlenecks exist in freeways, and they can be either permanent or temporary. Typical permanent bottlenecks are on-ramps and off-ramps, road restrictions and curves, uphill and downhill gradients. Examples of temporary bottlenecks are lane closures (or, in the worst case, road closures) due to accidents in the road. Bottlenecks may show two different states, active or inactive [28]. More specifically, a bottleneck is *activated* when breakdown occurs, i.e. the bottleneck has free-flow conditions downstream and congested conditions upstream. The performance of an active bottleneck is not affected by any bottlenecks occurring downstream [29, 30].

Figure 2.9 shows an example of time series of speed and flow in a given location where breakdown occurs. These series are referred to a time period of 5 h from 5 to 10 p.m., with a sample time of 5 min. The mean speed is higher than 100 km/h before 6.00, then it is between 90 and 100 km/h between 6.00 and 6.30, after which the sudden drop in speed occurs, until reaching a value lower than 40 km/h between 7.30 and 7.45. This drop in speed corresponds to the breakdown, i.e. to the transition from uncongested to congested conditions. After 7.45, speed starts increasing until

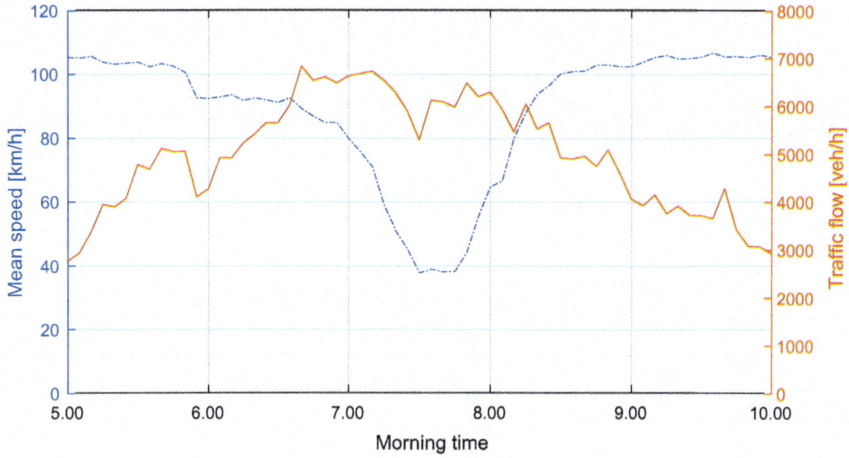

Fig. 2.9 Example of speed and flow time series during the morning

reaching an average value of about 105 km/h between 9.00 and 10.00. By analysing the time series of flow, it is possible to observe that the highest value of the flow is equal to 6836 veh/h and occurs at 6:40, much before the time of breakdown.

2.2.2 *Shock Waves*

A very important feature of traffic dynamics is the fact that traffic conditions change over space and time and it is often possible to observe discontinuities in flow and density. A *shock wave* is defined as the boundary in the time–space domain which constitutes a discontinuity in flow–density conditions [31]. Shock wave analysis is a key topic in traffic flow theory, dating back to the 50s [32], and is a quantitative study about the propagation of changes in flow or density. An important aspect which has been deeply investigated by researchers is the speed associated with such propagation. Shock waves are sometimes called also *shock fronts* or, simply, *shocks*.

Shock waves can be mild phenomena, for instance, when a platoon of high-speed vehicles reaches a platoon of slower vehicles, or significant events, for instance, when vehicles travelling fast approach a queue of stopped vehicles. An easy way to observe a shock wave is to look down the freeway traffic from a high point of view in the moment in which congestion is forming, for instance, for an accident: the wave of brake lights moving backwards (due to vehicles braking for the downstream congestion) is a shock wave, i.e. it represents the boundary between different traffic densities.

Shock waves were analysed in detail in [31], where a classification in six specific typologies is proposed and deeply discussed. Traffic shock waves can be classified

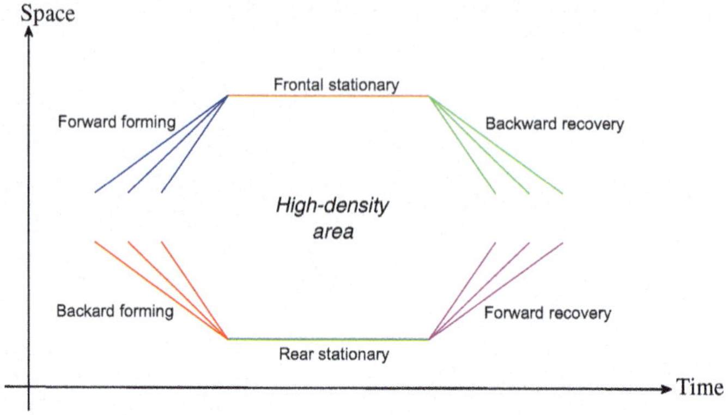

Fig. 2.10 Classification of shock waves

based on their direction and depending on the fact that they produce a tendency to congestion or to dissipation. Figure 2.10 sketches the six types of typical shock waves, which are backward forming, forward recovery, backward recovery, forward forming, rear stationary and frontal stationary shock waves.

The *backward forming* shock wave is the most obvious typology. It is normally associated with a bottleneck, which can be stationary, temporary (e.g. because of an accident or some maintenance works) or moving. It corresponds to a situation in which the demand is higher than the capacity and there is a queue formation. In this case, the shock wave front propagates backwards, i.e. in the opposite direction of traffic.

The *forward recovery* shock wave happens when there has been congestion but the demand tends to decrease below the bottleneck capacity, so that the length of congestion is going to reduce. In this case, the shock wave propagates forward, in the same direction of traffic, with free-flow conditions occurring farther and farther downstream over time.

The *backward recovery* shock waves occur when a congestion dissipates in such a way that the shock front propagates backwards. This happens in case of dissolving queue of a stationary or temporary bottleneck, for instance, when the highway original capacity is restored after an accident and the vehicles at the front of the congested region start to accelerate, progressively reducing the region density.

The *forward forming* shock waves are less frequent to observe and may occur in case of a moving bottleneck, for instance, when trucks are slowing down due to the grade of the roadway and the road geometry is such that there are limited overtaking opportunities for the other vehicles.

A *rear stationary* shock wave occurs when the flow of the arriving vehicles is equal to the capacity of the congested region. In this case, the shock wave with higher densities downstream and lower densities upstream is stationary, i.e. its position does not change over time. In other words, the tail of the queue remains at the same place as time elapses.

A *frontal stationary* shock wave is present at a bottleneck location where traffic demand exceeds capacity (both for recurrent and non-recurrent congestion situations). In that case, the front of the congested region is stationary, i.e. it does not change location over time. This stationary front, with lower densities downstream and higher densities upstream, normally happens to the head of a queue in case of stationary or temporary bottleneck or for a queue of vehicles stopped at a traffic light.

2.2.3 Phantom Traffic Jams

In Sect. 2.2.1, it has been discussed that breakdowns occur in bottleneck locations because of high traffic demands. However, the experience of drivers suggests that sometimes jams occur also without bottlenecks. These phenomena are often called *phantom traffic jams* and correspond to those situations, often experienced in highways, characterised by evident slowdowns for no apparent reason and without any obstacles in the road.

This situation creates *stop-and-go waves* in the traffic flows, which are caused by delays of the drivers in adapting their driving behaviours to the present traffic conditions. This phenomenon is particularly evident when traffic densities exceed a given threshold and is a consequence of finite acceleration and braking capabilities of the vehicles, as well as finite reaction times of the drivers. Let us consider a platoon of vehicles with a rather high density, i.e. with a rather short headway between vehicles: if the first vehicle brakes, the following one brakes as well and, if the distance from the vehicle in front is rather small, the following driver brakes more than requested for a sort of overreaction phenomenon. The following vehicles have a similar behaviour until the vehicle which stops completely, giving rise to stop-and-go situations. In these cases, there are local peaks of high traffic density in a freeway portion where the average traffic density is not so high. Of course, the frequency and intensity of phantom traffic jams depend on the type of drivers and on the traffic conditions.

A famous experiment showing real evidence for this phenomenon was carried out in Japan and described in [33]. The experiment was performed on a circular road 230 m long, with homogeneous lane conditions and flat ground. On this road, 22 vehicles were initially put equidistantly and the drivers were requested to drive at the same fixed speed of about 30 km/h, hence maintaining the same distance from the preceding vehicle. After some time in which vehicles were able to travel at a constant speed, small fluctuations appeared in the headway distances and grew until several vehicles were forced to stop completely for some time. This experiment showed that stop-and-go waves are generated in real traffic conditions and propagate backwards even in case drivers are asked to try to avoid them.

A more recent and very interesting experiment, inspired by the one described in [33], has been carried out in the U.S. in order to evaluate the impact of controlled autonomous vehicles on the dissipation of stop-and-go waves (see [34] and the references therein). Specifically, the experiment involved more than 20 vehicles in a circular road, among which one vehicle was controlled with simple speed control strategies. This allowed to alleviate stop-and-go phenomena, leading to consequent benefits in terms of lower speed variations and decreased fuel consumptions for all the vehicles involved. This experiment seems to suggest relevant future developments in traffic control techniques with autonomous vehicles as mobile actuators, even in case of low penetration rates.

2.2.4 Capacity Drop

Capacity drop phenomena are directly related to bottlenecks. As aforementioned, a bottleneck may be active or inactive. When it is active, it is characterised by free-flow conditions downstream and congested conditions upstream. When a bottleneck activates, a drop in the capacity occurs, i.e. the discharge flow of vehicles from the queue can be substantially lower than the flow measured before the queue formation. In other words, even though the demand upstream the bottleneck is higher than the capacity, the flow exiting the bottleneck is not the maximum supported capacity but is lower. The difference between the free-flow capacity of the bottleneck and the congested capacity of the bottleneck is often called *capacity drop*.

In the literature, different works have been published about the measurement and quantification of the capacity drop. Quantitatively, the measured capacity drops vary from place to place and generally range between very small (almost negligible) values up to more than 20%. The causes of this phenomenon are related by researchers to microscopic phenomena, such as lane-changing manoeuvres, low-speed merging behaviours and bounded acceleration capabilities (see, e.g. [35, 36]).

The first empirical studies on this phenomenon can be found in [17, 37], where a test case in the area of San Diego, U.S., is studied, and in [38], where the capacity drop phenomenon is measured and analysed in a test case in Toronto, Canada. In [18], capacity drops are measured by analysing time series of outputs or cumulative outputs at the bottleneck, considering two test cases in the metropolitan area of Toronto, Canada.

More recent works have been devoted to study how the bottleneck geometries can vary the mechanism of the capacity drop. In [39], real data of three different bottlenecks are analysed in detail, referring, respectively, to an on-ramp merge, a reduction of lanes and a horizontal curve. The authors of that work observe, in each of the bottlenecks, a strong relation between density and capacity drop, because capacity drop can be recovered once densities near the bottleneck diminish. As a consequence, it is clear that capacity drop phenomena may be avoided with specific control actions which maintain density under specified thresholds. A similar analysis was performed in [40], where the possibility of ramp metering control to increase maximum flow rates by preventing capacity drop is investigated. Analogously, in [28], the authors analyse whether ramp metering might increase the capacity of active freeway bottlenecks, reducing or even avoiding the drop near bottlenecks, by means of an extensive test referred to 27 active freeway bottlenecks in the Twin Cities, Minnesota, U.S., analysed for 7 weeks without ramp metering and 7 weeks with ramp metering control. Other empirical observations of capacity drops in freeway merges are reported and discussed in [41], where phase diagrams are adopted to quantitatively estimate the capacity drop.

Other research works dealt with the derivation of an analytical expression to estimate the capacity drop (see, e.g. [42, 43]). A recent study in this direction is [44], where previous works are extended by proposing a generic framework which takes into account the presence of heterogeneous vehicles with different characteristics and the physical interactions related to vehicles entering in merging areas.

2.3 Classification of Traffic Models

The development of models to describe the dynamics of vehicles and their interaction (including the traffic phenomena described above) is fundamental for the design, prediction and control of traffic systems. Since the first traffic model developed by Lighthill and Whitham in 1955 [45], there has been a growing interest of the scientific community for modelling the dynamic behaviour of traffic systems. This results in a high number of traffic models which have been developed during the years, with different levels of detail, and are now available to scientists and traffic managers. Of course, in this huge variety of models, the most appropriate to be chosen for a given real case basically depends on the type and scope of the considered application.

The adoption of dynamic traffic models can be useful in different contexts and can be applied for different purposes, such as

- *design of new road facilities and infrastructures*: the models can be used to simulate different design options, in different traffic scenarios, and to make what-if-analyses;
- *testing and evaluation of traffic control measures*: when new control rules are defined, they need to be evaluated to assess their performance, and this is done by making simulation tests in which the traffic system behaviour, under the application of these control rules, is simulated with appropriate traffic models;
- *prediction of short-term traffic conditions*: traffic models can be used to make a forecast of the traffic state over a short-term horizon, in order to provide the drivers with reliable information;
- *definition of prediction modules to be embedded in specific tools*: traffic models can be used to make predictions in model-based approaches, for monitoring, control and estimation purposes, which act in real time.

Several criteria may be applied for classifying traffic models, as highlighted, for instance, in the surveys [46, 47]. The most common classification of traffic models is related to their *level of detail*, distinguishing among microscopic, mesoscopic and macroscopic models. Another relevant distinction is associated with the continuous or discrete nature of the variables representing *space and time*. Of course, other classifications can be made, associated with other aspects. Figure 2.11 shows a classification of traffic models according to the two main criteria, i.e. the level of

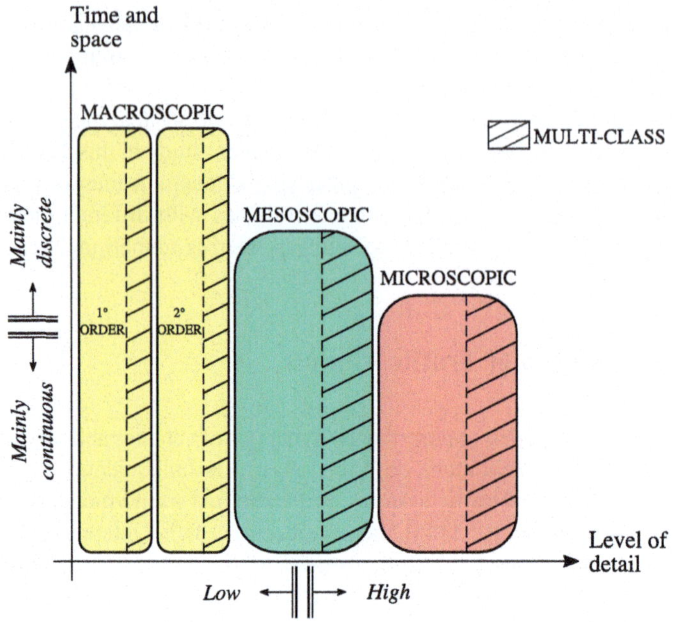

Fig. 2.11 Classification of traffic models according to their level of detail and their time and space representation

detail, as well as the time and space representation. It is worth noting that, for each class of models, it is possible to define a more detailed *multi-class* version, in which vehicles are categorised in different typologies.

2.3.1 Microscopic, Mesoscopic and Macroscopic Models

A possible classification of traffic models is made according to their level of detail, corresponding to the following cases:

- *microscopic models*, in which the dynamics of all vehicles and their interactions are represented in detail. Normally each vehicle is described with a dynamic model, with different parameters (representing, for instance, the desired speed or acceleration capabilities of the vehicle, as well as the aggressiveness and reaction times of the driver). These types of models are very detailed and, consequently, can be computationally intensive for representing large road networks. Microscopic models are often embedded in simulation software tools;
- *macroscopic models*, in which the traffic dynamics is represented at an aggregate level. Specifically, the flow of vehicles is seen as a unique stream, in analogy with the flow of fluids or gases, and its dynamics is described by means of aggregate

variables, such as density, mean speed and flow. These models are less computationally intensive than microscopic models and can allow fast simulations also for large-scale traffic networks. Macroscopic models are further classified according to the number of state variables, in first-order, second-order or higher-order models, and according to the number of represented vehicle types, in one-class or multi-class models;

- *mesoscopic models*, which present an intermediate level of detail. They do not distinguish individual vehicles but represent the heterogeneity of the drivers' choices in probabilistic terms.

In some cases, a further category is introduced, i.e. *submicroscopic models*, corresponding to a very high descriptive detail, in which also vehicle subunits are modelled.

As aforementioned, macroscopic models require a lower computational effort compared to microscopic models. In addition, it is normally easier to calibrate macroscopic models than microscopic models, since the former are characterised by fewer parameters. For these reasons, macroscopic models are surely the most suitable choice when they are used for control purposes, especially if the control law must be computed in real time, as, for instance, in predictive control schemes.

An interesting analysis of macroscopic traffic models is reported in [48] with reference to the general theory of model developments applied to scientific contexts. Generally speaking, it is possible to distinguish between deductive and inductive modelling approaches. On the one hand, purely *deductive* approaches (white box modelling) apply known physical laws, e.g. in Newtonian physics. On the other hand, *inductive* approaches (black box modelling) apply generic parametrised models, including no physical representations, which are fitted to real data. Among the two, there is an intermediate approach (grey box modelling), according to which a mathematical model with a physical interpretation is developed and the parameters of the model are calibrated with real data.

As specified in [48], even if most of macroscopic models are derived from microscopic observations and physical considerations, the resulting macroscopic models, due to the heterogeneity of drivers and vehicles, need to be calibrated using real data. Hence, in [48] it is argued that macroscopic traffic flow models cannot reach the same level of accuracy achieved in other science areas (e.g. Newtonian physics of thermodynamics), since the only accurate physical relation is represented by the conservation equation, while the other mathematical relations included in these models are the result of analogies and elaborations of empirical evidence. Nevertheless, macroscopic models are generally the most adequate for representing traffic dynamics in large systems, such as in freeways or extra-urban roads, while microscopic models can be more suitable for specific cases of urban traffic (such as road intersections).

The chapters of the book regarding traffic models follow this classification. In particular, Chaps. 3 and 4 are completely dedicated to macroscopic models, respectively, of first-order and second-order type, which are the most relevant class of models for the purposes of this book, mainly referred to freeway traffic control. Then, Chap. 5 includes a description of mesoscopic and microscopic models that are treated in a less specific way, since they are rarely used for control purposes in freeway traffic systems.

2.3.2 *Continuous and Discrete Models*

Another very relevant classification of traffic models regards the definition of the adopted independent variables. Indeed, all traffic models describe the evolution of the traffic system, both in space and in time, and these independent variables can be either continuous or discrete. Hence, it is possible to distinguish

- *continuous models*, in which space and time are considered as continuous variables and, consequently, the dynamics of the system is represented with differential equations;
- *discrete models*, in which space and time are discretised and the system dynamics is given by difference equations. Specifically, a road traffic system is divided into a set of road portions with finite length, and the time horizon is subdivided into a given number of time intervals.

Note that traffic engineers generally adopt discrete models instead of continuous models, considering a resolution in the order of 500 m for the space discretisation and 10–15 s for the time discretisation [48]. In particular, discrete models are normally adopted when embedded in traffic control systems. There are also some mixed models, such as the cluster models of mesoscopic time, described in Chap. 5, which are discrete in space and continuous in time.

In this book, Chaps. 3 and 4 describe macroscopic models both of continuous and discrete type, while in Chap. 5 mesoscopic and microscopic models are, again, either continuous or discrete.

2.3.3 *Other Classifications*

Besides the classifications previously reported, traffic models may be also categorised considering other criteria.

For instance, it is possible to distinguish between *deterministic* and *stochastic* models. In the former case, the relations among the variables are deterministic, i.e. if a given road stretch is simulated twice starting from the same initial conditions and considering the same boundary conditions, the evolution of the system is the same. The macroscopic traffic flow models described in Chaps. 3 and 4 are examples of deterministic models. Stochastic traffic models include some stochastic variables; hence, different simulations of a given road stretch with the same initial and boundary conditions will provide different results. Modelling stochastic processes can increase the descriptive power of traffic models, at the expense of higher computational times and increased difficulty in estimating the associated probability density functions. Some of the microscopic and mesoscopic models described in Chap. 5 have a stochastic nature.

References

1. Greenshields BD, Bibbins JR, Channing WS, Miller HH (1935) A study of traffic capacity. Highw Res Board 14:448–477
2. Greenberg H (1959) An analysis of traffic flow. Oper Res 7:79–85
3. Underwood RT (1961) Speed, volume and density relationships: quality and theory of traffic flow. Yale Bureau of Highway Traffic, Connecticut, pp 141–188
4. Newell GF (1961) Nonlinear effects in the dynamics of car following. Oper Res 9:209–229
5. Drake JS, Schofer JL, May AD (1967) A statistical analysis of speed-density hypotheses in vehicular traffic science. Highw Res Rec Proc 154:112–117
6. Chiabaut N, Buisson C, Leclercq L (2009) Fundamental diagram estimation through passing rate measurements in congestion. IEEE Trans Intell Transp Syst 10:355–359
7. Ji Y, Daamen W, Hoogendoorn S, Hoogendoorn-Lanser S, Qian X (2010) Investigating the shape of the macroscopic fundamental diagram using simulation data. Transp Res Rec 2161:40–48
8. Wang H, Li J, Chen QY, Ni D (2011) Logistic modeling of the equilibrium speed-density relationship. Transp Res Part A 45:554–566
9. Carey M, Bowers M (2012) A review of properties of flow-density functions. Transp Rev 32:49–73
10. Newell GF (1962) Theories of instability in dense highway traffic. J Oper Res Soc Jpn 5:9–54
11. Zhang HM (1999) A mathematical theory of traffic hysteresis. Transp Res Part B 33:1–23
12. Laval JA (2011) Hysteresis in traffic flow revisited: an improved measurement method. Transp Res Part B 45:385–391
13. Kerner BS (2004) The physics of traffic. Springer, Heidelberg
14. Kerner BS (2004) Three-phase traffic theory and highway capacity. Phys A 333:379–440
15. Schönhof M, Helbing D (2009) Criticism of three-phase traffic theory. Transp Res Part B 43:784–797
16. Treiber M, Kesting A, Helbing D (2010) Three-phase traffic theory and two-phase models with a fundamental diagram in the light of empirical stylized facts. Transp Res Part B 44:983–1000
17. Banks JH (1990) Flow processes at a freeway bottleneck. Transp Res Rec 1287:20–28
18. Cassidy MJ, Bertini RL (1999) Some traffic features at freeway bottlenecks. Transp Res Part B 33:25–42
19. Geroliminis N, Daganzo CF (2008) Existence of Urban-scale macroscopic fundamental diagrams: some experimental findings. Transp Res Part B 42:759–770
20. Geroliminis N, Sun J (2011) Properties of a well-defined macroscopic fundamental diagram for urban traffic. Transp Res Part B 45:605–617
21. Geroliminis N, Sun J (2011) Hysteresis phenomena of a macroscopic fundamental diagram in freeway networks. Transp Res Part A 45:966–979
22. Gayah VV, Daganzo CF (2011) Clockwise hysteresis loops in the macroscopic fundamental diagram: an effect of network instability. Transp Res Part B 45:643–655
23. Elefteriadou L (2014) An introduction to traffic flow theory. Springer, Berlin
24. HCM 2010 - Highway capacity manual. Transportation Research Board of the National Academy of Sciences, Washington D.C., 2010
25. Elefteriadou L (2003) Defining, measuring and estimating freeway capacity. In: Proceedings of the 82nd annual meeting of the transportation research board
26. Lorenz M, Elefteriadou L (2000) A probabilistic approach to defining freeway capacity and breakdown. In: Proceedings of the fourth international symposium on highway capacity, pp 84–95

27. Shiomi Y, Yoshii T, Kitamura R (2011) Platoon-based traffic flow model for estimating breakdown probability at single-lane expressway bottlenecks. Procedia Soc Behav Sci 17:591–610
28. Zhang L, Levinson D (2010) Ramp metering and freeway bottleneck capacity. Transp Res Part A 44:218–235
29. Daganzo CF (1997) Fundamentals of transportation and traffic operations. Elsevier Science New York
30. Bertini RL, Leal MT (2005) Empirical study of traffic features at a freeway lane drop. J Transp Eng 131:397–407
31. May AD (1990) Traffic flow fundamentals. Prenctice-Hall, Englewood Cliffs
32. Richards PI (1956) Shock waves on the highway. Oper Res 4:42–51
33. Sugiyama Y, Fukui M, Kikuchi M, Hasebe K, Nakayama A, Nishinari K, Tadaki S-I, Yukawa S (2008) Traffic jams without bottlenecks - experimental evidence for the physical mechanism of the formation of a jam. New J Phys 10:033001
34. Stern RE, Cui S, Delle Monache ML, Bhadani R, Bunting M, Churchill M, Hamilton N, Haulcy R, Pohlmann H, Wu F, Piccoli B, Seibold B, Sprinkle J, Work DB (2017) Dissipation of stop-and-go waves via control of autonomous vehicles: field experiments. arXiv:1705.01693
35. Cassidy MJ, Ahn S (1934) Driver turn-taking behavior in congested freeway merges. Transp Res Rec 2005:140–147
36. Treiber M, Kesting A, Helbing D (2006) Understanding widely scattered traffic flows, the capacity drop, and platoons as effects of variance-driven time gaps. Phys Rev E 74:1–10
37. Banks JH (1991) The two-capacity phenomenon: some theoretical issues. Transp Res Rec 1320:234–241
38. Hall F-L, Agyemang-Duah K (1991) Freeway capacity drop and the definition of capacity. Transp Res Rec 1320:91–98
39. Chung K, Rudjanakanoknad J, Cassidy MJ (2007) Relation between traffic density and capacity drop at three freeway bottlenecks. Transp Res Part B 41:82–95
40. Banks JH (1990) The two-capacity phenomenon at freeway bottlenecks: a basis for ramp metering? Transp Res Rec 1320:83–90
41. Srivastava A, Geroliminis N (2013) Empirical observations of capacity drop in freeway merges with ramp control and integration in a first-order model. Transp Res Part C 30:161–177
42. Laval JA (1988) Stochastic processes of moving bottlenecks: approximate formulas for highway capacity. Transp Res Rec 2006:86–91
43. Leclercq L, Laval JA, Chiabaut N (2011) Capacity drops at merges: an endogenous model. Transp Res Part B 45:1302–1313
44. Leclercq L, Laval JA, Chiabaut N (2016) Capacity drops at merges: new analytical investigations. Transp Res Part C 62:171–181
45. Lighthill MJ, Whitham GB (1955) On kinematic waves II: a theory of traffic flow on long crowded roads. Proc Royal Soc A 229:317–345
46. Hoogendoorn SP, Bovy PHL (2001) State-of-the-art of vehicular traffic flow modelling. Proc Inst Mech Eng Part I: J Syst Control Eng 215:283–303
47. van Wageningen-Kessels F, van Lint H, Vuik K, Hoogendoorn SP (2015) Genealogy of traffic flow models. EURO J Transp Logist 4:445–473
48. Papageorgiou M (1998) Some remarks on macroscopic traffic flow modelling. Transp Res Part A 32:323–329

Part II
Freeway Traffic Modelling

Chapter 3
First-Order Macroscopic Traffic Models

3.1 Macroscopic Modelling Aspects

All macroscopic models, both of first-order type and of higher orders, describe the evolution of aggregate quantities referred to the traffic system over time. This means that two independent variables are involved, i.e. space and time. In *continuous traffic models*, these independent variables are assumed to be continuous, while they are discretised in *discrete traffic models*. In this latter case, a freeway stretch is divided into a number of small road portions, and the time horizon is subdivided into a given number of time intervals.

Let us now introduce the proper notation of macroscopic traffic models, specifically differentiated for the continuous and the discrete case.

3.1.1 The Continuous Case

Referring to a generic location x (in a given road, possibly composed of several lanes) and time t, the main aggregate variables considered in continuous macroscopic traffic models are:

- $\rho(x, t)$ is the traffic density [veh/km];
- $v(x, t)$ is the average speed [km/h];
- $q(x, t)$ is the traffic flow [veh/h].

A first relation constituting the basis of every macroscopic model is the *hydrodynamic equation*, which computes the flow as the product of mean speed and density, i.e.

$$q(x, t) = \rho(x, t)v(x, t) \tag{3.1}$$

A second relation is the *continuity equation* or *conservation equation*, directly derived from the conservation law of vehicle flows and expressed as

A. Ferrara et al., *Freeway Traffic Modelling and Control*, Advances in Industrial Control, https://doi.org/10.1007/978-3-319-75961-6_3

$$\frac{\partial \rho(x,t)}{\partial t} + \frac{\partial q(x,t)}{\partial x} = 0 \tag{3.2}$$

All the continuous macroscopic traffic models are based on (3.1) and (3.2) and differ for the other equations which relate the variables $\rho(x,t)$, $v(x,t)$ and $q(x,t)$. This chapter and the following one will introduce the most important continuous macroscopic models, respectively, of first-order and second-order type (see in particular Sects. 3.2 and 4.1). The interested reader can find an overview on continuous traffic models in [1].

As already discussed in Sect. 2.1.3, the theoretical relation between density and flow in steady-state conditions is the so-called *Fundamental Diagram*. This is a relation $Q\left(\rho(x,t)\right)$, which has to satisfy the following conditions

$$Q(0) = 0, \quad Q(\rho^{\max}) = 0, \quad \left.\frac{dQ(\rho)}{d\rho}\right|_{\rho=\rho^{\mathrm{cr}}} = 0 \tag{3.3}$$

where ρ^{cr} is the *critical density* [veh/km] and $\rho^{\max}$ is the *jam density* [veh/km]. Moreover, $q^{\max}$ is the *capacity* [veh/h].

Analogously, the steady-state relation between mean traffic speed and density is denoted with $V\left(\rho(x,t)\right)$ and must satisfy the following conditions

$$V(0) = v^{\mathrm{f}}, \quad V(\rho^{\max}) = 0, \quad \frac{dV(\rho)}{d\rho} \leq 0 \tag{3.4}$$

where v^{f} indicates the *free-flow speed* [km/h].

Different shapes of these steady-state relations have been proposed in the literature. The first types of diagrams were introduced by Greenshields in 1935 [2] and correspond to a linear form for $V\left(\rho(x,t)\right)$ and a parabolic form for $Q\left(\rho(x,t)\right)$, i.e.

$$\begin{aligned} V\left(\rho(x,t)\right) &= v^{\mathrm{f}}\left[1 - \frac{\rho(x,t)}{\rho^{\max}}\right] \\ Q\left(\rho(x,t)\right) &= \rho(x,t)v^{\mathrm{f}}\left[1 - \frac{\rho(x,t)}{\rho^{\max}}\right] \end{aligned} \tag{3.5}$$

In case relations (3.5) are applied, it holds by definition that $\rho^{\mathrm{cr}} = \frac{1}{2}\rho^{\max}$ and $q^{\max} = \frac{1}{4}v^{\mathrm{f}}\rho^{\max}$. An example of steady-state relations of type (3.5) is shown in Fig. 3.1.

Other possible shapes, widely used in the literature and adopted in the next chapters of this book, are

$$\begin{aligned} V\left(\rho(x,t)\right) &= v^{\mathrm{f}} \exp\left[-\frac{1}{a}\left(\frac{\rho(x,t)}{\rho^{\mathrm{cr}}}\right)^{a}\right] \\ Q\left(\rho(x,t)\right) &= \rho(x,t)v^{\mathrm{f}} \exp\left[-\frac{1}{a}\left(\frac{\rho(x,t)}{\rho^{\mathrm{cr}}}\right)^{a}\right] \end{aligned} \tag{3.6}$$

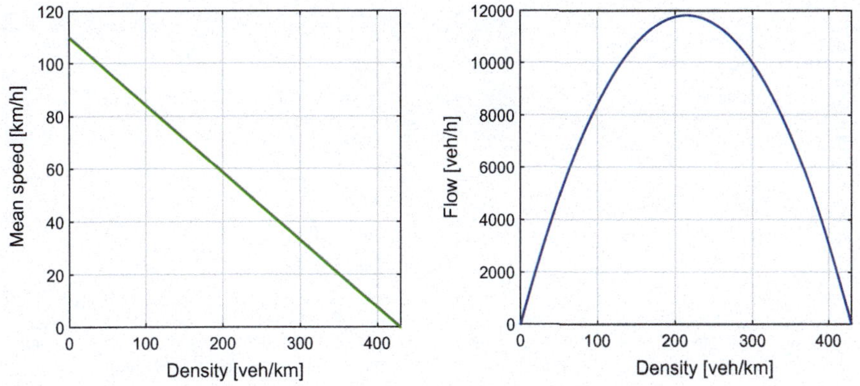

Fig. 3.1 Example of steady-state relations of type (3.5) with $v^{\mathrm{f}} = 110$ [km/h], $\rho^{\max} = 430$ [veh/km]

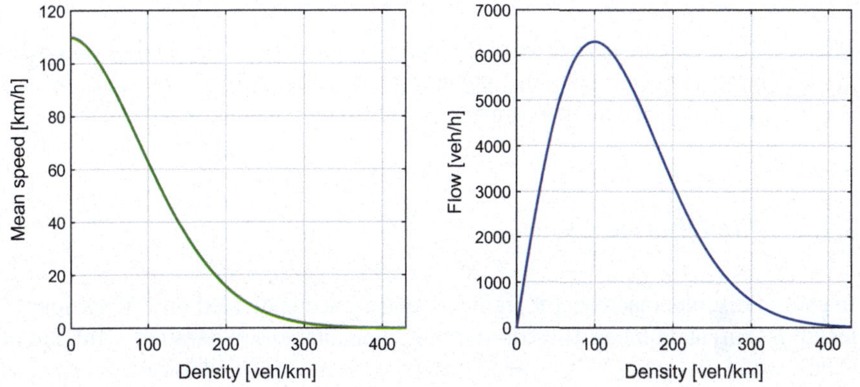

Fig. 3.2 Example of steady-state relations of type (3.6) with $v^{\mathrm{f}} = 110$ [km/h], $\rho^{\mathrm{cr}} = 100$ [veh/km], $a = 1.8$

where $a > 0$ is a suitable parameter. Note that these exponential relations do not meet the conditions $V(\rho^{\max}) = 0$ and $Q(\rho^{\max}) = 0$, respectively, in (3.4) and (3.3), but it holds that the values of $V(\rho^{\max})$ and $Q(\rho^{\max})$ in (3.6) are very small, hence in some way approximating conditions $V(\rho^{\max}) = 0$ and $Q(\rho^{\max}) = 0$. An example of steady-state relations of type (3.6) is reported in Fig. 3.2.

Finally, other common shapes of steady-state relations are

$$V\left(\rho(x,t)\right) = v^{\mathrm{f}}\left[1 - \left(\frac{\rho(x,t)}{\rho^{\max}}\right)^{l}\right]^{m}$$

$$Q\left(\rho(x,t)\right) = \rho(x,t)v^{\mathrm{f}}\left[1 - \left(\frac{\rho(x,t)}{\rho^{\max}}\right)^{l}\right]^{m} \tag{3.7}$$

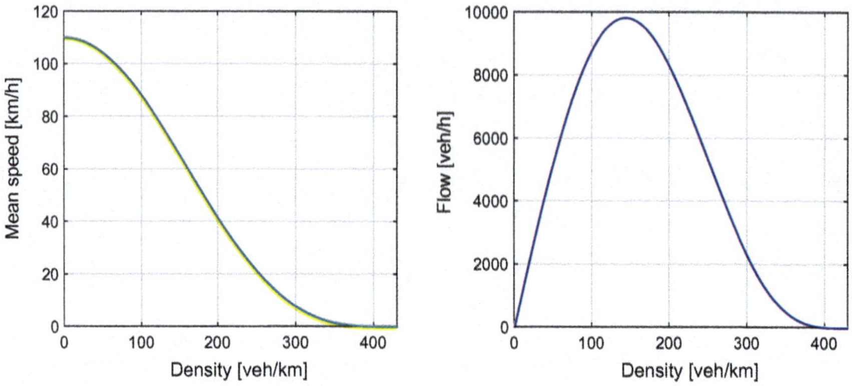

Fig. 3.3 Example of steady-state relations of type (3.7) with $v^{\mathrm{f}} = 110$ [km/h], $\rho^{\max} = 430$ [veh/km], $l = 2, m = 4$

where $l > 0, m > l$ are parameters. Note that (3.7) are very general and can represent most of the shapes reported in the literature, such as (3.6), for given values of l and m [3]. Figure 3.3 provides an example of steady-state relations of type (3.7).

3.1.2 *The Discrete Case*

In case of discrete macroscopic traffic models, space is divided into N portions of length L [km] and time is discretised into K time intervals of duration T [h]. Let us denote with $i = 1, \ldots, N$ the generic road portion (in some models called *cell* and in others called *section*), and with $k = 0, \ldots, K$ the generic time step. In some models, the space discretisation is not uniform, hence each portion i may have a different length $L_i, i = 1, \ldots, N$.

Referring to a generic portion i (in a road which can be composed of several lanes) and time step k, the main aggregate variables to be considered are:

- $\rho_i(k)$ is the traffic density at time kT [veh/km];
- $v_i(k)$ is the mean speed at time kT [km/h];
- $q_i(k)$ is the traffic flow during time interval $[kT, (k+1)T)$ [veh/h].

The hydrodynamic and continuity equations, in the discrete case, become

$$q_i(k) = \rho_i(k)v_i(k) \tag{3.8}$$

$$\rho_i(k+1) = \rho_i(k) + \frac{T}{L}[I_i(k) - O_i(k)] \tag{3.9}$$

where $I_i(k)$ is the traffic flow entering portion i during time interval $[kT, (k+1)T)$ [veh/h], and $O_i(k)$ is the traffic flow exiting portion i in the same time interval [veh/h].

Analogously to the continuous case, steady-state relations among flow, density and mean speed can be defined. In particular, relation (3.5) becomes

$$V\left(\rho_i(k)\right) = v^{\mathrm{f}}\left[1 - \frac{\rho_i(k)}{\rho^{\max}}\right]$$
$$Q\left(\rho_i(k)\right) = \rho_i(k) v^{\mathrm{f}}\left[1 - \frac{\rho_i(k)}{\rho^{\max}}\right] \tag{3.10}$$

Similarly, relation (3.6) can be written as

$$V\left(\rho_i(k)\right) = v^{\mathrm{f}} \exp\left[-\frac{1}{a}\left(\frac{\rho_i(k)}{\rho^{\mathrm{cr}}}\right)^{a}\right]$$
$$Q\left(\rho_i(k)\right) = \rho_i(k) v^{\mathrm{f}} \exp\left[-\frac{1}{a}\left(\frac{\rho_i(k)}{\rho^{\mathrm{cr}}}\right)^{a}\right] \tag{3.11}$$

and (3.7) as

$$V\left(\rho_i(k)\right) = v^{\mathrm{f}}\left[1 - \left(\frac{\rho_i(k)}{\rho^{\max}}\right)^{l}\right]^{m}$$
$$Q\left(\rho_i(k)\right) = \rho_i(k) v^{\mathrm{f}}\left[1 - \left(\frac{\rho_i(k)}{\rho^{\max}}\right)^{l}\right]^{m} \tag{3.12}$$

In some models, a different steady-state relation is considered for each road portion. In these cases, the parameters of the previous relations can be indexed with i, namely, $v_i^{\mathrm{f}}, \rho_i^{\max}, \rho_i^{\mathrm{cr}}, a_i, l_i, m_i, i = 1, \ldots, N$.

3.2 Continuous First-Order Models

The first macroscopic traffic model was developed by Lighthill and Whitham [4] and by Richards [5] in the 50s and is now known as the *Lighthill–Whitham–Richards* (LWR) model. The basic assumption of the LWR model is that vehicles adjust their speeds instantaneously to the value given by the steady-state relation depending on the present density. This model has been extended to consider boundary conditions, sources and inhomogeneities, as well as to represent traffic networks, as it will be described in the following subsections.

Most of the results on the LWR model have been obtained considering that it belongs to the class of conservation laws, for which a thorough theory has been developed by mathematicians (see, e.g. the books [6–10]). Considering more specifically the LWR model, especially its application for traffic networks, the interested reader can find more details in [11, 12], where all the related mathematical aspects are discussed in detail.

The LWR model presents several limitations. For instance, it does not contain any inertial effects, since it assumes that vehicles adjust their speeds instantaneously. This can produce unrealistically high accelerations or decelerations of vehicles. Moreover, it systematically predicts that the output flow from a congested area is equal to the capacity flow, if the portion of road downstream is not congested. This is in contrast with the capacity drop phenomenon observed in real-world traffic networks, as discussed in Sect. 2.2.4. Other qualitative considerations on first-order macroscopic models are included in [13].

3.2.1 The LWR Model

The LWR model is based on the assumption that the traffic flow instantaneously follows the density according to the Fundamental Diagram. The LWR model is then given by (3.1), (3.2), and the following relation:

$$v(x,t) = V\left(\rho(x,t)\right) \tag{3.13}$$

or, alternatively,

$$q(x,t) = Q\left(\rho(x,t)\right) \tag{3.14}$$

Hence, the LWR model can be rewritten as

$$\frac{\partial \rho(x,t)}{\partial t} + \frac{\partial (Q\left(\rho(x,t)\right))}{\partial x} = 0 \tag{3.15}$$

or

$$\frac{\partial \rho(x,t)}{\partial t} + \frac{\partial (\rho(x,t) V\left(\rho(x,t)\right))}{\partial x} = 0 \tag{3.16}$$

The LWR model belongs to the class of *first-order models*, in the sense that it captures the dynamics of a single variable, namely, the traffic density. Moreover, this model, in its original version, makes some assumptions on the shape of the Fundamental Diagram. Specifically, it assumes that $Q\left(\rho(x,t)\right)$ is a C^2 function, is strictly concave, and ensures that $Q(0) = 0$ and $Q(\rho^{\max}) = 0$, as in (3.3). According to these assumptions, (3.15) belongs to the class of *hyperbolic conservation laws*. The theory of systems of conservation laws has been extensively studied in the literature with particular attention to the problem of well-posedness, as done, for instance, in [14–18].

A very peculiar aspect associated with conservation laws is the generation of *discontinuities*. Referring in particular to the LWR model in the traffic case, the discontinuities resulting from the solution of the LWR model satisfactorily reproduce the shock waves which can be actually observed in traffic systems. These aspects were discussed in detail in [4, 5], where the wave theory is applied and the propagation of kinematic waves is discussed in detail, with reference to the real behaviour of road traffic systems. In [4], some preliminary comments to the traffic behaviour at road junctions are reported as well.

From a mathematical point of view, a relevant effort was put by researchers on solving the so-called LWR *initial value problem*, given by the conservation law (3.15) with a specified initial condition for the density $\rho(x, 0) = \rho_I(x)$. Referring to the general theory on conservation laws, it can be easily shown that the solution to this initial value problem can produce discontinuities in finite time, even in case of continuous initial conditions. This can be shown by applying the method of *characteristics*, allowing to rewrite the partial differential equation of the LWR model as a system of ordinary differential equations; the characteristics can be seen as lines in the (x, t) plane, starting from space–time points where initial conditions are known, along which the solution remains constant. If these lines do not intersect, the solution is unique; if instead they intersect, this means that there is a discontinuity (a shock) in the solution, and this is what normally happens with the LWR model. In this latter case, weak solutions must be dealt with (see, for instance, [11] for further mathematical details).

One of the most interesting cases to be analysed, especially when referring to traffic applications, is the solution of the LWR initial value problem in case the initial condition for the density $\rho_I(x)$ is piecewise constant. This corresponds, for instance, to the presence of vehicles waiting in front of a red traffic light: the density after the traffic light is low, while the density before is high. The opposite example is the case of queue formation for a red traffic light or for an accident in a freeway stretch: there is a point in space after which the density is very high and before which free-flow traffic conditions are present. The initial value problem in case of a discontinuous initial condition is called *Riemann problem*. Let us consider specifically the Riemann problem for (3.15) with the initial condition expressed as

$$\rho(x, 0) = \rho_I(x) = \begin{cases} \rho^- & \text{if } x < 0 \\ \rho^+ & \text{if } x > 0 \end{cases} \tag{3.17}$$

By applying the general results on conservation laws, it can be shown that this Riemann problem has not unique solution. The conventional mathematical approach to solve this problem is devoted to look for *entropy-admissible solutions* (see, for instance, [11] for a rigorous definition of this type of solutions), which present good properties, such as the uniqueness and the fact that they depend continuously on initial data. In [19], an interesting discussion about the choice of adopting entropy solutions for the LWR model is reported: the author explains, through an example, that the choice of the entropy solution is a mathematical sound choice, which guarantees existence, uniqueness and continuous dependency on initial conditions, but in some cases, these entropy solutions are not the best choice in order to provide a realistic behaviour of traffic.

The entropy-admissible solution for the Riemann problem for (3.15) with initial condition (3.17) can be written by distinguishing two cases:

1. if $\rho^- < \rho^+$, i.e. $\frac{dQ(\rho^-)}{d\rho} > \frac{dQ(\rho^+)}{d\rho}$ for the considered assumptions on $Q(\rho(x,t))$, the entropy-admissible solution is given by the *shock wave* expressed as

$$\rho(x,t) = \begin{cases} \rho^- & \text{if } x < \lambda t \\ \rho^+ & \text{if } x > \lambda t \end{cases} \tag{3.18}$$

where λ is obtained by applying the so-called Rankine–Hugoniot condition and is given by

$$\lambda = \frac{Q(\rho^+) - Q(\rho^-)}{\rho^+ - \rho^-} \tag{3.19}$$

This solution corresponds to a discontinuity, in which the density abruptly changes from ρ^- to ρ^+, propagating in space and time with speed λ, which then represents the *shock front propagation speed*;
2. if $\rho^- > \rho^+$, i.e. $\frac{dQ(\rho^-)}{d\rho} < \frac{dQ(\rho^+)}{d\rho}$, the entropy-admissible solution is given by the *rarefaction wave* expressed as

$$\rho(x,t) = \begin{cases} \rho^- & \text{if } x < \frac{dQ(\rho^-)}{d\rho} t \\ \left[\frac{dQ(\frac{x}{t})}{d\rho}\right]^{-1} & \text{if } \frac{dQ(\rho^-)}{d\rho} t < x < \frac{dQ(\rho^+)}{d\rho} t \\ \rho^+ & \text{if } x > \frac{dQ(\rho^+)}{d\rho} t \end{cases} \tag{3.20}$$

In this case, the solution is continuous, i.e. the propagation of the density in space and time occurs in a smooth way.

It is worth noting that, in case 1, the characteristics on the (x,t) plane overlap, as shown in the left graph of Fig. 3.4. Hence, the solution implies a discontinuity, which is highlighted in the right graph of Fig. 3.4, where the shock wave is represented by a dashed green line. Figure 3.5 shows instead the characteristics in case 2: they do not overlap; hence, there is a region in the (x,t) plane which appears to be empty. In that region, called expansion fan, characteristics are rays of constant density originating at $x = 0$ in order to guarantee continuity of the solution (see the right graph of Fig. 3.5).

The solution described so far holds for the case of strictly concave Fundamental Diagram. Nevertheless, also the case in which the Fundamental Diagram is *non-concave* can be interesting for real traffic applications. This case was treated in some research papers, such as in [20, 21].

Another version of the LWR model was introduced to overcome the fact that the LWR model produces discontinuities in finite time, leading to the so-called *LWR model with viscosity*. In this model, a viscosity term is added to (3.15), i.e.

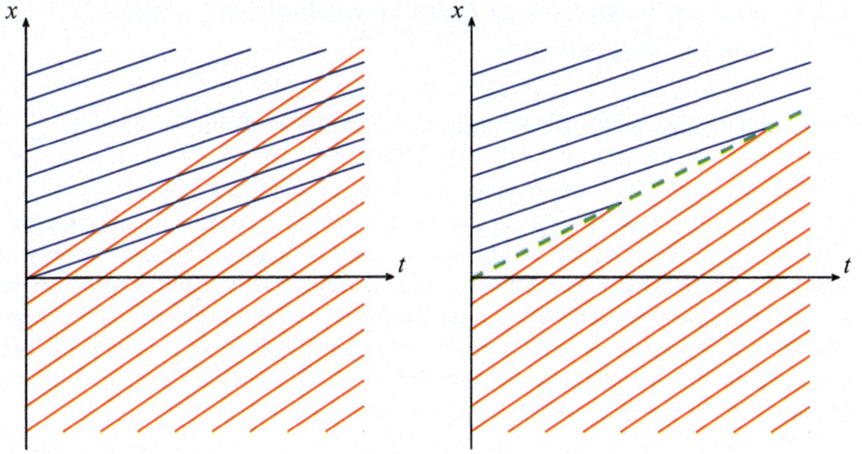

Fig. 3.4 Characteristics in case 1: shock wave

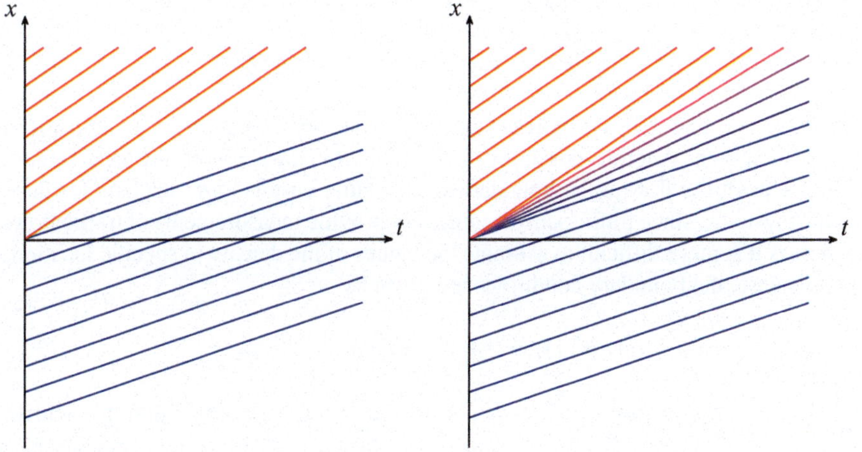

Fig. 3.5 Characteristics in case 2: rarefaction wave

$$\frac{\partial \rho(x,t)}{\partial t} + \frac{\partial (Q\left(\rho(x,t)\right))}{\partial x} = \mu \frac{\partial^2 \rho(x,t)}{\partial x^2} \tag{3.21}$$

and the discontinuities in the solution are eliminated. Nevertheless, in [11], it is shown that this model is not realistic to describe the traffic flow evolution.

3.2.2 *The LWR Model with Boundary Conditions, Sources and Inhomogeneities*

In the initial value problem described in Sect. 3.2.1, no boundary conditions are defined for the boundaries of the space domain. Clearly, this is somewhat unrealistic for a freeway traffic system, since normally a given road stretch is considered and the traffic conditions at the beginning and at the end of the stretch must be taken into account. To consider boundary conditions, an *initial-boundary value problem* must be addressed, in which the conservation law must satisfy, not only an initial condition but also the boundary conditions. Also, the initial-boundary value problem for the general class of hyperbolic conservation laws has been widely studied by mathematicians, with specific attention to the well-posedness of the problem, developing conditions for the existence and unicity of the solution (see, e.g. [18, 22, 23].

In the specific case of the LWR model for traffic systems, the initial-boundary value problem is given by (3.15), with the initial condition $\rho(x,0) = \rho_I(x)$ and boundary conditions that can be expressed in different ways. The boundary conditions can be related to the values of the density at the boundaries, i.e.

$$\rho(0,t) = \rho_0(t), \quad \rho(x_L,t) = \rho_L(t) \tag{3.22}$$

where $x = 0$ and $x = x_L$ indicate the initial and final location of the considered freeway stretch. Another possibility is that the boundary conditions are related to the values of the flow at the boundaries. This latter case is more realistic for many traffic systems, since traffic sensors generally provide measurements of traffic flows, whereas it is more difficult to estimate the values of the density in specific locations. In this case, the boundary conditions are given by

$$q\left(\rho(0,t)\right) = q_0(t), \quad q\left(\rho(x_L,t)\right) = q_L(t) \tag{3.23}$$

Some works in the literature deal with the initial-boundary value problem for the LWR model specifically referred to the case of freeway traffic. For instance, in [24], the boundary conditions are given in terms of traffic density, and the existence and uniqueness of a weak solution are proved. Also, the proposed numerical scheme is applied to a freeway scenario with data of the Interstate-80 Eastbound in West Berkeley and Emeryville, U.S. Another work dealing with the initial-boundary value problem for the LWR model is [25], referred to freeway stretches. In [25], the boundary conditions refer to the time-dependent flow entering a specific location, namely, $x = 0$. Moreover, some constraints on the flow at another specific location are included, namely, at $x = x_C$, modelling the presence of toll gates, construction sites or the occurrence of accidents, which limit the traffic flow. The boundary conditions are then expressed as

$$q\left(\rho(0,t)\right) = q_0(t), \quad q\left(\rho(x_C,t)\right) \leq \bar{q}_C(t) \tag{3.24}$$

where $\bar{q}_C(t)$ is the maximum flow allowed at $x = x_C$. The well-posedness result provided in [25] allows also to prove the existence of optimal management strategies for freeway traffic systems.

The LWR model described so far does not take into account the presence of on-ramps and off-ramps, which are instead a very important issue in modelling freeway stretches. In order to consider entrances of vehicles from on-ramps, exits from off-ramps, as well as local changes of the traffic flow due to inhomogeneities of the road, the LWR model must be written as a *conservation law with source* or *inhomogeneous conservation law*, i.e.

$$\frac{\partial \rho(x,t)}{\partial t} + \frac{\partial (Q\,(\rho(x,t)))}{\partial x} = s(x,t,\rho) \tag{3.25}$$

where $s(x, t, \rho)$ is the source term. Well-posedness results and numerical investigations for the inhomogeneous LWR model are presented in [26], where also second-order models with source terms are analysed.

Another interesting aspect to be included in the LWR model, relevant especially for real contexts, is related to consider the case in which the Fundamental Diagram depends explicitly and (sometimes discontinuously) on x and on t. This can allow to model intersections, sections with variable number of lanes, portions of the road with local and temporary variations of the parameters (e.g. capacity or free-flow speed). If the Fundamental Diagram only varies depending on time, the corresponding LWR model is said to present *time inhomogeneity*, whereas it has *space inhomogeneity* if the Fundamental Diagram only depends on space.

The LWR model with space inhomogeneity has been studied deeply and the well-posedness of the associated initial value problem (both for continuous and discontinuous dependence of the Fundamental Diagram on x) has been proven [19, 27]. The case of space–time inhomogeneity has been studied more recently, for example, in [28].

An alternative option for the solution of the LWR is given by the *Hamilton–Jacobi theory*, adopted, for instance, in [29, 30]. According to this theory, a *Lagrangian* approach, which is trajectory-based, is adopted, in contrast with the standard *Eulerian* framework used to solve conservation laws. This approach can assume particular relevance especially for the new type of sensors that are more and more widespread in freeway networks, i.e. mobile sensors which travel inside the domain along trajectories, providing *internal conditions* for the problem, in addition with boundary conditions provided by standard traffic sensors.

3.2.3 The LWR Model on Networks

In order to represent large-scale freeway systems, the LWR model has been extended to the case of *networks*, in which each road is modelled with the LWR model and specific conditions must be defined for the junctions where roads intersect. The first work in this direction was reported in [31], where a network of unidirectional roads is seen as a connected directed *graph*, with edges modelling the roads and vertices corresponding to junctions. The *junctions* play a key role in these network models, since at junctions the system is underdetermined even if the conservation of cars is taken into account; in other words, in order to obtain a well-defined solution, it is necessary to specify the distribution of vehicles at the junctions. In [31], the Riemann problem for the considered system is solved by *maximising the flow* at each intersection, and the existence of a solution to the general Cauchy problem is proven.

In [32], the road network is modelled as a graph, similarly to the case proposed in [31], but different conditions at junctions are taken into account. Specifically, it is assumed that there are some prescribed *preferences* of drivers, i.e. the traffic from incoming roads is distributed on outgoing roads according to fixed coefficients, and, by following these preferences, the drivers behave in order to maximise the traffic flows. Considering this model, the authors of [32] prove the existence of solutions to the Cauchy problem and show that the Lipschitz continuous dependence by initial data holds only under specific assumptions. Some other research works have dealt with developments of the LWR model on networks (see, e.g. [33, 34]), also considering specific types of junctions. For instance, the authors of [35] analyse the case of a T-junction, in which the interactions among incoming and outcoming flows are explicitly modelled. In [36], the specific case of freeways is addressed, and the considered junction is composed of the mainstream, an off-ramp and an on-ramp, this latter being modelled as a buffer of infinite capacity.

Some studies have also considered the case of *nodes with buffer*, i.e. the case in which there is a dynamics inside the junction, generally described by ordinary differential equations depending on incoming and outgoing flows. For instance, in [37], the storage capacity of the junctions is taken into account by using a reformulation of intersection models in terms of supply and demand functions. Similarly to [37], in [38], the solution of the Riemann problem at the node is provided and existence and well-posedness of solutions to the Cauchy problem are proven. A multi-buffer model is studied in [39], where a set of buffers, one for each outgoing road, is considered, allowing to correctly respect the preferences of drivers.

3.3 Discrete First-Order Models

Different numerical methods for non-linear conservation laws have been studied by researchers since many decades. While approximating a partial differential equation with a finite-difference equation, it is of course interesting to evaluate the error due to this approximation and to study relevant properties such as convergence and stability of the numerical method (see, e.g. [7, 40] for a detailed description of numerical methods for conservation laws). Moreover, to discretise partial differential equations, one can use both explicit and implicit numerical methods. With the *explicit* solution scheme, it is possible to explicitly express the dependence of each variable in the current time step on the variables in previous time steps. In the *implicit* case, instead, it is necessary to solve a system of equations involving both current and past values of the different variables (see, e.g. [41]). Implicit formulas are typically more stable than explicit ones, but harder to implement [42].

Referring to the specific case of traffic, i.e. to the LWR model, different finite-difference approximations have been proposed in the last decades. According to these numerical methods, the road space is divided into portions of finite length, time is discretised into time intervals of equal duration, and the partial differential equation of the LWR model is transformed into a finite-difference equation.

The most famous discretised version of the LWR model is the *Cell Transmission Model* (CTM), presented for the first time by Daganzo in [43, 44], making reference to a one-way road without any intermediate entrances or exits, and then extended in [45] for traffic networks with three-legged junctions, hence allowing to model on-ramps, off-ramps, freeway intersections and so on. According to the CTM, the discrete portions of the road are called *cells*, and two quantities are associated with the intersection between two cells, i.e. a *sending* function depending on the density before the intersection, and a *receiving* function depending on the density downstream the intersection.

In [43, 44], it is shown that the CTM is a discrete equivalent of the classical LWR model, both in case of continuous density and in presence of discontinuities. Moreover, the author of [43, 44] argues that the CTM could capture real-life features, such as stop-and-go phenomena, that the LWR theory is not able to model. This analysis is carried out by considering the specific case of triangular or trapezoidal Fundamental Diagram $Q(\rho)$, but it is asserted that the considerations reported in those papers can be generalised to other shapes of $Q(\rho)$. In [46], the propagation of disturbances of the CTM is also analysed and an asymptotic formula for the errors introduced by the finite difference approximation is presented.

A very interesting analysis of discretisation of first-order traffic flow models was conducted by Lebacque in [19], where he focuses on a specific numerical method, that is the so-called *Godunov scheme* [47]. This is a conservative finite-volume method which solves Riemann problems at each cell interface forward in time. In [19], it is shown that the CTM corresponds to the application of the Godunov scheme to the LWR model. In particular, the sending and receiving functions computed at the intersection between subsequent cells in the CTM are equivalent to the values of the flow at the singularity in the solution of the Riemann problem in the Godunov scheme. In [19], Lebacque introduces the terminology *demand* and *supply*, respectively, for the sending and receiving functions; this terminology is presently the most widespread when using the CTM and it is also the one adopted in this book.

By applying the Godunov scheme, a condition for the space discretisation L and the time discretisation T is also derived, which can be expressed as

$$T \max_{\rho \in [0, \rho^{\max}]} \left| \frac{dQ(\rho)}{d\rho} \right| \leq L \tag{3.26}$$

In [19], different shapes for the Fundamental Diagram $Q(\rho)$ (and consequently for the demand and supply functions) are investigated, also considering the case in which $Q(\rho)$ includes discontinuities, and it is argued that the concepts of supply and demand can provide an effective tool also for modelling intersections and networks.

Among the different discretisations of the LWR model proposed in the literature, it is worth mentioning also the *Link Transmission Model* (LTM), introduced for the first time in [48]. In the LTM, the evolution of traffic on a generic road network is represented in terms of the cumulative number of vehicles that pass the initial and final locations of each link at each time step. Hence, the numerical procedure characterising the LTM only requires calculations at the link boundaries, as in [49], instead of at each cell boundary, as in the CTM. This results in a computational advantage compared with the CTM. More efficient numerical schemes have been developed starting from the LTM, such as the iterative algorithm described in [50].

In the following subsections, we will focus on the CTM and its extensions, both for a freeway stretch and for a freeway network, since this is surely the most widespread first-order model in the traffic control engineering community, and, hence, of particular interest for the purposes of the present book.

3.3.1 The CTM for a Freeway Stretch

Let us start from the CTM for a freeway stretch including on-ramps and off-ramps. Note that the CTM described hereafter is derived from the original version proposed by Daganzo in [43–45], but it is presented with a different mathematical notation and nomenclature in order to conform to the notation and model classification adopted in this book.

As previously introduced in the general notation of a discrete traffic model, let N be the number of cells and K the number of time intervals. Let T denote the sample time [h] and L the length of each cell [km]. Moreover, in the CTM, on-ramps and off-ramps are assumed to be present at the interface between two subsequent cells.

For each cell $i = 1, \ldots, N$, and for each time step $k = 0, \ldots K$, let us define the following quantities:

- $\rho_i(k)$ is the traffic density of cell i at time kT [veh/km];
- $\Phi_i^+(k)$ is the total flow entering cell i during time interval $[kT, (k+1)T)$ [veh/h];
- $\Phi_i^-(k)$ is the total flow exiting cell i during time interval $[kT, (k+1)T)$ [veh/h];
- $\phi_i(k)$ is the mainstream (interface) flow entering cell i from cell $i-1$ during time interval $[kT, (k+1)T)$ [veh/h];
- $r_i(k)$ is the flow entering cell i from the on-ramp during time interval $[kT, (k+1)T)$ [veh/h];
- $s_i(k)$ is the flow exiting cell i through the off-ramp during time interval $[kT, (k+1)T)$ [veh/h];
- $\beta_i(k) \in [0, 1)$ is the split ratio of cell i during time interval $[kT, (k+1)T)$;

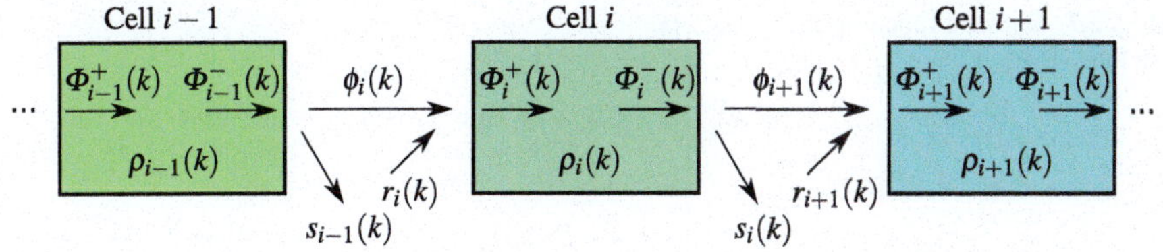

Fig. 3.6 Sketch of the division of the freeway stretch into cells and the relative notation in the CTM

- $D_i(k)$ is the demand of cell i (i.e. flow that can be sent from cell i to cell $i+1$) during time interval $[kT, (k+1)T)$ [veh/h];
- $S_i(k)$ is the supply of cell i (i.e. flow that can be received by cell i from cell $i-1$) during time interval $[kT, (k+1)T)$ [veh/h];
- $D_i^{\text{ramp}}(k)$ is the demand of the on-ramp of cell i (i.e. flow that can be sent from the on-ramp into cell i) during time interval $[kT, (k+1)T)$ [veh/h].

Figure 3.6 depicts a sketch of the subdivision of the freeway stretch into cells, with the main variables of the CTM.

The parameters of the CTM are as follows: v_i is the free-flow speed of cell i [km/h], w_i is the congestion wave speed of cell i [km/h], $q_i^{\max}$ is the capacity of cell i [veh/h], $\rho_i^{\max}$ is the jam density of cell i [veh/km], $p_i^{\text{ramp}} \in [0, 1]$ is the priority of the on-ramp flow with respect to the mainstream flow in cell i, $p_i \in [0, 1]$ is the priority of the mainstream flow with respect to the on-ramp flow in cell i, such that $p_i^{\text{ramp}} + p_i = 1, i = 1, \ldots, N$.

The CTM is characterised by the following equations describing the traffic density, this latter being the state variable of dimension N:

$$\rho_i(k+1) = \rho_i(k) + \frac{T}{L}\left[\Phi_i^{+}(k) - \Phi_i^{-}(k)\right] \tag{3.27}$$

where $i = 1, \ldots, N, k = 0, \ldots, K-1$, and the total flows entering and exiting cell i are, respectively, given by

$$\Phi_i^{+}(k) = \phi_i(k) + r_i(k) \tag{3.28}$$

$$\Phi_i^{-}(k) = \phi_{i+1}(k) + s_i(k) \tag{3.29}$$

The flow exiting through the off-ramp is computed as

$$s_i(k) = \frac{\beta_i(k)}{1-\beta_i(k)}\phi_{i+1}(k) \tag{3.30}$$

since $s_i(k) = \beta_i(k)\Phi_i^{-}(k) = \beta_i(k)[\phi_{i+1}(k) + s_i(k)]$.

Two important concepts of the CTM are the *demand* and the *supply*, associated with each cell. In particular, referring to the boundary between cell $i-1$ and cell i,

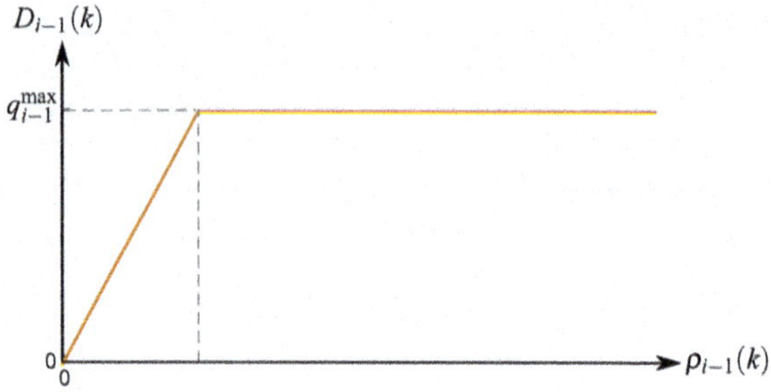

Fig. 3.7 Demand function in the CTM

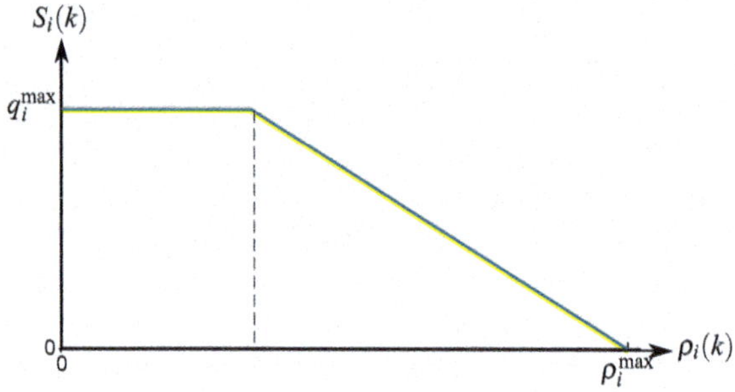

Fig. 3.8 Supply function in the CTM

let us introduce the demand of cell $i-1$, namely, $D_{i-1}(k)$, and the supply of cell i, namely, $S_i(k)$. The demand $D_{i-1}(k)$ is the flow that cell $i-1$ could send to the next cell i during time interval $[kT, (k+1)T)$, while the supply $S_i(k)$ is the flow that cell i could receive from cell $i-1$ in the same time interval. These two quantities are computed as

$$D_{i-1}(k) = \min\left\{(1-\beta_{i-1}(k))v_{i-1}\rho_{i-1}(k), q_{i-1}^{\max}\right\} \tag{3.31}$$

$$S_i(k) = \min\left\{w_i(\rho_i^{\max} - \rho_i(k)), q_i^{\max}\right\} \tag{3.32}$$

The demand and the supply are shown in Figs. 3.7 and 3.8, respectively, as functions of the density.

The *merge* between the on-ramp and the mainstream is analogous to the merge of two generic cells, as described in the original model proposed in [45]. In the generic case described in [45], a merge is given by two sending cells (characterised by two specific demands) and one receiving cell (characterised by a given supply); according to [45], the two sending cells send the maximum possible flow that the receiving cell is able to host. This merge model is adopted to compute the mainstream and on-ramp flows, since this latter case can be seen as a situation of two sending cells (the mainstream and the on-ramp) and one receiving cell downstream. In particular, for a given cell i during time interval $[kT, (k+1)T)$, the demands of the sending cells are $D_{i-1}(k)$ and $D_i^{\text{ramp}}(k)$, while the supply of the receiving cell is $S_i(k)$.

Two cases must be distinguished, corresponding, respectively, to free-flow and congested conditions.

Free-Flow Case This is the case in which there is enough space for the two flows that want to enter cell i, i.e.

$$\begin{aligned}&\text{If}\quad D_{i-1}(k)+D_i^{\text{ramp}}(k)\leq S_i(k)\\&\text{then}\quad \phi_i(k)=D_{i-1}(k),\quad r_i(k)=D_i^{\text{ramp}}(k)\end{aligned}\tag{3.33}$$

Congested Case The congested case is the opposite situation in which not all the flows that want to enter cell i can be received by it, i.e.

$$\begin{aligned}&\text{If}\quad D_{i-1}(k)+D_i^{\text{ramp}}(k)>S_i(k)\\&\text{then}\quad \phi_i(k)=\text{mid}\left\{D_{i-1}(k),\,S_i(k)-D_i^{\text{ramp}}(k),\,p_iS_i(k)\right\}\\&\qquad r_i(k)=\text{mid}\left\{D_i^{\text{ramp}}(k),\,S_i(k)-D_{i-1}(k),\,p_i^{\text{ramp}}S_i(k)\right\}\end{aligned}\tag{3.34}$$

where the function mid returns the middle value.

In order to better understand the merge model in the congested case, remind that the case $D_{i-1}(k)+D_i^{\text{ramp}}(k)\geq S_i(k)$ corresponds to a situation in which it is not possible to completely satisfy the demand $D_{i-1}(k)$ from the mainstream and the demand $D_i^{\text{ramp}}(k)$ from the on-ramp. Moreover, remind that parameters p_i and p_i^{ramp} model, respectively, the priority of the mainstream flow and the on-ramp flow in the merge and that $p_i^{\text{ramp}}+p_i=1$.

The basic idea of the merge model is that the demand $D_{i-1}(k)$ has a 'reserved' flow equal to $p_iS_i(k)$, while the demand $D_i^{\text{ramp}}(k)$ has a 'reserved' flow of $p_i^{\text{ramp}}S_i(k)$. Another important assumption of the merge model proposed in [45] is that if one of the two demands is lower than the corresponding 'reserved' flow, the complementary flow will saturate the supply of the receiving cell.

By rewriting (3.34) as

$$\begin{aligned}&\text{If}\quad S_i(k)-D_i^{\text{ramp}}(k)\leq p_iS_i(k)\leq D_{i-1}(k)\\&\text{then}\quad \phi_i(k)=p_iS_i(k),\quad r_i(k)=p_i^{\text{ramp}}S_i(k)\\&\text{If}\quad p_iS_i(k)\leq S_i(k)-D_i^{\text{ramp}}(k)\leq D_{i-1}(k)\\&\text{then}\quad \phi_i(k)=S_i(k)-D_i^{\text{ramp}}(k),\quad r_i(k)=D_i^{\text{ramp}}(k)\\&\text{If}\quad S_i(k)-D_i^{\text{ramp}}(k)\leq D_{i-1}(k)\leq p_iS_i(k)\\&\text{then}\quad \phi_i(k)=D_{i-1}(k),\quad r_i(k)=S_i(k)-D_{i-1}(k)\end{aligned}\tag{3.35}$$

it is possible to distinguish three sub-cases:

- if $D_{i-1}(k)\geq p_iS_i(k)$ and $D_i^{\text{ramp}}(k)\geq p_i^{\text{ramp}}S_i(k)$, then the 'reserved' flows are guaranteed, resulting in $\phi_i(k)=p_iS_i(k)$ and $r_i(k)=p_i^{\text{ramp}}S_i(k)$;
- if $D_{i-1}(k)\geq p_iS_i(k)$ and $D_i^{\text{ramp}}(k)\leq p_i^{\text{ramp}}S_i(k)$, i.e. the demand from the on-ramp is lower than the 'reserved' flow, all the demand from the on-ramp enters the cell

and the flow entering from cell $i-1$ is obtained in order to saturate the supply $S_i(k)$, resulting in $\phi_i(k) = S_i(k) - D_i^{\text{ramp}}(k)$ and $r_i(k) = D_i^{\text{ramp}}(k)$;
- if $D_{i-1}(k) \leq p_i S_i(k)$ and $D_i^{\text{ramp}}(k) \geq p_i^{\text{ramp}} S_i(k)$, i.e. the mainstream demand is lower than the 'reserved' flow, all the mainstream demand enters the cell and the flow entering from the on-ramp is obtained in order to saturate the supply $S_i(k)$, resulting in $\phi_i(k) = D_{i-1}(k)$ and $r_i(k) = S_i(k) - D_{i-1}(k)$.

Note that, in any case, in congested situations the total flow entering cell i is given by $\Phi_i^+(k) = \phi_i(k) + r_i(k) = S_i(k)$.

Summarising, the CTM for a freeway with off-ramps and on-ramps in all the cells is given by (3.27)–(3.34). The *boundary conditions* are the demand in the cell before the first one, i.e. $D_0(k)$, the supply of the cell after the last one, i.e. $S_{N+1}(k)$, the on-ramp demands, i.e. $D_i^{\text{ramp}}(k)$, and the split ratios, i.e. $\beta_i(k)$, $i = 1, \ldots, N$, $k = 0, \ldots, K$.

Finally, let us consider the case in which some cells have *no off-ramps and no on-ramps*. To adapt the CTM previously described to the case in which some cells do not present any ramps, it is possible to fix $\beta_{i-1}(k) = 0$, $D_i^{\text{ramp}}(k) = 0$ and $p_i^{\text{ramp}} = 0$ in case there are not on-ramps and off-ramps between cell $i-1$ and cell i, $i = 1, \ldots, N$. In this way, it is assured that $r_i(k) = 0$ and $s_{i-1}(k) = 0$, $k = 0, \ldots, K$. Note that in this case the interface flow can be computed as

$$\phi_i(k) = \min\{D_{i-1}(k), S_i(k)\} \tag{3.36}$$

according to the first CTM proposed in [43].

3.3.2 The CTM with On-Ramp Queue Dynamics

The CTM described in Sect. 3.3.1 considers a freeway stretch with on-ramps and off-ramps and models the dynamic evolution of the traffic density. In the literature, this version of the CTM has been extended to consider also the dynamics of the queue lengths present at the on-ramps and the possibility to regulate the flow entering from the on-ramp via ramp metering control. This augmented version is adopted especially when ramp metering control approaches are designed, for instance, in [51–53].

In this case, the following dynamic quantities are added to the model (see Fig. 3.9):

- $l_i(k)$ is the queue length of vehicles waiting in the on-ramp of cell i at time kT [veh];
- $d_i(k)$ is the flow accessing the on-ramp of cell i during time interval $[kT, (k+1)T)$ [veh/h];
- $r_i^{\text{C}}(k)$ is the *ramp metering control variable*, i.e. the flow determined by the ramp metering controller to enter cell i from the on-ramp during time interval $[kT, (k+1)T)$ [veh/h].

The parameter $r_i^{\max}$ is also considered, representing the capacity of the on-ramp of section i, i.e. the maximum flow that can enter from that on-ramp, $i = 1, \ldots, N$.

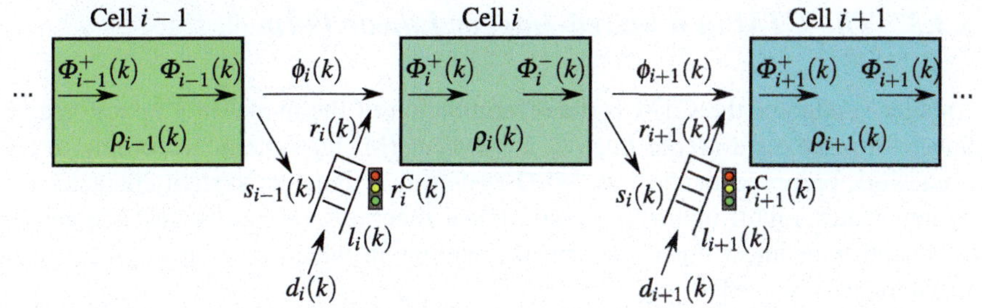

Fig. 3.9 Sketch of freeway stretch in case of on-ramp queues and the relative notation in the CTM

The dynamic equation of the on-ramp queue length, for $i = 1, \ldots, N$, $k = 0, \ldots, K-1$, is given by

$$l_i(k+1) = l_i(k) + T\,[d_i(k) - r_i(k)] \tag{3.37}$$

In this case, the on-ramp demand $D_i^{\text{ramp}}(k)$ to be used in (3.33) and (3.34) is no more a boundary condition but it is computed by taking into account the queue length evolution and the flow accessing the on-ramp. It is possible to distinguish between two cases, corresponding, respectively, to uncontrolled and controlled on-ramp flows.

Uncontrolled On-Ramps If the on-ramp in section i is not controlled, the on-ramp demand of cell i during time interval $[kT, (k+1)T)$ is given by

$$D_i^{\text{ramp}}(k) = \min\left\{d_i(k) + \frac{l_i(k)}{T}, r_i^{\max}\right\} \tag{3.38}$$

Controlled On-Ramps If the on-ramp in section i is controlled, the on-ramp demand of cell i is given by

$$D_i^{\text{ramp}}(k) = \min\left\{d_i(k) + \frac{l_i(k)}{T}, r_i^{\text{C}}(k), r_i^{\max}\right\} \tag{3.39}$$

The augmented CTM to include the on-ramp queue dynamics, for a freeway with off-ramps and on-ramps in all the cells, is given by (3.27)–(3.34), (3.37), together with (3.38) for the uncontrolled on-ramps and (3.39) for the controlled on-ramps. The *boundary conditions* are now the demand in the cell before the first one, i.e. $D_0(k)$, the supply of the cell after the last one, i.e. $S_{N+1}(k)$, the flows accessing the on-ramp queues, i.e. $d_i(k)$, and the split ratios, i.e. $\beta_i(k)$, $i = 1, \ldots, N$, $k = 0, \ldots, K$.

3.3.3 *The CTM in a Mixed-Integer Linear Form*

Another version of the CTM is the reformulation of the model in a mixed-integer linear form, i.e. as a *Mixed Logical Dynamical* (MLD) system. According to the framework proposed in [54], an MLD system is a dynamic system characterised by logic rules, on/off inputs, piecewise linear functions, discrete states, and can be expressed with linear equalities and inequalities in which continuous and binary variables are involved.

The CTM in MLD form has been first proposed in [51, 55], where it has been used as prediction model in Model Predictive Control (MPC) schemes. The advantage of using the CTM in MLD form is related to computational issues, since the non-linearities present in the original model are avoided, resulting in a mixed-integer linear model which is equivalent to the original one. This is obtained by adding some equalities and inequalities, as well as some auxiliary variables, both binary and continuous.

The non-linear relations present in the CTM are the minimum functions in (3.31) and (3.32), as well as the relations (3.33) and (3.34). Let us start from equation (3.31) and let us introduce a binary variable $\delta_{i-1}^{\mathrm{D}}(k)$ such that $[\delta_{i-1}^{\mathrm{D}}(k) = 1]$ iff $[(1-\beta_{i-1}(k))v_{i-1}\rho_{i-1}(k) \le q_{i-1}^{\max}]$. Exploiting the transformations of propositional logic in linear inequalities reported in [54], this latter relation can be transformed as

$$\begin{aligned} (1-\beta_{i-1}(k))v_{i-1}\rho_{i-1}(k) - q_{i-1}^{\max} &\le D_{i-1}^{\max}(1-\delta_{i-1}^{\mathrm{D}}(k)) \\ (1-\beta_{i-1}(k))v_{i-1}\rho_{i-1}(k) - q_{i-1}^{\max} &\ge \varepsilon + (D_{i-1}^{\min} - \varepsilon)\delta_{i-1}^{\mathrm{D}}(k) \end{aligned} \tag{3.40}$$

where ε is a small tolerance, $D_{i-1}^{\max}$ and $D_{i-1}^{\min}$ are the maximum and minimum value of function $(1-\beta_{i-1}(k))v_{i-1}\rho_{i-1}(k) - q_{i-1}^{\max}$, respectively, i.e. $D_{i-1}^{\max} = v_{i-1}\rho_{i-1}^{\max}$ and $D_{i-1}^{\min} = -q_{i-1}^{\max}$. Now (3.31) can be substituted by the following equation:

$$D_{i-1}(k) = \delta_{i-1}^{\mathrm{D}}(k)[(1-\beta_{i-1}(k))v_{i-1}\rho_{i-1}(k)] + (1-\delta_{i-1}^{\mathrm{D}}(k))q_{i-1}^{\max} \tag{3.41}$$

which is still non-linear, since it contains a multiplication between variables. This non-linearity can be overcome by introducing another auxiliary variable $z_{i-1}^{\mathrm{D}}(k)$, such that $z_{i-1}^{\mathrm{D}}(k) = \delta_{i-1}^{\mathrm{D}}(k)\rho_{i-1}(k)$. Then, (3.41) becomes

$$D_{i-1}(k) = (1-\beta_{i-1}(k))v_{i-1}z_{i-1}^{\mathrm{D}}(k) + (1-\delta_{i-1}^{\mathrm{D}}(k))q_{i-1}^{\max} \tag{3.42}$$

The definition $z_{i-1}^{\mathrm{D}}(k) = \delta_{i-1}^{\mathrm{D}}(k)\rho_{i-1}(k)$ can be obtained with the following set of inequalities

$$\begin{aligned} &R_{i-1}^{\min}\delta_{i-1}^{\mathrm{D}}(k) \le z_{i-1}^{\mathrm{D}}(k) \le R_{i-1}^{\max}\delta_{i-1}^{\mathrm{D}}(k) \\ &z_{i-1}^{\mathrm{D}}(k) \ge \rho_{i-1}(k) - R_{i-1}^{\max}(1-\delta_{i-1}^{\mathrm{D}}(k)) \\ &z_{i-1}^{\mathrm{D}}(k) \le \rho_{i-1}(k) - R_{i-1}^{\min}(1-\delta_{i-1}^{\mathrm{D}}(k)) \end{aligned} \tag{3.43}$$

in which $R_{i-1}^{\max}$ and $R_{i-1}^{\min}$ can be estimated as the maximum and minimum value of function $\rho_{i-1}(k)$, i.e. $R_{i-1}^{\max} = \rho_{i-1}^{\max}$ and $R_{i-1}^{\min} = 0$.

Analogously, it is possible to consider Eq. (3.32), for which it is necessary to introduce a binary variable $\delta_i^{\mathrm{S}}(k)$ with the following meaning: $[\delta_i^{\mathrm{S}}(k) = 1]$ iff $[w_i(\rho_i^{\max} - \rho_i(k)) \leq q_i^{\max}]$. Such relation can be transformed as follows:

$$\begin{aligned} w_i(\rho_i^{\max} - \rho_i(k)) - q_i^{\max} &\leq S_i^{\max}(1 - \delta_i^{\mathrm{S}}(k)) \\ w_i(\rho_i^{\max} - \rho_i(k)) - q_i^{\max} &\geq \varepsilon + (S_i^{\min} - \varepsilon)\delta_i^{\mathrm{S}}(k) \end{aligned} \tag{3.44}$$

where $S_i^{\max}$ and $S_i^{\min}$ are the maximum and minimum value of function $w_i(\rho_i^{\max} - \rho_i(k)) - q_i^{\max}$, respectively, i.e. $S_i^{\max} = w_i\rho_i^{\max}$ and $S_i^{\min} = -q_i^{\max}$. Now (3.32) can be written as

$$S_i(k) = \delta_i^{\mathrm{S}}(k)[w_i(\rho_i^{\max} - \rho_i(k))] + (1 - \delta_i^{\mathrm{S}}(k))q_i^{\max} \tag{3.45}$$

which is still non-linear; to overcome this, another variable $z_i^{\mathrm{S}}(k)$ is defined as $z_i^{\mathrm{S}}(k) = \delta_i^{\mathrm{S}}(k)\rho_i(k)$. Then, (3.45) becomes

$$S_i(k) = \delta_i^{\mathrm{S}}(k)w_i\rho_i^{\max} - w_i z_i^{\mathrm{S}}(k) + (1 - \delta_i^{\mathrm{S}}(k))q_i^{\max} \tag{3.46}$$

The relation $z_i^{\mathrm{S}}(k) = \delta_i^{\mathrm{S}}(k)\rho_i(k)$ can be replaced by the following set of inequalities:

$$\begin{aligned} R_i^{\min}\delta_i^{\mathrm{D}}(k) &\leq z_i^{\mathrm{S}}(k) \leq R_i^{\max}\delta_i^{\mathrm{S}}(k) \\ z_i^{\mathrm{S}}(k) &\geq \rho_i(k) - R_i^{\max}(1 - \delta_i^{\mathrm{S}}(k)) \\ z_i^{\mathrm{S}}(k) &\leq \rho_i(k) - R_i^{\min}(1 - \delta_i^{\mathrm{S}}(k)) \end{aligned} \tag{3.47}$$

The CTM in MLD form considers a simplified version of the merge model, i.e. of relations (3.33), (3.34) and (3.39). In particular, the CTM in MLD form considers the following simplified merge model:

$$\begin{aligned} &\text{If} \quad D_{i-1}(k) + r_i^{\mathrm{C}}(k) \leq S_i(k) \\ &\text{then} \quad \phi_i(k) = D_{i-1}(k) \\ &\text{else} \quad \phi_i(k) = S_i(k) - r_i^{\mathrm{C}}(k) \end{aligned} \tag{3.48}$$

Since (3.48) is non-linear, it is necessary to introduce a binary variable $\delta_i^{\mathrm{M}}(k)$ defined as $[\delta_i^{\mathrm{M}}(k) = 1]$ iff $[D_{i-1}(k) + r_i^{\mathrm{C}}(k) \leq S_i(k)]$, corresponding to the following inequalities:

$$\begin{aligned} D_{i-1}(k) + r_i^{\mathrm{C}}(k) - S_i(k) &\leq M_i^{\max}(1 - \delta_i^{\mathrm{M}}(k)) \\ D_{i-1}(k) + r_i^{\mathrm{C}}(k) - S_i(k) &\geq \varepsilon + (M_i^{\min} - \varepsilon)\delta_i^{\mathrm{M}}(k) \end{aligned} \tag{3.49}$$

where $M_i^{\max}$ and $M_i^{\min}$ are the maximum and minimum value of function $D_{i-1}(k) + r_i^{\mathrm{C}}(k) - S_i(k)$, respectively, i.e. $M_i^{\max} = q_{i-1}^{\max} + r_i^{\max}$ and $M_i^{\min} = -q_i^{\max}$. It is now possible to write (3.48) as

$$\phi_i(k) = \delta_i^{\mathrm{M}}(k) D_{i-1}(k) + (1 - \delta_i^{\mathrm{M}}(k))[S_i(k) - r_i^{\mathrm{C}}(k)] \tag{3.50}$$

which is still non-linear because of the products between variables. Then, other three variables should be defined. First of all, the auxiliary variable $z_i^{\mathrm{M_D}}(k)$ is defined as $z_i^{\mathrm{M_D}}(k) = \delta_i^{\mathrm{M}}(k) D_{i-1}(k)$ and corresponds to

$$\begin{aligned}
&M_{\mathrm{D},i}^{\min} \delta_i^{\mathrm{M}}(k) \le z_i^{\mathrm{M_D}}(k) \le M_{\mathrm{D},i}^{\max} \delta_i^{\mathrm{M}}(k) \\
&z_i^{\mathrm{M_D}}(k) \ge D_{i-1}(k) - M_{\mathrm{D},i}^{\max}(1 - \delta_i^{\mathrm{M}}(k)) \\
&z_i^{\mathrm{M_D}}(k) \le D_{i-1}(k) - M_{\mathrm{D},i}^{\min}(1 - \delta_i^{\mathrm{M}}(k))
\end{aligned} \tag{3.51}$$

in which $M_{\mathrm{D},i}^{\max}$ and $M_{\mathrm{D},i}^{\min}$ are the maximum and minimum value of function $D_{i-1}(k)$, i.e. $M_{\mathrm{D},i}^{\max} = q_{i-1}^{\max}$ and $M_{\mathrm{D},i}^{\min} = 0$.

Then, the auxiliary variable $z_i^{\mathrm{M_S}}(k)$ is defined as $z_i^{\mathrm{M_S}}(k) = \delta_i^{\mathrm{M}}(k) S_i(k)$ and given by

$$\begin{aligned}
&M_{\mathrm{S},i}^{\min} \delta_i^{\mathrm{M}}(k) \le z_i^{\mathrm{M_S}}(k) \le M_{\mathrm{S},i}^{\max} \delta_i^{\mathrm{M}}(k) \\
&z_i^{\mathrm{M_S}}(k) \ge S_i(k) - M_{\mathrm{S},i}^{\max}(1 - \delta_i^{\mathrm{M}}(k)) \\
&z_i^{\mathrm{M_S}}(k) \le S_i(k) - M_{\mathrm{S},i}^{\min}(1 - \delta_i^{\mathrm{M}}(k))
\end{aligned} \tag{3.52}$$

in which $M_{\mathrm{S},i}^{\max}$ and $M_{\mathrm{S},i}^{\min}$ are the maximum and minimum value of function $S_i(k)$, i.e. $M_{\mathrm{S},i}^{\max} = q_i^{\max}$ and $M_{\mathrm{S},i}^{\min} = 0$.

Finally, the auxiliary variable $z_i^{\mathrm{M_R}}(k)$ is defined as $z_i^{\mathrm{M_R}}(k) = \delta_i^{\mathrm{M}}(k) r_i^{\mathrm{C}}(k)$ and corresponds to

$$\begin{aligned}
&M_{\mathrm{R},i}^{\min} \delta_i^{\mathrm{M}}(k) \le z_i^{\mathrm{M_R}}(k) \le M_{\mathrm{R},i}^{\max} \delta_i^{\mathrm{M}}(k) \\
&z_i^{\mathrm{M_R}}(k) \ge r_i(k) - M_{\mathrm{R},i}^{\max}(1 - \delta_i^{\mathrm{M}}(k)) \\
&z_i^{\mathrm{M_R}}(k) \le r_i(k) - M_{\mathrm{R},i}^{\min}(1 - \delta_i^{\mathrm{M}}(k))
\end{aligned} \tag{3.53}$$

in which $M_{\mathrm{R},i}^{\max}$ and $M_{\mathrm{R},i}^{\min}$ can be estimated as the maximum and minimum value of function $r_i(k)$, i.e. $M_{\mathrm{R},i}^{\max} = r_i^{\max}$ and $M_{\mathrm{R},i}^{\min} = 0$.

Then, (3.50) becomes

$$\phi_i(k) = z_i^{\mathrm{M_D}}(k) + S_i(k) - r_i^{\mathrm{C}}(k) - z_i^{\mathrm{M_S}}(k) + z_i^{\mathrm{M_R}}(k) \tag{3.54}$$

Moreover, the following inequalities must be verified:

$$r_i^{\mathrm{C}}(k) \le r_i^{\max} \tag{3.55}$$

$$r_i^{\mathrm{C}}(k) \le d_i(k) + \frac{l_i(k)}{T} \tag{3.56}$$

The CTM in MLD form is given by (3.27)–(3.30), (3.40), (3.42)–(3.44), (3.46)–(3.47), (3.49), (3.51)–(3.56). Note that the CTM in MLD form is characterised by three sets of auxiliary binary variables, namely, $\delta_i^{\mathrm{D}}(k)$, $\delta_i^{\mathrm{S}}(k)$, $\delta_i^{\mathrm{M}}(k)$, $i = 1, \ldots, N$, $k = 0, \ldots, K$, and five sets of auxiliary continuous variables, namely, $z_i^{\mathrm{D}}(k)$, $z_i^{\mathrm{S}}(k)$, $z_i^{\mathrm{M_D}}(k)$, $z_i^{\mathrm{M_S}}(k)$, $z_i^{\mathrm{M_R}}(k)$, $i = 1, \ldots, N$, $k = 0, \ldots, K$.

3.3.4 The CTM Including Capacity Drop Phenomena

First-order traffic models, both of continuous and discrete type, are not able to capture the capacity drop, which is a common traffic phenomenon detected in real cases (see Sect. 2.2.4). In the literature, some research works have been devoted to include capacity drop phenomena in first-order traffic models (see, e.g. [56–60]). In the following, two interesting extended versions of the CTM are reported, respectively, obtained by changing the demand function and by changing both the demand and the supply according to a 5-step piecewise linear Fundamental Diagram.

CTM with Capacity Drop: Change in the Demand Function A possibility of modelling capacity drop phenomena in the CTM has been proposed in [60, 61]. In this model, the drop is represented by simply modifying the demand function, so that in case of congestion the demand function is linearly decreasing, as shown in Fig. 3.10. More specifically, the demand of cell $i-1$, for $i = 1, \ldots, N$ and $k = 0, \ldots K$, instead of being represented by (3.31), is given by

$$D_{i-1}(k) = \min\left\{(1-\beta_{i-1}(k))v_{i-1}\rho_{i-1}(k),\, q_{i-1}^{\max} + w'_{i-1}(\rho_{i-1}^{\mathrm{cr}} - \rho_{i-1}(k))\right\} \tag{3.57}$$

where w'_i is the decreasing capacity rate due to the capacity drop phenomenon referred to cell i ($w'_i < w_i$), while ρ_i^{cr} is the critical density of cell i causing a breakdown in capacity.

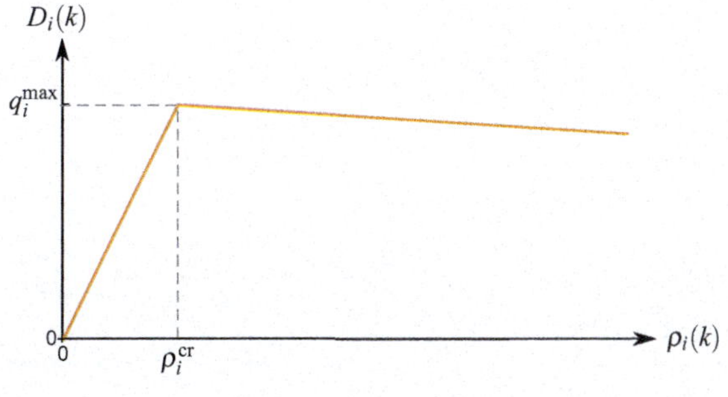

Fig. 3.10 Modified demand function (CTM with capacity drop)

This simple modification of the demand function allows to model capacity drop by maintaining a simple linear formulation of the model that is useful especially for control purposes. Of course, this simple modification is not sufficient to create a capacity drop at the head of a congestion under all circumstances [60].

A very recent extension of the CTM to include the capacity drop phenomenon has been proposed in [62]. This model has been used in a model-based predictive control scheme in [63], in which it has been extended to consider the application of variable speed limits. In [63], the proposed modified CTM is described in comparison with the one reported in [60]. The model discussed in [63] is validated in [64], where it is calibrated with real traffic data from a Dutch freeway and compared with a second-order traffic flow model.

CTM with Capacity Drop: a 5-step Piecewise Linear Fundamental Diagram
Another possibility of including the capacity drop in the CTM has been proposed in [59]. In that work, a 5-step piecewise linear Fundamental Diagram is defined, based on empirical data, in which two values of capacity are explicitly considered. Then, the capacity drop (between these two values of capacity) is modelled by introducing a memory-state binary variable which determines whether the bottleneck is active or inactive.

Let us start from the 5-step piecewise linear Fundamental Diagram. In a given location the steady-state relation between traffic flow and density is assumed to be a 5-step piecewise linear function, which can be written as follows:

$$Q(\rho_i(k)) = \begin{cases} v\rho_i(k) & \text{if } \; 0 \le \rho_i(k) \le \rho^a \wedge \sigma_i(k) = 0 \\ \kappa + v'\rho_i(k) & \text{if } \; \rho^a \le \rho_i(k) \le \rho^b \wedge \sigma_i(k) = 0 \\ q^{\max,\mathrm{H}} & \text{if } \; \rho^b \le \rho_i(k) \le \rho^c \wedge \sigma_i(k) = 0 \\ q^{\max,\mathrm{L}} & \text{if } \; \rho^b \le \rho_i(k) \le \rho^d \wedge \sigma_i(k) = 1 \\ w(\rho^{\max} - \rho_i(k)) & \text{if } \; \rho^d \le \rho_i(k) \le \rho^{\max} \wedge \sigma_i(k) = 1 \end{cases} \tag{3.58}$$

A representation of the piecewise linear relation (3.58) is given in Fig. 3.11. Each block of this function is defined by the density boundaries ρ^a, ρ^b, ρ^c, ρ^d, the jam density $\rho^{\max}$ and the congestion state $\sigma_i(k) \in \{0, 1\}$. This latter is a binary quantity,

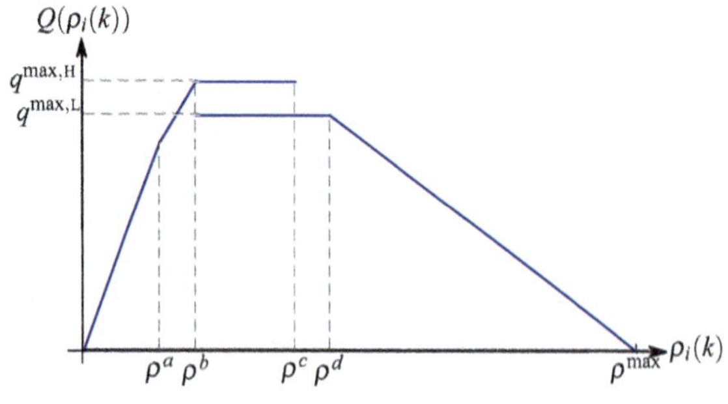

Fig. 3.11 5-step piecewise linear approximation of the Fundamental Diagram

equal to 0 when the state is uncongested and equal to 1 when it is congested. Note that the density boundaries must verify $0 < \rho^a < \rho^b < \rho^c < \rho^d < \rho^{\max}$.

The first two blocks represent the uncongested phase of traffic. The first block is for *light conditions* in which vehicles move at free-flow speed v. The second block represents the *undersaturated* state of traffic, in which the interactions among vehicles decrease the mean speed (that is equal to $v' < v$). The third and fourth blocks represent, respectively, the *pre-congestion* and *post-congestion* situations. Indeed, there is a time interval in which, despite the high density, the freeway works at the maximum capacity $q^{\max,\text{H}}$. After that time, a breakdown occurs and capacity decreases to a lower value, that is $q^{\max,\text{L}}$. Finally, the fifth block represents the behaviour in the *congested* phase; therefore, the (negative) slope is assumed to be equal to the congestion wave speed w. Moreover, if the density value is equal to the maximum value $\rho^{\max}$, the flow is equal to zero.

According to this 5-step piecewise linear Fundamental Diagram, the standard CTM has been modified in [59], by changing the demand and supply functions, as well as by introducing a relation to update the value of the congestion state variable. Further parameters are added to the standard ones in the CTM. Such parameters, referred to cell i, $i = 1, \ldots, N$, are the undersaturated speed v_i' [km/h], the constant κ_i [veh/h], the high and low capacity values $q_i^{\max,\text{H}}$ and $q_i^{\max,\text{L}}$ [veh/h], the density boundaries ρ_i^a, ρ_i^b, ρ_i^c, ρ_i^d [veh/km].

Taking into account (3.58) and Fig. 3.11, it is possible to split the graph into two parts: the left part of the graph (from the first to the third block) is related to the demand function, while the right part (fourth and fifth blocks) is associated with the supply function. Specifically, the demand of cell $i-1$ and the supply of cell i, instead of being given by (3.31) and (3.32), are, respectively, defined as

$$D_{i-1}(k) = \min\left\{(1-\beta_{i-1}(k))v_{i-1}\rho_{i-1}(k),\, (1-\beta_{i-1}(k))\left[\kappa_{i-1} + v_{i-1}'\rho_{i-1}(k)\right],\, q_{i-1}^{\max,\text{H}}\right\} \tag{3.59}$$

$$S_i(k) = \begin{cases} \min\left\{w_i(\rho_i^{\max} - \rho_i(k)),\, q_i^{\max,\text{H}}\right\} & \text{if } \sigma_i(k-1) = 0 \\ \min\left\{w_i(\rho_i^{\max} - \rho_i(k)),\, q_i^{\max,\text{L}}\right\} & \text{if } \sigma_i(k-1) = 1 \end{cases} \tag{3.60}$$

The congested state variable $\sigma_i(k)$ indicates if the state of cell i at time kT is uncongested or congested and is given by

$$\sigma_i(k) = \begin{cases} 1 \text{ if } \left(\rho_i(k) \geq \rho_i^c\right) \vee \left(\rho_i(k) \geq \rho_i^b \wedge \sigma_i(k-1) = 1\right) \\ 0 \text{ otherwise} \end{cases} \tag{3.61}$$

According to [59], in the 5-step piecewise linear Fundamental Diagram, there is not one value for the critical density, but two values, i.e. ρ_i^b and ρ_i^c. These values of density are responsible for changing $\sigma_i(k)$ to 0 or to 1.

3.3.5 *The CTM for a Freeway Network*

The CTM for a freeway network has been first introduced in [45], in which only three-legged junctions are modelled. According to this assumption, the cells are classified into three types (see Fig. 3.12):

- a *diverge* cell is characterised by only one entering link and two leaving links;
- a *merge* cell presents two entering links and one exiting link;
- an *ordinary* cell has just one entering and one leaving link.

With these three types of cells, any freeway networks with three-legged junctions can be modelled. Nevertheless, no generality is lost because the case of junctions with more than three legs can be easily represented as combinations of three-legged junctions, as discussed in [45].

The dynamics of ordinary cells has already been described in Sect. 3.3.1. As for merge and diverge cells, the state equation for traffic density (3.27) still holds, but the definition of entering and exiting flows should be modified. In particular, for a merge cell i, the total flow $\Phi_i^+(k)$ entering cell i during time interval $[kT, (k+1)T)$ depends on the flows coming from the preceding cells. Conversely, for a diverge cell i, the total flow $\Phi_i^-(k)$ exiting cell i during time interval $[kT, (k+1)T)$ depends on the flows going to the following cells. Let us analyse these two cases separately.

Merge Cell Let us consider that cell i is a merge cell and let us denote with j and l the two preceding cells. Let us denote with $\phi_{j,i}(k)$ and $\phi_{l,i}(k)$ the flows entering cell i during time interval $[kT, (k+1)T)$, from cell j and l, respectively, as shown in Fig. 3.13.

For cells j and l it is possible to define the demand, i.e. the flow that can be sent from cell j and l, respectively, to cell i during time interval $[kT, (k+1)T)$. Analogously to (3.31), these demands can be computed as

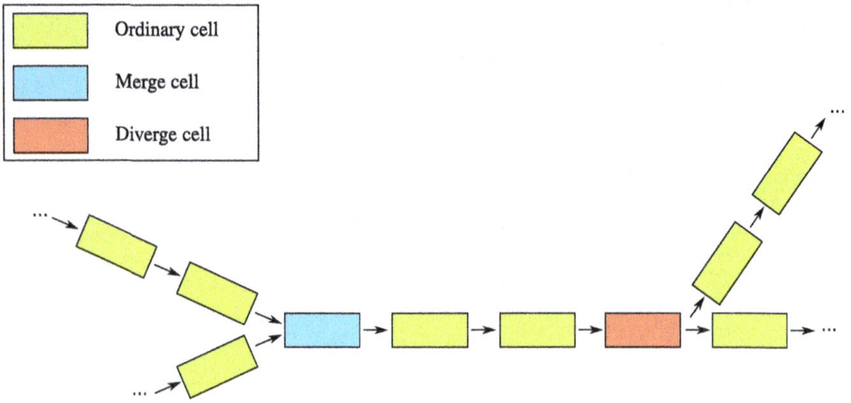

Fig. 3.12 Different types of cells in a freeway network according to the CTM

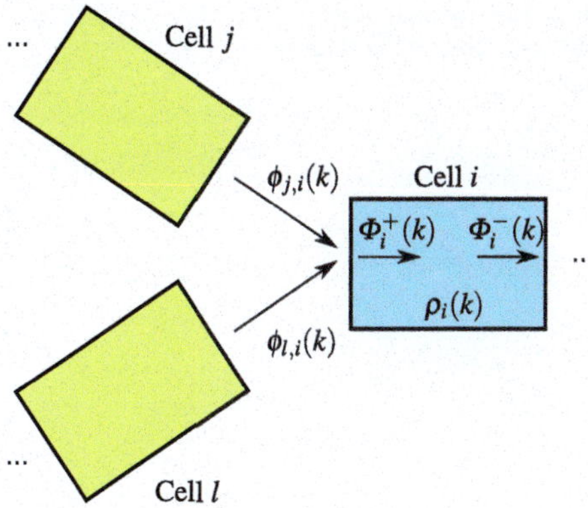

Fig. 3.13 Sketch of a merge cell and the relative notation

$$D_j(k) = \min\left\{(1 - \beta_j(k))v_j\rho_j(k), q_j^{\max}\right\} \tag{3.62}$$

$$D_l(k) = \min\left\{(1 - \beta_l(k))v_l\rho_l(k), q_l^{\max}\right\} \tag{3.63}$$

The supply of cell i represents the flow that can be received by cell i from cells j and l during time interval $[kT, (k+1)T)$ and is still given by (3.32).

The total flow $\Phi_i^+(k)$ entering cell i is computed as the sum of the flows entering from cells j and l, i.e.

$$\Phi_i^+(k) = \phi_{j,i}(k) + \phi_{l,i}(k) \tag{3.64}$$

As already analysed in Sect. 3.3.1 for the mainstream flow and the on-ramp flow, we are again in the situation in which there are two sending cells and one receiving cell. Two cases are distinguished, corresponding to free-flow and congested conditions. In the free-flow case, in cell i there is enough space for the two flows coming from cells j and l, and a condition analogous to (3.33) can be written, i.e.

$$\begin{aligned} &\text{If} \quad D_j(k) + D_l(k) \le S_i(k) \\ &\text{then} \quad \phi_{j,i}(k) = D_j(k), \quad \phi_{l,i}(k) = D_l(k) \end{aligned} \tag{3.65}$$

If the previous condition is not satisfied, this means that not all the flows coming from cells j and l can be received by cell i, and, analogously to (3.34), the following conditions for the congested case hold:

$$\begin{aligned} &\text{If} \quad D_j(k) + D_l(k) > S_i(k) \\ &\text{then} \quad \phi_{j,i}(k) = \text{mid}\left\{D_j(k), S_i(k) - D_l(k), p_j S_i(k)\right\} \\ &\qquad\quad \phi_{l,i}(k) = \text{mid}\left\{D_l(k), S_i(k) - D_j(k), p_l S_i(k)\right\} \end{aligned} \tag{3.66}$$

where p_j and p_l are the priorities of cell j and l in the merge, with $p_j + p_l = 1$.

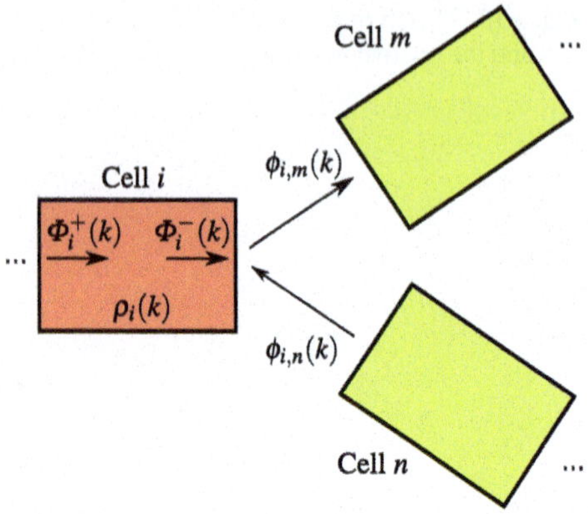

Fig. 3.14 Sketch of a diverge cell and the relative notation

Diverge Cell Let us consider that cell i is a diverge cell and let us denote with m and n the two following cells. Let us denote with $\phi_{i,m}(k)$ and $\phi_{i,n}(k)$ the flows exiting cell i during time interval $[kT, (k+1)T)$ and going to cells m and n, respectively, as shown in Fig. 3.14.

For cell i, the demand $D_i(k)$ is the flow that can be sent from cell i to cells m and n during time interval $[kT, (k+1)T)$, respectively, and it is computed as in (3.31), i.e.

$$D_i(k) = \min\left\{(1 - \beta_i(k))v_i\rho_i(k), q_i^{\max}\right\} \tag{3.67}$$

The supply of cells m and n is instead the flow that can be received by cells m and n, respectively, from cell i during time interval $[kT, (k+1)T)$. These quantities are computed analogously to (3.32), i.e.

$$S_m(k) = \min\left\{w_m(\rho_m^{\max} - \rho_m(k)), q_m^{\max}\right\} \tag{3.68}$$

$$S_n(k) = \min\left\{w_n(\rho_n^{\max} - \rho_n(k)), q_n^{\max}\right\} \tag{3.69}$$

The total flow $\Phi_i^-(k)$ exiting cell i is computed by taking into account the assumptions of the diverge model. The basic idea is that this total flow is restricted in case at least one of the two diverging branches cannot receive its allocated flow. According to this assumption, vehicles which cannot go to the next cell prevent all the other vehicles behind them to continue, supposing that vehicles at the diverge area are served according to a first-in-first-out rule. Of course, this is not completely true in real cases, especially for low exit percentages, but it is worth noting that some blockage phenomena can occur in reality for high exit percentages and in specific traffic conditions, in some way motivating this assumption.

By denoting with $\beta_m(k)$ and $\beta_n(k)$ the portions of traffic flow present in cell i going, respectively, to cell m and n (supposing that these quantities are exogenously determined), the total flow exiting cell i is computed as

$$\Phi_i^-(k) = \min \left\{ D_i(k), \frac{S_m(k)}{\beta_m(k)}, \frac{S_n(k)}{\beta_n(k)} \right\} \tag{3.70}$$

The flows exiting cell i and going to cells m and n are then computed as

$$\phi_{i,m}(k) = \beta_m(k)\Phi_i^-(k) \tag{3.71}$$

$$\phi_{i,n}(k) = \beta_n(k)\Phi_i^-(k) \tag{3.72}$$

3.3.6 Other CTM Versions

Other modifications of the original CTM have been proposed in the literature in the last two decades. Hereafter some of them are briefly commented for the reader's convenience, while others can be addressed with the relevant references in [65]. Note that some of the modifications regard the extension of the CTM to include the case of a freeway in which variable speed limits or route guidance strategies are applied. These modifications to the CTM are not reported in this book, whereas these types of control have been included in second-order models (see Chap. 4), being this latter the most common choice in the scientific literature.

Asymmetric Cell Transmission Model The Asymmetric Cell Transmission Model (ACTM) is a modification of the CTM proposed in [52]. The relevant difference between the two models is the treatment of traffic merges. More specifically, merges in the ACTM are considered as asymmetric connections, such as the junctions of the on-ramps into the mainstream. According to the logic of the standard CTM, the merge is oriented to move as much of the demand as possible from the two merging cells into the receiving cell. The ACTM, instead, makes separate allocations of supply for each merging flow. The flows can then be computed separately as the minimum among the demand, the allocated supply, and the capacity. This modification is justified by the fact that the non-concave/non-convex mid functions of the CTM in (3.34) are replaced with concave min functions, which is an advantage when this model is used as a basis to solve model-based traffic control problems. Moreover, in [52] it is proved that the ACTM, as the CTM, ensures not to predict unrealistic behaviours such as backward moving traffic, negative densities and densities exceeding the jam density.

Link-Node Cell Transmission Model The Link-Node Cell Transmission Model (LN-CTM) is an extension of the CTM to simulate traffic in road networks [66]. In this model, the traffic network is represented with a directed graph, in which links represent road segments and nodes are the junctions among links. Normal links are used to connect two nodes, source links are used to introduce traffic in the network, whereas sink links are used to receive traffic moving out of the network. According to this logic, the on-ramps are represented as source links, while the off-ramps are sinks. The LN-CTM uses a more accurate model of the merging phenomena compared with the ACTM. In particular, in congested conditions, the available supply is shared by the incoming flows proportionally to the demands. However, this more detailed representation of the merge comes at an additional cost of added non-linearity [67], and therefore the results proved in [52] cannot be applied for the LN-CTM.

Lagged Cell Transmission Model In order to improve the accuracy of the original CTM, in [68] a modification of the model has been proposed, based on the fact that the downstream density, used to calculate the supply, is measured at an earlier time instant compared to the current time step, i.e. it is lagged. The introduction of lags can be justified because traffic information travels more slowly in the upstream than in the downstream direction. An improved version of the Lagged CTM has been proposed in [69] to avoid the occurrence of negative densities and of densities larger than the maximum value.

Variable-Length Cell Transmission Model The Variable-Length Cell Transmission Model (VLM) has been proposed in [70] and differs from the standard CTM for the fact that a limited number of cells (of variable length) are used. A road network is subdivided into several sections which are assumed to be composed of a downstream congested cell followed by a free upstream cell. Both cells have variable lengths and are described by two lumped densities (one congested, the other free). The model includes one more state describing the length variation for each cell.

Switched Interpretation of CTM In the literature, some switched interpretations of the CTM have also appeared. Indeed, the CTM is a piecewise linear model and can be regarded as a hybrid system that switches among different sets of linear difference equations. Each set describes a specific operation mode of the freeway traffic system. Since the number of modes can become very high [71, 72], some assumptions can be made to reduce the number of modes. A typical assumption is to consider at most one wave front in the considered freeway stretch. The presence of a single wave front is an assumption reasonable for short freeway stretches with only one on-ramp and one off-ramp. The switched interpretation of the CTM with the single-wave front assumption is called in the literature *Switching-Mode Model* (SMM) [73]. The reduced set of modes can be associated with a graph, since the transition between modes has to follow specific rules, also dictated by the fact that the congestion moves upwards or downwards. The switched model with the associated graph is regarded as a *Graph Constrained CTM* [74].

3.4 Multi-class First-Order Models

Multi-class traffic models have been developed by researchers in order to distinguish different classes of vehicles travelling in the same road system. Depending on the objective of the model, the vehicle classes can be referred to different types of vehicles (e.g. cars, trucks, public transport vehicles and so on) or to specific features of the drivers (such as driving behaviours, travel purposes and so on). In recent applications, it is becoming more and more relevant to distinguish vehicles according to the driver information level, specifically representing the class of 'intelligent vehicles', i.e. vehicles equipped with innovative technology enabling the exchange of data with other vehicles and the traffic infrastructure.

3.4.1 Motivations for Multi-class Models

Regardless of the considered vehicle typologies, multi-class models are characterised by a higher descriptive capability than single-class models, allowing to more realistically represent the dynamic behaviour of a real traffic system. Multi-class models may allow the description of relevant traffic phenomena that can not be captured by models representing only one class of vehicles, in particular all the phenomena related to the interaction of different groups of vehicles which have to share the same infrastructure.

Referring specifically to macroscopic traffic models, a multi-class macroscopic model assumes that the traffic behaviour is represented as the interaction of different traffic flows corresponding to different vehicle categories, whereas a single-class model assumes that the whole traffic is a homogeneous fluid. Let us consider in particular the easiest and better known example of multi-class traffic, i.e. a freeway traffic system in which both cars and trucks travel. In this case, it is easy to observe that trucks have a strong impact on the overall traffic flow for many reasons (because of their high dimensions and low operating capabilities, because their presence has a psychological impact on the drivers of nearby vehicles and so on). Also, these two classes of vehicles have different behaviours and, in many traffic scenarios, can be seen as two different flows sharing the same infrastructure. In particular considering roads with multiple lanes, as it normally happens in freeways, fast vehicles can overtake slow vehicles, so that the traffic behaviour is given by the dynamics of two different flows which influence each other. Representing explicitly the two flows and the interaction between them, instead of modelling the whole traffic as a unique flow, allows to better describe the real traffic system.

A further advantage of the multi-class modelling framework is related to the possibility of designing multi-class controllers, providing different control actions for different vehicle classes. This aspect will be further investigated in Chap. 10.

3.4.2 An Overview of Multi-class First-Order Models

Most of the multi-class first-order traffic models present in the literature are multi-class versions of the LWR model, while only few are multi-class extensions of the CTM.

Multi-class Versions of the LWR Model In some cases, the heterogeneous properties of the traffic flow are represented through *multi-lane* models, as in [75], where a general multi-lane rule is introduced, or in [76], where two types of vehicles and a set of dedicated lanes are modelled. In particular, in [76], the vehicles of the first class can use all the lanes, whereas those belonging to the second class are compelled to travel in a subset of lanes usually located on the right side of the freeway.

Another multi-class first-order model is reported in [77], where the macroscopic model is derived from mesoscopic principles, i.e. from gas-kinetic equations, in order to model the fact that drivers accelerate/decelerate not only according to the desired speed of their class but also due to interactions with other vehicles, both belonging to their class and to other classes. The model proposed in [77] is extended in [78], where a multi-class multi-lane model is proposed for explicitly representing the presence of vehicles moving in platoons. A macroscopic behavioural theory of traffic dynamics for homogeneous and multi-lane freeways is proposed in [79, 80]. Taking into account that drivers can be distinguished in timid and aggressive, this theory can be used to make predictions for separate groups of lanes and is shown to be consistent with experimental observations.

A kinematic wave model of multi-commodity network traffic flow is presented in [81] and, then, extended in [82] to the lane-changing case. In these works, it is assumed that all vehicles have predefined paths and each commodity is represented by vehicles using the same path.

Another work developed in order to consider heterogeneous groups of drivers in the traffic flow is [83], where an extension of the LWR model is formulated with different speed distributions for each class of road users. Specifically, that model describes the dynamic behaviour of heterogeneous users in the traffic flow, in which faster vehicles can overtake the slower ones, both under uncongested and congested conditions, whereas slower vehicles behave in order to slow down the faster ones. In [84], the authors present a homogenised hyperbolic traffic flow model to take into account the presence of several types of vehicles (such as cars, trucks, buses, and so on). An n-population generalisation of the LWR model is proposed in [85], allowing to mathematically explain some practical traffic phenomena, such as overtaking among vehicles. In [86], a multi-class first-order model is presented to explain non-linear traffic phenomena, such as hysteresis and capacity drop. In that model, in free-flow conditions the different vehicle classes are characterised by specific desired speeds and overtaking is allowed; in congested conditions, instead, all the vehicles must travel at the same congested speed and it is not possible to overtake.

In [87], a new model is proposed for vehicle classes interacting in a non-cooperative way: slow vehicles can be seen as *moving bottlenecks* for the fast vehicles, which maximise their speed without influencing slower vehicles. In this model, each class represents a homogeneous group of vehicles, which interacts with the other vehicle classes within the traffic flow. According to this concept, each class is characterised by a different Fundamental Diagram. Note that the specific case of moving bottlenecks, i.e. the presence of slow vehicles moving in the traffic flow, resulting in a reduction of the capacity, is studied in a number of research works based on the LWR model (see, e.g. [88–92]). From a mathematical point of view, moving bottlenecks are normally represented as models in which the partial differential equations of the LWR model are coupled with ordinary differential equations describing the motion of slower vehicles.

Based on the same logic applied in [87], a model to describe a *disordered traffic system* is presented in [93]. In a disordered traffic system, there is no lane discipline, i.e. drivers of smaller vehicles exploit their manoeuvrability to move into lateral gaps at low speeds, whereas at high speeds larger vehicles exploit their greater power to move forward in the traffic flow. These types of systems are very common in developing countries.

A more recent development of macroscopic first-order models for the multi-class case is the Fastlane model, which was first developed in [94]. Fastlane was then successively extended in [95] to be applied for developing multi-class ramp metering in order to control separately the different vehicle classes. Fastlane is based on the LWR model and differs from earlier multi-class first-order macroscopic traffic models for the fact that it models the system dynamics in terms of state-dependent (instead of constant) passenger car equivalents. According to the Highway Capacity Manual, the *Passenger Car Equivalents* (PCE) are defined as the number of passenger cars displaced by a single heavy vehicle of a particular type under prevailing roadway, traffic and control conditions [96]. This factor depends on the considered freeway portion and the traffic conditions present in it, as discussed, for instance, in [97].

A recent work on multi-class traffic models is [98], where two types of vehicles are considered. The model is able to capture overtaking dynamics and creeping phenomena, these latter representing overtaking actions by small vehicles in highly congested situations when larger vehicles have completely stopped. In [98], it is shown that this two-class homogeneous model is equivalent to the ARZ model, of second-order type (see Sect. 4.1.2).

Multi-class Versions of the CTM In [99], the conventional single-class CTM is extended to a more generalised multi-class model in order to take into account the mixed composition of vehicle classes. In the experimental results reported in the paper, the multi-class model is compared with the single-class one and is proven to be significantly more accurate in representing real traffic scenarios.

In [100], a multi-class CTM is developed with the aim of distinguishing two specific classes of vehicles, i.e. *autonomous vehicles* and conventional vehicles. Indeed, autonomous vehicles may entail reduced headways and an increased capacity. Of course, this impact depends on the proportion of autonomous vehicles in the entire traffic flow. The idea of explicitly modelling the presence of Vehicle Automation and Communication Systems (VACS) in the traffic network can be also found in [60], where the CTM is modified to consider lane-changing and capacity drop phenomena, by specifically computing lateral and longitudinal flows.

A recent multi-class version of the CTM can be found in [101], where a unified framework to model heterogeneous traffic flows for large-scale networks is proposed. This model considers the interaction of different vehicle classes, each of which is characterised by homogeneous car-following behaviours and vehicle attributes, and represents three traffic states, corresponding, respectively, to free-flow, semi-congested, and full congested conditions. This model also allows the computation of travel times for each vehicle class.

A multi-class version of the CTM, specifically modelling the presence of cars and buses in the traffic flow, is presented in [102]. The proposed model is called BUS-CTM and tries to replicate the phenomenon of moving bottlenecks, caused by buses moving in the traffic flow. Specifically, buses and cars are considered as heterogeneous vehicles with different characteristics, such as free-flow speed, acceleration and size.

References

1. Piccoli B, Tosin A (2009) Vehicular traffic: a review of continuum mathematical models. In: Meyers RA (ed) Encyclopedia of complexity and systems science. Springer, Berlin, pp 9727–9749
2. Greenshields BD, Bibbins JR, Channing WS, Miller HH (1935) A study of traffic capacity. In: Highway research board proceedings, vol 14, pp 448–477
3. Papageorgiou M, Blosseville J-M, Hadj-Salem H (1989) Macroscopic modelling of traffic flow on the Boulevard Périphérique in Paris. Transp Res Part B 23:29–47
4. Lighthill MJ, Whitham GB (1955) On kinematic waves II: a theory of traffic flow on long crowded roads. Proc R Soc A 229:317–345
5. Richards PI (1956) Shock waves on the highway. Oper Res 4:42–51
6. Lax PD (1973) Hyperbolic systems of conservation laws and the mathematical theory of shock waves. SIAM, Philadelphia
7. Leveque RJ (1992) Numerical methods for conservation laws. Birkhäuser Verlag, Basel
8. Evans L (1998) Partial differential equations. American Mathematical Society, Providence
9. Serre D (1996) Systems of conservation laws. Cambridge University Press, Cambridge
10. Bressan A (2000) Hyperbolic systems of conservation laws. Oxford University Press, Oxford
11. Garavello M, Piccoli B (2016) Traffic flow on networks. American Institute of Mathematical Sciences
12. Garavello M, Han K, Piccoli B (2016) Models for vehicular traffic on networks. American Institute of Mathematical Sciences
13. Papageorgiou M (1998) Some remarks on macroscopic traffic flow modelling. Transp Res Part A 32:323–329
14. Lax P (1953) Nonlinear hyperbolic equations. Commun Pure Appl Math 6:231–258
15. Glimm J (1965) Solutions in the large for nonlinear hyperbolic systems of equations. Commun Pure Appl Math 18:697–715
16. Kruzkov S (1970) First order quasilinear equations in several independent variables. Math USSR-Sbornik 10:217–243
17. DiPerna RJ (1976) Global existence of solutions to nonlinear hyperbolic systems of conservation laws. J Differ Equ 20:187–212
18. Bardos C, Leroux A, Nedelec J (1979) First order quasilinear equations with boundary conditions. Commun Part Differ Equ 4:1017–1034
19. Lebacque J (1996) The Godunov scheme and what it means for first order traffic flow models. In: Proceedings of the 13th international symposium on transportation and traffic theory, pp 647–677
20. Li T (2003) Global solutions of nonconcave hyperbolic conservation laws with relaxation arising from traffic flow. J Differ Equ 190:131–149
21. Li T, Liu H (2005) Stability of a traffic flow model with nonconvex relaxation. Commun Math Sci 3:101–118

22. Dubois F, Le Floch P (1988) Boundary conditions for nonlinear hyperbolic systems of conservations laws. J Differ Equ 71:93–122
23. Le Floch P (1988) Explicit formula for scalar non-linear conservation laws with boundary condition. Math Methods Appl Sci 10:265–287
24. Strub IS, Bayen AM (2006) Weak formulation of boundary conditions for scalar conservation laws: an application to highway traffic modelling. Int J Robust Nonlinear Control 16:733–748
25. Colombo RM, Goatin P, Rosini MD (2011) On the modelling and management of traffic. ESAIM Math Model Numer Anal 45:853–872
26. Bagnerini P, Colombo RM, Corli A (2006) On the role of source terms in continuum traffic flow models. Math Comput Model 44:917–930
27. Jin W-L, Zhang HM (2003) The inhomogeneous kinematic wave traffic flow model as a resonant nonlinear system. Transp Sci 37:294–311
28. Li J, Zhang HM (2013) Modeling space-time inhomogeneities with the kinematic wave theory. Transp Res Part B 54:113–125
29. Claudel CG, Bayen AM (2010) Lax-Hopf based incorporation of internal boundary conditions into Hamilton–Jacobi equation, part I: theory. IEEE Trans Autom Control 55:1142–1157
30. Claudel CG, Bayen AM (2010) Lax-Hopf based incorporation of internal boundary conditions into Hamilton–Jacobi equation, part II: computational methods. IEEE Trans Autom Control 55:1158–1174
31. Holden H, Risebro NH (1995) A mathematical model of traffic flow on a network of unidirectional roads. SIAM J Math Anal 26:999–1017
32. Coclite GM, Garavello M, Piccoli B (2005) Traffic flow on a road network. SIAM J Math Anal 36:1862–1886
33. Bretti G, Natalini R, Piccoli B (2006) Numerical approximations of a traffic flow model on networks. Netw Heterog Media 1:57–84
34. Helbing D, Lämmer S, Lebacque J-P (2005) Self-organized control of irregular or perturbed network traffic. In: Deissenberg C, Hartl RF (eds) Opt Control Dyn Games. Springer, Dortrecht, pp 239–274
35. Marigo A, Piccoli B (2008) A fluid dynamic model for T-junctions. SIAM J Math Anal 39:2016–2032
36. Delle Monache ML, Reilly J, Samaranayake S, Krichene W, Goatin P, Bayen AM (2014) A PDE-ODE model for a junction with ramp buffer. SIAM J Appl Math 74:22–39
37. Herty M, Lebacque J-P, Moutari S (2009) A novel model for intersections of vehicular traffic flow. Netw Heterog Media 4:813–826
38. Garavello M, Goatin P (2012) The Cauchy problem at a node with buffer, discrete and continuous dynamical systems, Series A. American Institute of Mathematical Sciences
39. Garavello M, Piccoli B (2013) A multibuffer model for LWR road networks. In: Ukkusuri S, Ozbay K (eds) Advances in dynamic network modeling in complex transportation systems. Springer, New York, pp 143–161
40. Leveque RJ (2002) Finite volume methods for hyperbolic problems. Cambridge University Press, Cambridge
41. Caligaris C, Sacone S, Siri S (2011) On an implicit and stable resolution scheme for the Payne–Whitham model. Math Comput Model 54:378–387
42. Trefethen LN (1996) Finite difference and spectral methods for ordinary and partial differential equations. Cornell University
43. Daganzo CF (1993) The cell transmission model part I: a simple dynamic representation of highway traffic. University of California, Berkeley
44. Daganzo CF (1994) The cell transmission model: a dynamic representation of highway traffic consistent with the hydrodynamic theory. Transp Res Part B 28:269–287
45. Daganzo CF (1995) The cell transmission model, part II: network traffic. Transp Res Part B 29:79–93
46. Daganzo CF (1995) A finite difference approximation of the kinematic wave model of traffic flow. Transp Res Part B 29:261–276

47. Godunov SK (1959) A difference method for numerical calculation of discontinuous solutions of the equations of hydrodynamics. Matematicheskii Sbornik 47:271–306
48. Yperman I, Logghe S, Immers B (2005) The link transmission model: an efficient implementation of the kinematic wave theory in traffic networks. Advanced OR and AI methods in transportation, pp 122–127
49. Newell GF (1993) A simplified theory of kinematic waves in highway traffic. Transp Res Part B 27:281–313
50. Himpe W, Corthout R, Tampère MJC (2016) An efficient iterative link transmission model. Transp Res Part B 92:170–190
51. Ferrara A, Sacone S, Siri S (2015) Event-triggered model predictive schemes for freeway traffic control. Transp Res Part C 58:554–567
52. Gomes G, Horowitz R (2006) Optimal freeway ramp metering using the asymmetric cell transmission model. Transp Res Part C 14:244–262
53. Gomes G, Horowitz R, Kurzhanskiy AA, Varaiya P, Kwon J (2008) Behavior of the cell transmission model and effectiveness of ramp metering. Transp Res Part C 16:485–513
54. Bemporad A, Morari M (1999) Control of systems integrating logic, dynamics, and constraints. Automatica 35:407–427
55. Ferrara A, Nai Oleari A, Sacone S, Siri S (2015) Freeways as systems of systems: a distributed model predictive control scheme. IEEE Syst J 9:312–323
56. Lebacque J (1852) Two-phase bounded-acceleration traffic flow model: analytical solutions and applications. Transp Res Rec 2003:220–230
57. Laval JA, Daganzo CF (2006) Lane-changing in traffic streams. Transp Res Part B 40:251–264
58. Leclercq L, Laval JA, Chiabaut N (2011) Capacity drops at merges: an endogenous model. Procedia Soc Behav Sci 17:12–26
59. Srivastava A, Geroliminis N (2013) Empirical observations of capacity drop in freeway merges with ramp control and integration in a first-order model. Transp Res Part C 30:161–177
60. Roncoli C, Papageorgiou M, Papamichail I (2015) Traffic flow optimisation in presence of vehicle automation and communication systems - part I: a first-order multi-lane model for motorway traffic. Transp Res Part C 57:241–259
61. Roncoli C, Papageorgiou M, Papamichail I (2014) Optimal control for multi-lane motorways in presence of vehicle automation and communication systems. In: Proceedings of the 19th IFAC world congress, pp 4178–4183
62. Han Y, Yuan Y, Hegyi A, Hoogendoorn S (2016) New extended discrete first-order model to reproduce propagation of jam waves. Transp Res Rec 2560:108–118
63. Han Y, Hegyi A, Yuan Y, Hoogendoorn S, Papageorgiou M, Roncoli C (2017) Resolving freeway jam waves by discrete first-order model-based predictive control of variable speed limits. Transp Res Part C 77:405–420
64. Han Y, Hegyi A, Yuan Y, Hoogendoorn S (2017) Validation of an extended discrete first-order model with variable speed limits. Transp Res Part C 83:1–17
65. Alecsandru C, Quddus A, Huang KC, Rouhieh B, Khan AR, Zeng Q (2011) An assessment of the cell-transmission traffic flow paradigm: development and applications. Transportation Research Board
66. Muralidharan A, Dervisoglu G, Horowitz R (2009) Freeway traffic flow simulation using the link node cell transmission model. In: Proceedings of the American control conference, pp 2916–2921
67. Muralidharan A, Horowitz R (2012) Optimal control of freeway networks based on the link node cell transmission model. In: Proceedings of the American control conference, pp 5769–5774
68. Daganzo CF (1999) The lagged cell-transmission model. In: Proceedings of international symposium on transportation and traffic theory, pp 81–104
69. Szeto WY (2008) Enhanced lagged cell-transmission model for dynamic traffic assignment. Transp Res Rec 76–85
70. Canudas-de-Wit C, Ferrara A (2016) A variable-length cell road traffic model: application to ring road speed limit optimization. In: Proceedings of the IEEE 55th conference on decision and control, pp 6745–6752

71. Ferrara A, Sacone S, Siri S, Vivas C, Rubio FR (2016) Switched observer-based ramp metering controllers for freeway systems. In: Proceedings of the 55th IEEE conference on decision and control, pp 6777–6782
72. Ferrara A, Sacone S, Siri S (2015) A switched ramp-metering controller for freeway traffic systems. In: Proceedings of the 5th IFAC conference on analysis and design of hybrid systems, pp 105–110
73. Muñoz L, Sun X, Horowitz R, Alvarez L (2003) Traffic density estimation with the cell transmission model. In: Proceedings of the American control conference, pp 3750–3755
74. Morbidi F, Leon Ojeda L, Canudas de Wit C, Bellicot I (2014) A new robust approach for highway traffic density estimation. In: Proceedings of the European control conference, pp 2575–2580
75. Holland EN, Woods AW (1997) A continuum model for the dispersion of traffic on two-lane roads. Transp Res Part B 31:473–485
76. Daganzo CF (1997) A continuum theory of traffic dynamics for freeways with special lanes. Transp Res Part B 31:83–102
77. Hoogendoorn SP, Bovy PHL (2000) Continuum modeling of multiclass traffic flow. Transp Res Part B 34:123–146
78. Hoogendoorn SP, Bovy PHL (2001) Platoon-based multiclass modeling of multilane traffic flow. Netw Spat Econ 1:137–166
79. Daganzo CF (2002) A behavioral theory of multi-lane traffic flow part I: long homogeneous freeway sections. Transp Res Part B 36:131–158
80. Daganzo CF (2002) A behavioral theory of multi-lane traffic flow, part II: merges and the onset of congestion. Transp Res Part B 36:159–169
81. Jin W-L (2012) A kinematic wave theory of multi-commodity network traffic flow. Transp Res Part B 46:1000–1022
82. Jin W-L (2013) A multi-commodity Lighthill–Whitham–Richards model of lane-changing traffic flow. Transp Res Part B 57:361–377
83. Wong GCK, Wong SC (2002) A multi-class traffic flow model - an extension of LWR model with heterogeneous drivers. Transp Res Part A 36:827–841
84. Bagnerini P, Rascle M (2003) A multiclass homogenized hyperbolic model of traffic flow. SIAM J Math Anal 35:949–973
85. Benzoni-Gavage S, Colombo RM (2003) An n-populations model for traffic flow. Eur J Appl Math 14:587–612
86. Ngoduy D, Liu R (2007) Multiclass first-order simulation model to explain non-linear traffic phenomena. Phys A 385:667–682
87. Logghe S, Immers LH (2008) Multi-class kinematic wave theory of traffic flow. Transp Res Part B 42:523–541
88. Lebacque JP, Lesort JB, Giorgi F (1998) Introducing buses into first-order macroscopic traffic flow models. Transp Res Rec 1644:70–79
89. Leclercq L (2007) Hybrid approaches to the solutions of the "Lighthill–Whitham–Richards" model. Transp Res Part B 41:701–709
90. Lattanzio C, Maurizi A, Piccoli B (2011) Moving bottlenecks in car traffic flow: a PDE-ODE coupled model. SIAM J Math Anal 43:50–67
91. Delle Monache ML, Goatin P (2014) Scalar conservation laws with moving constraints arising in traffic flow modeling: an existence result. J Differ Equ 257:4015–4029
92. Simoni MD, Claudel CG (2017) A fast simulation algorithm for multiple moving bottlenecks and applications in urban freight traffic management. Transp Res Part B 104:238–255
93. Nair R, Mahmassani HS, Miller-Hooks E (2011) A porous flow approach to modeling heterogeneous traffic in disordered systems. Transp Res Part B 45:13311–345
94. van Lint JWC, Hoogendoorn SP, Schreuder M (2008) Fastlane: new multiclass first-order traffic flow model. Transp Res Rec 2088:177–187
95. Schreiter T, van Lint H, Hoogendoorn S (2011) Multi-class ramp metering: concepts and initial results. In: Proceedings of the 14th international IEEE conference on intelligent transportation systems, pp 885–889

96. Special Report 209: highway capacity manual. (1994) 3rd edn. Transportation Research Board, Washington DC
97. Al-Kaisy AF, Hall FL, Reisman ES (2002) Developing passenger car equivalents for heavy vehicles on freeways during queue discharge flow. Transp Res Part A 36:725–742
98. Fan S, Work D (2015) A heterogeneous multiclass traffic flow model with creeping. SIAM J Appl Math 75:813–835
99. Tuerprasert K, Aswakul C (2010) Multiclass cell transmission model for heterogeneous mobility in general topology of road network. J Intell Transp Syst 14:68–82
100. Levin MW, Boyles SD (2016) A multiclass cell transmission model for shared human and autonomous vehicle roads. Transp Res Part C 62:103–116
101. Qian Z, Li J, Li X, Zhang M, Wang H (2017) Modeling heterogeneous traffic flow: a pragmatic approach. Transp Res Part B 99:183–204
102. Liu H, Wang J, Wijayaratna K, Dixit VV, Travis Waller S (2015) Integrating the bus vehicle class into the cell transmission model. IEEE Trans Intell Transp Syst 16:2620–2630

Chapter 4
Second-Order Macroscopic Traffic Models

4.1 Continuous Second-Order Models

In order to overcome the weaknesses of first-order models of continuous type (see Sect. 3.2), second-order traffic flow models were developed and appeared approximately 20 years later. These models, besides considering the dynamics of the traffic density, explicitly introduce a dynamic equation for the mean speed. The first continuous second-order traffic flow model was proposed by Payne [1] and Whitham [2], in the 70s and is generally known as the *Payne–Whitham* (PW) model. This model received some critiques, the major one being formulated by Daganzo [3], showing that classical second-order models can exhibit non-physical solutions. This critique led to the development of new second-order models, such as those developed by Aw and Rascle [4], on the one side, and Zhang [5], on the other side. This latter model is often known as *Aw–Rascle–Zhang* (ARZ) model. These models are briefly described in the following subsections, the interested reader can find more mathematical details in books specifically dedicated to continuous traffic models, for example, in [6, 7].

4.1.1 The PW Model

The PW model is a continuous traffic flow model of macroscopic type, i.e. it represents the dynamics of aggregate variables referred to the traffic flow. As described in Sect. 3.1.1, the main variables considered in continuous macroscopic models are the traffic density $\rho(x, t)$ [veh/km], the mean speed $v(x, t)$ [km/h], and the traffic flow $q(x, t)$ [veh/h], with x representing the location and t indicating time.

The PW model is based on the two basic equations of traffic flow models, i.e. the hydrodynamic equation and the continuity equation, described in Sect. 3.1.1 and reported in the following:

$$q(x, t) = \rho(x, t)v(x, t) \tag{4.1}$$

A. Ferrara et al., *Freeway Traffic Modelling and Control*, Advances in Industrial Control, https://doi.org/10.1007/978-3-319-75961-6_4

$$\frac{\partial \rho(x,t)}{\partial t} + \frac{\partial q(x,t)}{\partial x} = 0 \tag{4.2}$$

In the PW model, (4.1) and (4.2) are coupled with a partial differential equation describing the dynamics of the mean speed, analogously to the momentum equation of fluid dynamics. This equation is derived from a car-following rule, by applying Taylor expansion, and it yields

$$\frac{\partial v(x,t)}{\partial t} + v(x,t)\frac{\partial v(x,t)}{\partial x} = \frac{1}{\tau}\left[V(\rho(x,t)) - v(x,t)\right] + \frac{1}{2\tau\,\rho(x,t)}\frac{dV(\rho)}{d\rho}\frac{\partial \rho(x,t)}{\partial x} \tag{4.3}$$

where $\tau > 0$ is a constant called *speed adaptation time*. The speed equation (4.3) is composed of convection, relaxation and anticipation terms, which are now analysed in detail.

The *convection* term given by

$$v(x,t)\frac{\partial v(x,t)}{\partial x} \tag{4.4}$$

describes the fact that the vehicles travelling along the freeway do not adjust their speed instantaneously. More specifically, let us consider the case in which vehicles are travelling very fast and need to decrease their speed to adapt to a lower downstream traffic mean speed. They do this gradually, which implies that a higher upstream speed tends to increase the traffic speed downstream (and the opposite holds in case of lower upstream speed). In other words, this term describes how the upstream speed influences the downstream one.

The *relaxation* term expressed as

$$\frac{1}{\tau}\left[V(\rho(x,t)) - v(x,t)\right] \tag{4.5}$$

models the fact that all the vehicles tend to adjust their speed to the steady-state speed $V(\rho(x,t))$. The speed relaxation time τ is related to the reaction times of the drivers.

The *anticipation* term given by

$$\frac{1}{2\tau\,\rho(x,t)}\frac{dV(\rho)}{d\rho}\frac{\partial \rho(x,t)}{\partial x} \tag{4.6}$$

describes the capability of the drivers to look ahead and to adjust their actual speed to the speed compatible with the density downstream. Note that this term can also be written as

$$-\frac{1}{\rho(x,t)}\frac{dp(\rho)}{d\rho}\frac{\partial \rho(x,t)}{\partial x} \tag{4.7}$$

where $p(\rho) = -\frac{1}{2\tau}V(\rho)$ is the *pressure* term, in analogy with fluid dynamics. By virtue of the non-increasing nature of the steady-state relation between mean speed and density (see Sect. 3.1.1), the traffic pressure is a non-decreasing function of density.

It is worth noting that sometimes a *diffusive acceleration* term, or *viscosity* term, is added at the second member of (4.3). Such term is given by

$$\upsilon \frac{\partial^2 v(x,t)}{\partial x^2} \tag{4.8}$$

where $\upsilon \geq 0$ represents a *diffusion* coefficient, again by analogy with the fluid theory.

4.1.2 The ARZ Model

The continuous macroscopic models, and in particular the PW model described in Sect. 4.1.1, rely on the equivalence between traffic and fluids. Yet, as observed in [3], there are major differences between them, which need to be correctly captured by traffic models. For instance, in contrast with fluids, vehicles are *anisotropic* particles that mostly respond to frontal stimuli, i.e. they are influenced mainly (or only) by the traffic dynamics ahead of them. Moreover, differently from molecules, drivers have their own personality. These differences motivate the presence of inconsistencies in the PW model, corresponding to an unrealistic behaviour, such as negative speeds, the violation of the anisotropy principle, and the propagation of the information faster than the speed of vehicles, as highlighted in [3].

Aw and Rascle in [4] proposed a simple modification of the PW model in order to overcome its inconsistencies. Specifically, they consider a version of the PW model in which both the relaxation term and the diffusive term are neglected and the traffic pressure is defined as a smooth increasing function of the density ρ, i.e.

$$p(\rho) = \rho^{\gamma} \tag{4.9}$$

with $\gamma > 0$.

Then, instead of adopting the *Eulerian* point of view, i.e. the one of an external observer placed in a fixed spatial position x, the Aw–Rascle model rely on the *Lagrangian* point of view, i.e. the one of an internal observer flowing through x with speed v, as a single vehicle in the traffic flow does. Hence, in the Aw–Rascle model, the authors suggest to correct the anticipation factor involving the derivative of the pressure with respect to x with the so-called *convective derivative* (or material derivative) of the pressure term. Mathematically, this implies to use the convective derivative operator $D_t := \partial_t + v\partial_x$, where v is the actual fluid speed, to derive the anticipation term of the model. This term can be expressed as

$$D_t\left(p(\rho)\right) = \frac{\partial p(\rho)}{\partial t} + v(x,t)\frac{\partial p(\rho)}{\partial t} \tag{4.10}$$

so that the Aw–Rascle model is given by (4.1), (4.2) and

$$\frac{\partial}{\partial t}\left(v(x,t) + p(\rho(x,t))\right) + v(x,t)\frac{\partial}{\partial x}\left(v(x,t) + p(\rho(x,t))\right) = 0 \tag{4.11}$$

with the pressure $p(\rho(x,t))$ expressed as in (4.9). Note that, as highlighted by the same authors in [4], the model can create some difficulties from the mathematical point of view when the density is close to zero, since the model is not well-posed near the vacuum.

An interesting paper dealing with the controversy on Daganzo's criticism against second-order models and the proposal by Aw and Rascle to overcome such drawbacks is [8]. In this paper, the linear stability of these traffic models is analysed, by mainly focusing on the characteristic speeds. One of the theoretical inconsistencies associated with second-order macroscopic traffic models is related to the fact that they predict two characteristic speeds, one of which is faster than the average speed. In [8], arguments for and against this view are discussed, by comparing the PW model with the Aw–Rascle model.

The Aw–Rascle model was extended by introducing relaxation terms, as can be found, for example, in [9, 10]. Moreover, a model similar to the one proposed by Aw and Rascle was developed independently and following a different rationale by Zhang [5]. To correctly consider all the contributors, this model is now often referred to as the Aw–Rascle–Zhang model or by its acronym ARZ, as for instance in [11].

Analogously to the application of the LWR model to networks (see Sect. 3.2.3), also the ARZ model has been considered for modelling *road networks*. In this case, the traffic dynamics on roads is given by the ARZ model, while specific conditions or rules must be defined for junctions, in order to determine a unique solution. One of the first works considering the second-order ARZ model applied to networks is [12], where the Riemann problem at junctions is solved by specifying suitable rules on traffic distributions and the maximisation of flows and other quantities. In [13], a road network is considered as well, with the roads modelled by the ARZ model, in which a different model for the junctions is taken into account in order to ensure the conservation of all moments. A further extension of these two junction models can be found in [14], where the solutions guarantee that all the moments are conserved and, at the same time, the total flow at the junction is maximised.

4.1.3 Phase-Transition Models

As discussed in Sect. 2.1.3, a very relevant peculiarity of traffic flow is associated with the form of the Fundamental Diagram, which is obtained from experimental data. While the left side of this relation (corresponding to the free-flow case) can be easily approximated with a straight line, the left part of the curve (corresponding to the congested regime) is more difficult to be approximated with a single line, since real data are often very sparse. First-order traffic flow models, which assume that the speed of vehicles instantaneously adapts to its steady-state value, cannot capture the aforementioned phenomenon, and this is one of the reasons that have motivated researchers to introduce second-order models.

By following the three-flow phase theory developed by Kerner [15], reporting that three different behaviours can be observed in traffic flow (free-flow, synchronised flow, and wide moving jams), some *phase-transition models* appeared in the literature. In [16], free-flow and congestion are seen as two different phases, governed by different dynamic equations. In particular, in free-flow conditions, a classical LWR model, of first-order type, is used, while the congested case is represented through a second-order model with dynamic equations defined for the density and for the linearised momentum.

A generalisation of the model proposed in [16] can be found in [17], where a different Fundamental Diagram form is used for the free-flow phase and a variety of possible Fundamental Diagrams is allowed for the congested case, depending on the shape resulting from real data. The accuracy and practicality of this phase-transition model were assessed in [18]. Another phase-transition model was proposed in [19], where the first-order LWR model is coupled with the second-order ARZ model and a transition dynamics from the free-flow to the congested behaviour is introduced. Such model well fits the experimental data and is able to overcome some of the drawbacks of the ARZ model.

The extension of phase-transition models to road networks was addressed for the first time in [20], where, specifically, the phase-transition model presented in [16] is taken into account. In [20], the existence of solutions is proved, without any restriction on the network geometry. The same phase-transition model was also adopted in [21], where a specific model for junctions is considered, also including the presence of precedences among different incoming and outcoming flows.

4.2 Discrete Second-Order Models

The first discretised versions of the PW model appeared in the literature in the late 80s [22, 23], with applications to the Boulevard Périphérique in Paris. In particular, in [22, 23], the PW model is discretised in space and time, considering new terms to model the influence of on-ramp and off-ramp flows on the mainstream dynamic behaviour. This model was then extended to consider a freeway network [24, 25], by means of the simulation program called *METANET*, which is an acronym for 'Modèle d'Écoulement de Trafic sur Autoroute NETworks'. Even if the name METANET was firstly associated with the simulation tool for the freeway network, it is now normally used to indicate the second-order traffic flow model in the discretised version. This latter is the meaning of METANET adopted in this book.

In the remainder of this chapter different versions of the METANET model are reported, for a freeway stretch and for a freeway network, both in the single-class case and in the multi-class version.

4.2.1 METANET for a Freeway Stretch

Let us consider the METANET model for a freeway stretch with on-ramps and off-ramps, as proposed in [22, 23] and reported here with a slightly different mathematical notation in order to adapt to the notation adopted in this book.

As aforementioned, this model is discrete in space and time, i.e. the freeway stretch is divided into a given number of road portions, called *sections*, and the time horizon is partitioned into time intervals of equal length. Let N be the number of sections, each one having length L_i [km], $i = 1, \ldots, N$, and K be the number of time intervals, with sample time T [h]. In the METANET model, on-ramps and off-ramps are assumed to be present within the sections, differently from the case considered in the CTM, where ramps are assumed to be at the interface between subsequent cells (see Sect. 3.3).

Figure 4.1 shows a sketch of the subdivision of the freeway stretch into sections, with the main variables of the METANET model. For each section $i = 1, \ldots, N$, and for each time step $k = 0, \ldots K$, the following quantities are defined:

- $\rho_i(k)$ is the traffic density in section i at time kT [veh/km];
- $v_i(k)$ is the mean traffic speed in section i at time kT [km/h];
- $q_i(k)$ is the traffic flow leaving section i during time interval $[kT, (k+1)T)$ [veh/h];
- $r_i(k)$ is the on-ramp traffic flow entering section i during time interval $[kT, (k+1)T)$ [veh/h];
- $s_i(k)$ is the off-ramp traffic flow exiting section i during time interval $[kT, (k+1)T)$ [veh/h].

The parameters of the model are as follows: v_i^{f} is the free-flow speed [km/h] of section i, ρ_i^{cr} is the critical density [veh/km] of section i, $\rho_i^{\max}$ is the jam density

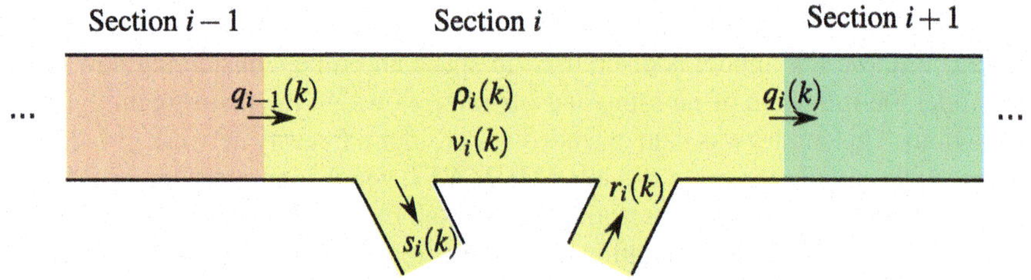

Fig. 4.1 Sketch of the division of the freeway stretch into sections and the relative notation in METANET

[veh/km] of section i, $i = 1, \ldots, N$, τ, ν, χ, δ_{on} are model parameters present in the speed equation, while a is a parameter present in the steady-state speed–density relation.

The METANET model is given by the following finite difference equations:

$$\rho_i(k+1) = \rho_i(k) + \frac{T}{L_i}\left[q_{i-1}(k) - q_i(k) + r_i(k) - s_i(k)\right] \tag{4.12}$$

$$\begin{aligned} v_i(k+1) = v_i(k) &+ \frac{T}{\tau}\left[V\left(\rho_i(k)\right) - v_i(k)\right] + \frac{T}{L_i} v_i(k)\left[v_{i-1}(k) - v_i(k)\right] \\ &- \frac{\nu T\left[\rho_{i+1}(k) - \rho_i(k)\right]}{\tau L_i\left[\rho_i(k) + \chi\right]} - \delta_{\text{on}} T \frac{v_i(k) r_i(k)}{L_i\left[\rho_i(k) + \chi\right]} \end{aligned} \tag{4.13}$$

where $i = 1, \ldots, N$, $k = 0, \ldots, K-1$, while the traffic flow to be used in (4.12) is

$$q_i(k) = \rho_i(k) v_i(k) \tag{4.14}$$

and the steady-state speed–density relation adopted in (4.13) is given by

$$V\left(\rho_i(k)\right) = v_i^{\text{f}} \exp\left[-\frac{1}{a}\left(\frac{\rho_i(k)}{\rho_i^{\text{cr}}}\right)^a\right] \tag{4.15}$$

Equation (4.12) represents the conservation of vehicles, while (4.13) is a discretisation of the speed equation of the PW model, with an additional term. Hence, as in the speed equation (4.3) of the PW model, relaxation, convection and anticipation terms can be identified, as well as a fourth term to model the influence of cars entering from the on-ramp.

The *relaxation term*, i.e. $\frac{T}{\tau}\left[V\left(\rho_i(k)\right) - v_i(k)\right]$, models the fact that vehicles tend to reach the steady-state speed depending on the experienced density $\rho_i(k)$, according to a parameter τ, which represents the swiftness of drivers. Hence, vehicles accelerate if their actual speed is lower than the steady-state value, and they decelerate otherwise.

The *convection term*, i.e. $\frac{T}{L_i} v_i(k)\left[v_{i-1}(k) - v_i(k)\right]$, represents the fact that vehicles arriving in section i from section $i-1$ cannot adapt immediately their speed. If vehicles travel in section $i-1$ at a higher speed than in section i, they decelerate when they reach section i but this change of speed is not instantaneous. A similar argument applies in case of acceleration from section $i-1$ to section i.

The *anticipation term*, i.e. $-\frac{\nu T\left[\rho_{i+1}(k) - \rho_i(k)\right]}{\tau L_i\left[\rho_i(k) + \chi\right]}$, takes into account that drivers adjust their speed also on the basis of the situation they see downstream, hence there is a deceleration if a higher density is seen ahead, and an acceleration in the opposite case.

Finally, the fourth term $-\delta_{\text{on}} T \frac{v_i(k) r_i(k)}{L_i\left[\rho_i(k) + \chi\right]}$ was introduced in [22] to model the direct impact of the on-ramp entering flow $r_i(k)$ on the speed dynamics (note that a similar term was also proposed in [26]). Indeed, vehicles entering from the on-ramps normally have a lower speed than vehicles in the mainstream, inducing a deceleration on these latter vehicles, which is more relevant if the entering flows are high. In [22], the authors suggest to use a similar term with $s_i(k)$ replacing $r_i(k)$ and δ_{off} replacing δ_{on}, to model the speed reduction due to the exit of vehicles through the off-ramps.

In the METANET model for a freeway stretch, given by (4.12)–(4.15), the *boundary conditions* are the traffic flow entering the first road section, i.e. $q_0(k)$, the on-ramp and off-ramp traffic flows $r_i(k)$ and $s_i(k)$, $i = 1, \ldots, N$, the mean traffic speed in the section before the first one, i.e. $v_0(k)$, the traffic density in the section after the last one, i.e. $\rho_{N+1}(k)$, $k = 0, \ldots, K$.

Note that the variables referred to on-ramps and off-ramps are defined for all the sections and are imposed to be equal to 0, i.e. $r_i(k) = 0$, $s_i(k) = 0$, $k = 0, \ldots, K$, in case section $i \in \{1, \ldots, N\}$ is not equipped with ramps.

4.2.2 *METANET with On-Ramp Queue Dynamics*

The METANET model reported in Sect. 4.2.1 describes the dynamic evolution of the traffic density and the mean speed in a freeway stretch with on-ramps and off-ramps, but it does not model the possible queues at the on-ramps. This latter aspect is particularly relevant when ramp metering control approaches are studied, as, for instance, in [27, 28].

As shown in Fig. 4.2, the following dynamic quantities are added to the model in order to include the dynamics of the queues at the on-ramps and the possible presence of a ramp metering controller:

- $l_i(k)$ is the queue length of vehicles waiting in the on-ramp of section i at time kT [veh];
- $d_i(k)$ is the flow accessing the on-ramp of section i during time interval $[kT, (k+1)T)$ [veh/h];

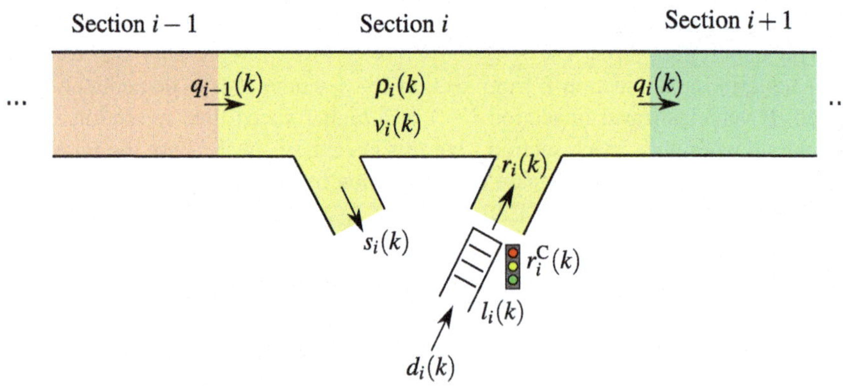

Fig. 4.2 Sketch of freeway stretch in case of on-ramp queues and the relative notation in METANET

- $r_i^{\mathrm{C}}(k)$ is the *ramp metering control variable*, i.e. the flow computed by the ramp metering controller that should enter section i from the on-ramp during time interval $[kT, (k+1)T)$ [veh/h].

Besides the parameters of the model described in Sect. 4.2.1, another parameter is considered. This parameter is $r_i^{\max}$, which represents the capacity of the on-ramp of section i, $i = 1, \ldots, N$.

The dynamic equation of the on-ramp queue length, for $i = 1, \ldots, N$, $k = 0, \ldots, K-1$, is given by

$$l_i(k+1) = l_i(k) + T\,[d_i(k) - r_i(k)] \tag{4.16}$$

In this model, the flow $r_i(k)$ entering the mainstream from the on-ramp is not a boundary condition, as in the model described in Sect. 4.2.1, since the boundary condition is now given by the demand $d_i(k)$. The flow $r_i(k)$ is computed in a different way depending on the fact that the on-ramp in section i is uncontrolled or controlled with ramp metering policies. Let us distinguish these two cases.

Uncontrolled On-Ramps In case the on-ramp of section i is uncontrolled, the flow $r_i(k)$ entering the mainstream from the on-ramp of section i during time interval $[kT, (k+1)T)$ is computed as

$$r_i(k) = \min\left\{d_i(k) + \frac{l_i(k)}{T},\, r_i^{\max},\, r_i^{\max}\frac{\rho_i^{\max} - \rho_i(k)}{\rho_i^{\max} - \rho_i^{\mathrm{cr}}}\right\} \tag{4.17}$$

Equation (4.17) computes the on-ramp flow as the minimum between three values: the flow corresponding to the vehicles in the on-ramp (waiting in the queue or reaching it), the on-ramp capacity, and the maximum flow that should enter the mainstream due to the traffic conditions. Note that this third term is computed as a reduction of the on-ramp capacity in case traffic conditions in the mainstream become congested, i.e. if $\rho_i(k) > \rho_i^{\mathrm{cr}}$.

Controlled On-Ramps If the on-ramp of section i is controlled, the flow $r_i(k)$ entering the mainstream from the on-ramp of section i during time interval $[kT, (k+1)T)$ is given by

$$r_i(k) = \min\left\{d_i(k) + \frac{l_i(k)}{T},\, r_i^{\max},\, r_i^{\mathrm{C}}(k),\, r_i^{\max}\frac{\rho_i^{\max} - \rho_i(k)}{\rho_i^{\max} - \rho_i^{\mathrm{cr}}}\right\} \tag{4.18}$$

in which the flow computed by the ramp metering controller $r_i^{\mathrm{C}}(k)$ is added as a fourth term in the minimum function.

In some cases, it is preferable to represent the controlled case in a slightly different way, i.e. instead of considering the flow $r_i^{\mathrm{C}}(k)$ as ramp metering control variable, the *metering rate* $\mu_i(k) \in [\mu_i^{\min}, 1]$ is adopted as control variable, $\mu_i^{\min}$ being the minimum on-ramp metering rate. If $\mu_i(k) = 1$, no ramp metering policy is applied, while ramp metering becomes active if $\mu_i(k) < 1$. In this case, (4.18) is replaced by the following:

$$r_i(k) = \mu_i(k) \min \left\{ d_i(k) + \frac{l_i(k)}{T}, r_i^{\max}, r_i^{\max} \frac{\rho_i^{\max} - \rho_i(k)}{\rho_i^{\max} - \rho_i^{\text{cr}}} \right\} \tag{4.19}$$

The augmented METANET model to include the on-ramp queue dynamics for a freeway with off-ramps and on-ramps is given by (4.12)–(4.16), with (4.17) if the on-ramps are not controlled, and (4.18) or (4.19) for the controlled on-ramps. The *boundary conditions* are, in this case, the traffic flow entering the first road section, i.e. $q_0(k)$, the off-ramp traffic flows $s_i(k)$, the on-ramp demands $d_i(k), i = 1, \ldots, N$, the mean traffic speed in the section before the first one, i.e. $v_0(k)$, the traffic density in the section after the last one, i.e. $\rho_{N+1}(k), k = 0, \ldots, K$.

As in Sect. 4.2.1, the variables referred to on-ramps and off-ramps are defined for all the sections. They are then fixed to 0, i.e. $r_i(k) = 0$, $d_i(k) = 0$, $l_i(k) = 0$, $s_i(k) = 0, k = 0, \ldots, K$, if section $i \in \{1, \ldots, N\}$ is not equipped with ramps.

Note that the equation for the queue dynamics can be also adopted to consider the possible queue which is created to enter the considered freeway stretch. In that case, $l_0(k)$ denotes this queue length and its dynamics is modelled similarly to (4.16), where the demand is $d_0(k)$ and the flow entering the mainstream is $q_0(k)$.

4.2.3 *METANET for a Freeway Network*

The METANET model described in the previous sections for a freeway stretch has been extended to consider a freeway network of arbitrary topology, including freeway stretches, bifurcations, on-ramps and off-ramps, in all types of traffic conditions, and also in case of events causing capacity reduction [24, 25].

According to this model, the freeway network is represented by means of a directed graph (see Fig. 4.3) composed of:

- M freeway links, i.e. freeway stretches with homogeneous geometric characteristics (number of lanes, curvatures and so on);
- O origin links, i.e. links which forward traffic flows from outside into the considered freeway network (they can represent either on-ramps or other freeway stretches merging in the considered network);
- N nodes, representing junctions, bifurcations, merging on-ramps or diverging off-ramps, connecting no more than three links.

Note that the assumptions aforementioned are not restrictive and allow to represent any type of freeway network. As a matter of fact, in case a freeway stretch presents inhomogeneous characteristics, it can be represented by two or more consecutive links separated by nodes positioned where the road geometry changes. Moreover, in case of a complex node connecting more than three links, it can be easily decomposed into more nodes meeting such condition, by introducing dummy links and dummy

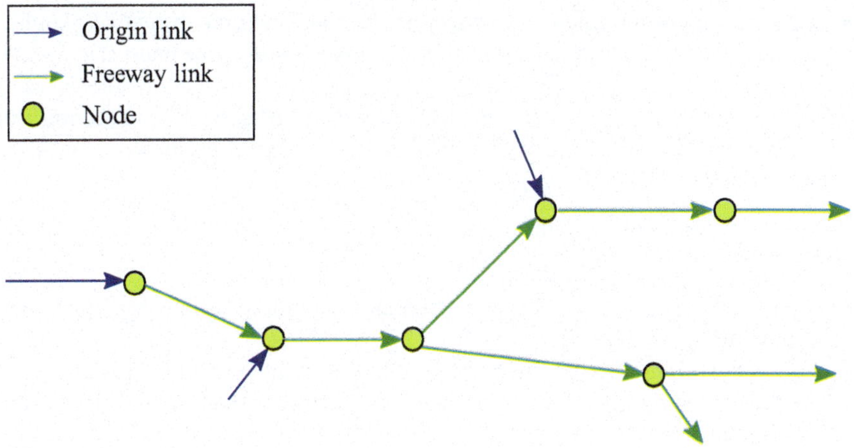

Fig. 4.3 Links and nodes in a freeway network according to METANET

nodes. Finally, note that the two directions of a freeway stretch should be represented as separate links with opposite directions.

Each freeway link $m = 1, \ldots, M$ is further divided into N_m sections which have a length denoted with L_m [km] and a number of lanes indicated with λ_m. Also, for each node $n = 1, \ldots, N$, O_n is the set of exiting links, and I_n, $\bar{I}_n$ are the set of entering freeway links and entering origin links, respectively.

The METANET model may be used to describe the traffic behaviour in a freeway network in two different ways: in a *non-destination-oriented mode*, when the traffic assignment problem is not considered and the destination of vehicles travelling in the network is neglected, or in a *destination-oriented mode*, when instead the drivers' route choice behaviour is considered and the choice of road users among alternative paths is explicitly modelled. In this section, the destination-oriented model is reported, since it is more general and particularly useful in case route guidance control is applied to the traffic network. The model in the non-destination-oriented mode is similar to the destination-oriented one, but it does not include the variables which depend on the destination of drivers (see [25] for further details).

In the destination-oriented model, for each link and for each node, it is necessary to specify the set of reachable destinations. Let us denote with J_m, $\bar{J}_o$, $\bar{\bar{J}}_n$, respectively, the sets of destinations reachable from freeway link $m = 1, \ldots, M$, from origin link $o = 1, \ldots, O$, and from node $n = 1, \ldots, N$.

The time horizon is divided into K time intervals, with sample time interval T [h]. The variables referring to the *freeway links*, for each freeway link $m = 1, \ldots, M$, for each section $i = 1, \ldots, N_m$, and for each time step $k = 0, \ldots, K$, are:

- $\rho_{m,i,j}(k)$ is the partial traffic density in section i of link m at time instant kT with destination $j \in J_m$ [veh/km/lane];
- $\rho_{m,i}(k)$ is the traffic density in section i of link m at time instant kT [veh/km/lane];

- $v_{m,i}(k)$ is the mean traffic speed in section i of link m at time instant kT [km/h];
- $q_{m,i}(k)$ is the traffic flow leaving section i of link m during time interval $[kT, (k+1)T)$ [veh/h];
- $\gamma_{m,i,j}(k) \in [0, 1]$ is the *composition rate*, i.e. the portion of flow in section i of link m at time instant kT having destination $j \in J_m$; the values of the composition rates must verify that $\sum_{j \in J_m} \gamma_{m,i,j}(k) = 1$.

The variables referring to the *origin links*, for each origin link $o = 1, \ldots, O$ and for each time step $k = 0, \ldots, K$, are:

- $d_{o,j}(k)$ is the partial origin demand entering origin link o at time instant kT with destination $j \in \bar{J}_o$ [veh/h];
- $d_o(k)$ is the origin demand entering origin link o at time instant kT [veh/h];
- $l_{o,j}(k)$ is the partial queue length at origin link o at time instant kT with destination $j \in \bar{J}_o$ [veh];
- $l_o(k)$ is the queue length at origin link o at time instant kT [veh];
- $\gamma_{o,j}(k) \in [0, 1]$ is the *composition rate*, i.e. the portion of flow leaving origin link o at time instant kT having destination $j \in \bar{J}_o$; the values of the composition rates must verify that $\sum_{j \in \bar{J}_o} \gamma_{o,j}(k) = 1$;
- $\theta_{o,j}(k) \in [0, 1]$ is the portion of the demand originating in origin link o at time instant kT having destination $j \in \bar{J}_o$; analogously to the composition rates, it holds that $\sum_{j \in \bar{J}_o} \theta_{o,j}(k) = 1$;
- $q_o(k)$ is the traffic flow leaving origin link o during time interval $[kT, (k+1)T)$ [veh/h];
- $r_o^{\mathrm{C}}(k)$ is the *ramp metering control variable*, i.e. the flow computed by the ramp metering controller that should enter from the origin link o during time interval $[kT, (k+1)T)$ [veh/h].

The variables referring to the *nodes*, for each node $n = 1, \ldots, N$ and for each time step $k = 0, \ldots, K$, are:

- $Q_{n,j}(k)$ is the flow entering node n during time interval $[kT, (k+1)T)$ with destination $j \in \bar{\bar{J}}_n$ [veh/h];
- $\beta_{m,n,j}(k) \in [0, 1]$ is the *splitting rate*, i.e. the portion of flow present in node n at time instant kT which chooses link m to reach destination $j \in \bar{\bar{J}}_n$; the values of the splitting rates must verify that $\sum_{\mu \in O_n} \beta_{\mu,n,j}(k) = 1$.

The model parameters are: v_m^{f} is the free-flow speed [km/h] in each section of link m, ρ_m^{cr} is the critical density [veh/km/lane] in each section of link m, $\rho_m^{\max}$ is the jam density [veh/km/lane] in each section of link m, $m = 1, \ldots, M$, $q_o^{\max}$ is the capacity of origin link o, $o = 1, \ldots, O$, whereas $\tau, \nu, \chi, \delta_{\mathrm{on}}, \phi$ are model parameters present in the speed equation, and a_m, $m = 1, \ldots, M$, is a parameter characterising the steady-state speed–density relation.

Let us now distinguish the model equations of the freeway links, the origin links and the nodes, in case of possible controlled on-ramps and route guidance control actions. Moreover, an additional description is added to show how METANET has been extended to include the application of mainstream control actions in terms of variable speed limits.

Freeway Links The equations characterising freeway links are an extension of (4.12)–(4.15), taking into account that in the network model the traffic densities are expressed per lane and that the destinations are taken into account. In particular, the conservation equation is here written for the partial traffic density, i.e.

$$\rho_{m,i,j}(k+1) = \rho_{m,i,j}(k) + \frac{T}{L_m \lambda_m} \left[\gamma_{m,i-1,j}(k) q_{m,i-1}(k) - \gamma_{m,i,j}(k) q_{m,i}(k)\right] \tag{4.20}$$

where $m = 1, \dots, M, i = 1, \dots, N_m, j \in J_m, k = 0, \dots, K-1$, and the following relations hold:

$$\rho_{m,i}(k) = \sum_{j \in J_m} \rho_{m,i,j}(k) \tag{4.21}$$

$$\gamma_{m,i,j}(k) = \frac{\rho_{m,i,j}(k)}{\rho_{m,i}(k)} \tag{4.22}$$

The speed dynamic equation is

$$v_{m,i}(k+1) = v_{m,i}(k) + \frac{T}{\tau}\left[V\left(\rho_{m,i}(k)\right) - v_{m,i}(k)\right]$$
$$+ \frac{T}{L_m} v_{m,i}(k)\left[v_{m,i-1}(k) - v_{m,i}(k)\right] - \frac{\nu T\left[\rho_{m,i+1}(k) - \rho_{m,i}(k)\right]}{\tau L_m\left[\rho_{m,i}(k) + \chi\right]} \tag{4.23}$$

where $m = 1, \dots, M, i = 1, \dots, N_m, k = 0, \dots, K-1$. The traffic flow in (4.20) and the steady-state speed–density relation in (4.23) are given by

$$q_{m,i}(k) = \rho_{m,i}(k) v_{m,i}(k) \lambda_m \tag{4.24}$$

$$V\left(\rho_{m,i}(k)\right) = v_m^{\mathrm{f}} \exp\left[-\frac{1}{a_m}\left(\frac{\rho_{m,i}(k)}{\rho_m^{\mathrm{cr}}}\right)^{a_m}\right] \tag{4.25}$$

In (4.23), an additional term can be added to take into account the speed reduction caused by merging phenomena near on-ramps, analogous to the fourth term in (4.13). In particular, let us consider a node in which an origin link o enters and let us denote with m the link exiting that node; in the first section of link m there is a speed reduction given by

$$-\delta_{\mathrm{on}} T \frac{v_{m,1}(k) q_o(k)}{L_m \lambda_m \left[\rho_{m,1}(k) + \chi\right]} \tag{4.26}$$

In (4.23), it is possible to add a further additional term to model the speed reduction due to weaving phenomena in case of lane reductions in the mainstream. By denoting with $\Delta\lambda$ the number of lanes dropped between link m and the following one, the speed reduction in the last section of link m is given by

$$-\phi T \Delta\lambda \frac{v_{m,N_m}(k)^2 \rho_{m,N_m}(k)}{L_m \lambda_m \rho_m^{\mathrm{cr}}} \tag{4.27}$$

The boundary conditions in (4.23) are the virtual downstream density $\rho_{m,N_m+1}(k)$ at the end of the link and the virtual upstream speed $v_{m,0}(k)$ at the beginning of the link. In case of nodes with one input link and output link, these values are obtained directly from adjacent links, but in case of nodes with more than two links, these values must be computed as suitable weighted sums. In particular, if node n (at the end of link m) has more than one leaving link, the virtual downstream density can be computed as in (4.65), i.e.

$$\rho_{m,N_m+1}(k) = \frac{\sum_{\mu \in O_n} \rho_{\mu,1}(k)^2}{\sum_{\mu \in O_n} \rho_{\mu,1}(k)} \tag{4.28}$$

where the quadratical relation is used to represent the fact that a highly loaded link contributes to the spillback more than proportionally.

In case node n (at the beginning of link m) has more than one entering link, the virtual upstream speed may be computed as

$$v_{m,0}(k) = \frac{\sum_{\mu \in I_n} v_{\mu,N_\mu}(k) q_{\mu,N_\mu}(k)}{\sum_{\mu \in I_n} q_{\mu,N_\mu}(k)} \tag{4.29}$$

Origin Links The equations of the origin links are analogous to (4.16)–(4.19), adapted to the network model. In particular, the dynamic evolution of the partial queue length is calculated as

$$l_{o,j}(k+1) = l_{o,j}(k) + T\left[d_{o,j}(k) - \gamma_{o,j}(k) q_o(k)\right] \tag{4.30}$$

where $o = 1, \ldots, O$, $j = 1, \ldots, \bar{J}_o$, $k = 0, \ldots, K-1$, and the following relations hold:

$$l_o(k) = \sum_{j \in \bar{J}_o} l_{o,j}(k) \tag{4.31}$$

$$\gamma_{o,j}(k) = \frac{l_{o,j}(k)}{l_o(k)} \tag{4.32}$$

$$d_{o,j}(k) = \theta_{o,j}(k) d_o(k) \tag{4.33}$$

The traffic flow leaving origin link o and entering the mainstream, i.e. entering the downstream link m, is given by

$$q_o(k) = \min\left\{d_o(k) + \frac{l_o(k)}{T}, q_o^{\max}, q_o^{\max}\frac{\rho_m^{\max} - \rho_{m,1}(k)}{\rho_m^{\max} - \rho_m^{\text{cr}}}\right\} \tag{4.34}$$

If, instead, the considered origin link o is a controlled on-ramp, the traffic flow leaving origin link o and entering link m is computed as

$$q_o(k) = \min\left\{d_o(k) + \frac{l_o(k)}{T}, q_o^{\max}, r_o^{\text{C}}(k), q_o^{\max}\frac{\rho_m^{\max} - \rho_{m,1}(k)}{\rho_m^{\max} - \rho_m^{\text{cr}}}\right\} \tag{4.35}$$

As already mentioned in Sect. 4.2.2, in some cases the considered control variable is the metering rate $\mu_o(k) \in [\mu_o^{\min}, 1]$, $\mu_o^{\min}$ being the minimum on-ramp metering rate. In these cases, (4.35) can be substituted by

$$q_o(k) = \mu_o(k)\min\left\{d_o(k) + \frac{l_o(k)}{T}, q_o^{\max}, q_o^{\max}\frac{\rho_m^{\max} - \rho_{m,1}(k)}{\rho_m^{\max} - \rho_m^{\text{cr}}}\right\} \tag{4.36}$$

Equations (4.30)–(4.35) can be slightly modified to represent the so-called *store-and-forward* links, which are links characterised not only by a capacity and a queue length but also by constant travel times. These links are useful to consider urban zones or motorway-to-motorway control [25].

Nodes The model of the nodes does not represent any dynamic behaviour, but only conservation of flows. The total traffic flow entering node $n = 1, \ldots, N$ with destination $j \in \bar{\bar{J}}_n$, referred to time step $k = 0, \ldots, K$, is computed as the sum of the entering flows with destination j, i.e.

$$Q_{n,j}(k) = \sum_{\mu \in I_n} q_{\mu,N_\mu}(k)\gamma_{\mu,N_\mu,j}(k) + \sum_{o \in \bar{I}_n} q_o(k)\gamma_{o,j}(k) \tag{4.37}$$

The traffic flow exiting node $n = 1, \ldots, N$ and entering the first section of link $m = 1, \ldots, M$, referred to time step $k = 0, \ldots, K$, is calculated as the sum of flows choosing link m in the bifurcation, i.e.

$$q_{m,0}(k) = \sum_{j \in J_m} \beta_{m,n,j}(k)Q_{n,j}(k) \tag{4.38}$$

Equation (4.38) is used to set the boundary conditions $q_{m,0}(k)$ in (4.20), where other boundary conditions are $\gamma_{m,0,j}(k)$, which are computed as

$$\gamma_{m,0,j}(k) = \frac{\beta_{m,n,j}(k)Q_{n,j}(k)}{q_{m,0}(k)} \tag{4.39}$$

In presence of *route guidance* control, the splitting rates become the control variables, but it is in this case important to distinguish among different variables representing splitting rates. The following quantities are added to the model:

- $\beta^{\mathrm{C}}_{m,n,j}(k) \in [0, 1]$ is the *route guidance control variable*, i.e. the splitting rate defined by a suitable traffic controller to be actuated at node n and representing the portion of flow present in node n at time instant kT which should choose link m to reach destination $j \in \bar{\bar{J}}_n$;
- $\beta^{\mathrm{N}}_{m,n,j}(k) \in [0, 1]$ is the nominal splitting rate, i.e. the portion of flow present in node n at time instant kT which would spontaneously choose link m to reach destination $j \in \bar{\bar{J}}_n$.

Note that $\beta^{\mathrm{C}}_{m,n,j}(k)$ is the splitting rate defined with a suitable control approach and communicated to drivers through the visualisation of proper recommendations on Variable Message Signs (VMSs). If, for instance, $\beta^{\mathrm{C}}_{m,n,j}(k) = 1$, this means that it is recommended to drivers to choose link m to reach destination j.

The effective splitting rates depend both on these splitting rates suggested through VMSs and the natural and spontaneous route choice of the drivers, according to a compliance rate $\varepsilon_{m,n} \in [0, 1]$, which is a model parameter. In the considered model, the effective splitting rates $\beta_{m,n,j}(k)$ are obtained as a weighing sum of the suggested rates $\beta^{\mathrm{C}}_{m,n,j}(k)$ and the nominal rates $\beta^{\mathrm{N}}_{m,n,j}(k)$ resulting in absence of route recommendations, i.e.

$$\beta_{m,n,j}(k) = (1 - \varepsilon_{m,n})\beta^{\mathrm{N}}_{m,n,j}(k) + \varepsilon_{m,n}\beta^{\mathrm{C}}_{m,n,j}(k) \tag{4.40}$$

Freeway links controlled with variable speed limits The original METANET model [25] does not describe the effect of variable speed limits applied in freeway links through VMSs. There are many different ways in which researchers have modelled this aspect but, up to now, there is not one model that is universally known as a suitable representation of variable speed limits in freeways. We will report two possible developments of METANET, widely adopted by researchers, which model the presence of variable speed limits in a freeway link in terms of a variation of the steady-state speed–density relation $V\left(\rho_{m,i}(k)\right)$ given by (4.25).

The model proposed in [29, 30] was developed in contrast with early models [31] in which the effect of speed limits was considered by scaling down the desired speed, consequently changing the shape of the whole Fundamental Diagram and reducing the capacity. According to the authors of [29, 30], that approach was not realistic and, then, a more realistic model was introduced, by assuming that the steady-state speed in case of variable speed limits is the minimum between the usual steady-state speed and the speed caused by the limit imposed through VMSs. According to this view, let us consider the following additional variable:

- $v^{\mathrm{C}}_{m,i}(k)$ is the *variable speed limit control variable* representing the traffic speed to display in section i of link m during time interval $[kT, (k+1)T)$ [km/h].

Then, the steady-state speed–density relationship becomes

$$V\left(\rho_{m,i}(k)\right) = \min\left\{v^{\mathrm{f}}_m \exp\left[-\frac{1}{a_m}\left(\frac{\rho_{m,i}(k)}{\rho^{\mathrm{cr}}_m}\right)^{a_m}\right], (1+\alpha)v^{\mathrm{C}}_{m,i}(k)\right\} \tag{4.41}$$

where α models the compliance rate of drivers. In (4.41), the control variable $v_{m,i}^{\mathrm{C}}(k)$ is multiplied for $(1+\alpha)$ because drivers normally do not follow completely the speed limits and their desired speed is usually higher than the imposed speed limit (see [29, 30] for more details on this model).

Another development of METANET to consider variable speed limits has been proposed more recently [32, 33]. In [32, 33], again, the steady-state speed–density relationship depends on the speed displayed on VMSs but the dependence is different from the one provided in (4.41), as well as the meaning of the control variable. Specifically, let us introduce the following additional variable:

- $b_m(k) \in [b^{\min}, 1]$ is the *variable speed limit control variable*, representing the variable speed limit rate to display in each section of link m during time interval $[kT, (k+1)T)$.

Note that $b^{\min}$ is a lower admissible bound for the control variable. This latter can be interpreted as a rate limiting the speed of vehicles, hence $b_m(k) = 1$ means that no variable speed limits are applied, while the control case corresponds to $b_m(k) < 1$. In this model, the steady-state speed–density relationship is written analogously to (4.25) but with parameters dependent on $b_m(k)$, i.e.

$$V\left(\rho_{m,i}(k), b_m(k)\right) = v_m^{\mathrm{f}}\left(b_m(k)\right) \exp\left[-\frac{1}{a_m\left(b_m(k)\right)}\left(\frac{\rho_{m,i}(k)}{\rho_m^{\mathrm{cr}}\left(b_m(k)\right)}\right)^{a_m(b_m(k))}\right] \tag{4.42}$$

where the dependence of the parameters on $b_m(k)$ is of affine type, as follows:

$$v_m^{\mathrm{f}}\left(b_m(k)\right) = v_m^{\mathrm{f}} b_m(k) \tag{4.43}$$

$$\rho_m^{\mathrm{cr}}\left(b_m(k)\right) = \rho_m^{\mathrm{cr}}\left[1 + A_m(1 - b_m(k))\right] \tag{4.44}$$

$$a_m\left(b_m(k)\right) = a_m\left[E_m - (E_m - 1)b_m(k)\right] \tag{4.45}$$

in which A_m and E_m are model parameters. Note that, when $b_m(k) = 1$, (4.42) is equal to (4.25).

To summarise, the METANET model for a freeway network is given by (4.20)–(4.40), with (4.41) or (4.42)–(4.45) instead of (4.25) in case of variable speed limits. The *boundary conditions* are the demands of the origin links $d_o(k)$, with the ratios of these demands for each destination, i.e. $\theta_{o,j}(k), o = 1, \ldots, O, j \in \bar{J}_o, k = 0, \ldots, K$, and the traffic density in the sections downstream the considered freeway network, i.e. $\rho_{m,N_m+1}(k), k = 0, \ldots, K$, for links $m \in \{1, \ldots, M\}$ which are destination links.

4.3 Multi-class Second-Order Models

As discussed in Sect. 3.4, there are many motivations for explicitly modelling the presence of multiple classes of vehicles in the traffic flow. These motivations not only apply for first-order traffic flow models but also for second-order models. Nevertheless, less research studies have dealt with the developments of second-order macroscopic models for the multi-class context, compared with first-order models, and these studies are rather recent.

In particular, few works in the multi-class literature deal with *continuous* second-order traffic flow models. For instance, in [34], starting from a car-following model for heterogeneous traffic flow and exploiting the relationship between micro and macro variables, a macroscopic traffic model is developed to represent the flow dynamics of cars and buses. In [35], the Aw–Rascle model is extended to represent heterogeneous traffic flow: in that paper, this model is calibrated using data from an arterial section in India and the results are compared with those obtained from other multi-class traffic models.

The literature on *discrete* second-order traffic flow models extended to the multi-class case is limited as well. In [36], the METANET model is adapted to represent a heterogeneous flow, considering an interpolation among the different Fundamental Diagrams of each class of vehicles, and is exploited within a Model Predictive Control (MPC) approach. A different multi-class second-order traffic model is proposed in [37], extending the approach proposed in [38]. In [37], each vehicle class is subject to its own single-class Fundamental Diagram, and is limited within an assigned space.

Another multi-class extension of the METANET model was proposed in [39], then slightly modified in [40, 41] for a freeway stretch and in [42] for a freeway network. In these latter models, the interaction among the different classes of vehicles is modelled through a Fundamental Diagram, different for each class, in which the flow of each class depends on the total density. In the following subsections, this latter multi-class model is analysed more in detail, respectively for a freeway stretch and for a network.

4.3.1 *A Multi-class Second-Order Model for a Freeway Stretch*

The model reported in this section was proposed in [40, 41], for the case of a multi-class ramp metering strategy. It is worth noting that considering a multi-class ramp metering policy implies that separate lanes and separate traffic lights are present at the on-ramps for different vehicles classes; the most realistic case is surely the one in which cars and trucks are distinguished and the on-ramps are divided in two different lanes.

The considered model extends the METANET model for a freeway stretch, described in Sects. 4.2.1 and 4.2.2, to the case in which different classes of vehicles are taken into account. Even though some notation of the multi-class model is common to the one-class model, for the reader's convenience the entire nomenclature of the multi-class model is reported and described in the following.

The considered multi-class macroscopic traffic flow model is based on the division of the freeway stretch into N sections and the discretisation of the time horizon into K time intervals. Moreover, C classes of vehicles are considered. Let T indicate the sample time interval and L_i the length [km] of section i, $i = 1, \dots, N$.

In order to correctly model the presence of different types of vehicles, let us introduce the parameter η^c, $c = 1, \dots, C$, which represents a conversion factor of vehicles of class c into cars. This parameter has a meaning analogous to the definition of *Passenger Car Equivalents* (PCE), that is the number of passenger cars displaced by a single heavy vehicle of a particular type under specific traffic and control conditions [43]. This parameter can vary depending on the traffic conditions in a road portion [44], but in this multi-class traffic model, η^c, $c = 1, \dots, C$, is assumed to be a constant value.

For each section $i = 1, \dots, N$, and for each time step $k = 0, \dots K$, the main aggregate variables of the model are defined for each class $c = 1, \dots, C$:

- $\rho_i^c(k)$ is the traffic density of class c in section i at time kT [veh^c/km];
- $v_i^c(k)$ is the mean traffic speed of class c in section i at time kT [km/h];
- $q_i^c(k)$ is the traffic flow of class c leaving section i during time interval $[kT, (k+1)T)$ [veh^c/h];
- $r_i^c(k)$ is the on-ramp traffic flow of class c entering section i during time interval $[kT, (k+1)T)$ [veh^c/h];
- $s_i^c(k)$ is the off-ramp traffic flow of class c exiting section i during time interval $[kT, (k+1)T)$ [veh^c/h];
- $l_i^c(k)$ is the queue length of vehicles of class c waiting in the on-ramp of section i at time kT [veh^c];
- $d_i^c(k)$ is the flow of class c accessing the on-ramp of section i during time interval $[kT, (k+1)T)$ [veh^c/h];
- $r_i^{\mathrm{C},c}(k)$ is the *ramp metering control variable*, i.e. the flow of class c, computed by the ramp metering controller, that should enter section i from the on-ramp during time interval $[kT, (k+1)T)$ [veh^c/h].

To correctly define the multi-class model, some variables referred to the total flow of vehicles are also required, as follows:

- $\rho_i(k)$ is the traffic density in section i at time kT [PCE/km];
- $r_i(k)$ is the total on-ramp traffic flow entering section i during time interval $[kT, (k+1)T)$ [PCE/h].

The considered model includes some traffic parameters. Specifically, $v_i^{\mathrm{f},c}$ is the free-flow speed [km/h] referred to class c and section i, ρ_i^{cr} is the critical density [PCE/km] of section i, $\rho_i^{\max}$ is the jam density [PCE/km] of section i, $r_i^{\max,c}$ is the on-ramp capacity for class c and section i [veh^c/h], $c = 1, \dots, C$, $i = 1, \dots, N$, while τ^c, ν^c, χ^c, δ_{on}^c are model parameters present in the speed equation, and l^c, m^c are parameters of the steady-state speed–density relation, $c = 1, \dots, C$.

On the basis of the single-class model (4.12)–(4.18), the multi-class traffic model for a freeway stretch is given by the following dynamic equations:

$$\rho_i^c(k+1) = \rho_i^c(k) + \frac{T}{L_i}\left[q_{i-1}^c(k) - q_i^c(k) + r_i^c(k) - s_i^c(k)\right] \tag{4.46}$$

$$\begin{aligned} v_i^c(k+1) = v_i^c(k) &+ \frac{T}{\tau^c}\left[V^c(\rho_i(k)) - v_i^c(k)\right] + \frac{T}{L_i} v_i^c(k)\left[v_{i-1}^c(k) - v_i^c(k)\right] \\ &- \frac{\nu^c T\left[\rho_{i+1}(k) - \rho_i(k)\right]}{\tau^c L_i\left[\rho_i(k) + \chi^c\right]} - \delta_{\text{on}}^c T \frac{v_i^c(k) r_i(k)}{L_i\left[\rho_i(k) + \chi^c\right]} \end{aligned} \tag{4.47}$$

$$l_i^c(k+1) = l_i^c(k) + T\left[d_i^c(k) - r_i^c(k)\right] \tag{4.48}$$

where $c = 1, \ldots, C$, $i = 1, \ldots, N$, $k = 0, \ldots, K-1$. Note that, in the speed equation (4.47) for vehicles of class c, the anticipation term depends on the total density $\rho_i(k)$ downstream, as well as the fourth term depends on the total on-ramp flow $r_i(k)$ merging in the mainstream, since the acceleration or deceleration of class c depends on the total flow of vehicles seen ahead.

The traffic flow in (4.46) is obtained as

$$q_i^c(k) = \rho_i^c(k) v_i^c(k) \tag{4.49}$$

whereas the total density and the total on-ramp traffic flow used in (4.47) can be computed, respectively, as

$$\rho_i(k) = \sum_{c=1}^{C} \eta^c \rho_i^c(k) \tag{4.50}$$

$$r_i(k) = \sum_{c=1}^{C} \eta^c r_i^c(k) \tag{4.51}$$

and the steady-state speed–density relation in (4.47) is given by

$$V^c(\rho_i(k)) = v_i^{\text{f},c}\left[1 - \left(\frac{\rho_i(k)}{\rho_i^{\max}}\right)^{l^c}\right]^{m^c} \tag{4.52}$$

In (4.52) the steady-state speed of class c depends on the total density $\rho_i(k)$ and on parameters that are specific of class c, i.e. the free-flow speed $v_i^{\text{f},c}$, and parameters l^c and m^c. Note that, in the multi-class version of the freeway traffic model, the steady-state relation (4.52) has been used, instead of considering a multi-class version of (4.15), since this type of relation presents a more general form, as already discussed in Sect. 3.1.2.

If the on-ramps are not controlled, the on-ramp traffic flow is computed as

$$r_i^c(k) = \min\left\{d_i^c(k) + \frac{l_i^c(k)}{T}, r_i^{\max,c}, r_i^{\max,c}\frac{\rho_i^{\max} - \rho_i(k)}{\rho_i^{\max} - \rho_i^{\mathrm{cr}}}\right\} \tag{4.53}$$

whereas, in the controlled case, this flow is given by

$$r_i^c(k) = \min\left\{d_i^c(k) + \frac{l_i^c(k)}{T}, r_i^{\mathrm{C},c}(k), r_i^{\max,c}, r_i^{\max,c}\frac{\rho_i^{\max} - \rho_i(k)}{\rho_i^{\max} - \rho_i^{\mathrm{cr}}}\right\} \tag{4.54}$$

If, as mentioned in Sect. 4.2.2, the considered control variable is the metering rate $\mu_i^c(k) \in [\mu_i^{\min,c}, 1]$, $\mu_i^{\min,c}$ being the minimum on-ramp metering rate, (4.54) can be substituted by

$$r_i^c(k) = \mu_i^c(k)\min\left\{d_i^c(k) + \frac{l_i^c(k)}{T}, r_i^{\max,c}, r_i^{\max,c}\frac{\rho_i^{\max} - \rho_i(k)}{\rho_i^{\max} - \rho_i^{\mathrm{cr}}}\right\} \tag{4.55}$$

Note that, in the last term in (4.53)–(4.55), the total density $\rho_i(k)$ is considered, since the reduction of the on-ramp entering flow due to congestion in the mainstream is related to the density of all the vehicles present in the mainstream.

4.3.2 A Multi-class Second-Order Model for a Freeway Network

The multi-class second-order model for a freeway network presented here is the multi-class extension of the network model described in Sect. 4.2.3, taking into account the multi-class concepts already described in Sect. 4.3.1. This model was proposed in [42] for a freeway network in which the on-ramps are controlled and route guidance policies are applied.

Even though some notation and some definitions are common to the models described in the previous sections, all the notation of this model is described for the reader's convenience. On the other hand, the repeated information is only briefly described, and the reader can find more details in the aforementioned sections.

The time horizon is divided into K time intervals, with sample time interval T [h], and C classes of vehicles are considered, with η^c representing a conversion factor of vehicles of class c into cars, $c = 1, \ldots, C$. The freeway network is represented with a directed graph composed of M freeway links, O origin links, and N nodes. Each freeway link $m = 1, \ldots, M$ is further divided into N_m sections with length L_m [km] and number of lanes λ_m. For each node $n = 1, \ldots, N$, O_n is the set of exiting links, and I_n, $\bar{I}_n$ are the set of entering freeway links and entering origin links, respectively. The sets of destinations reachable from freeway link $m = 1, \ldots, M$, from origin link $o = 1, \ldots, O$, and from node $n = 1, \ldots, N$ are denoted with J_m, $\bar{J}_o$, $\bar{\bar{J}}_n$, respectively.

The main variables referring to the freeway links, for each vehicle class $c = 1, \ldots, C$, for each freeway link $m = 1, \ldots, M$, for each section $i = 1, \ldots, N_m$, and for each time step $k = 0, \ldots, K$, are:

- $\rho^c_{m,i,j}(k)$ is the partial traffic density of class c in section i of link m at time instant kT with destination $j \in J_m$ [veh^c/km/lane];
- $\rho^c_{m,i}(k)$ is the traffic density of class c in section i of link m at time instant kT [veh^c/km/lane];
- $\rho_{m,i}(k)$ is the traffic density in section i of link m at time instant kT [PCE/km/lane];
- $v^c_{m,i}(k)$ is the mean traffic speed of class c in section i of link m at time instant kT [km/h];
- $q^c_{m,i}(k)$ is the traffic flow of class c leaving section i of link m during time interval $[kT, (k+1)T)$ [veh^c/h];
- $\gamma^c_{m,i,j}(k) \in [0, 1]$ is the composition rate, i.e. the portion of the traffic flow of class c in section i of link m at time instant kT having destination $j \in J_m$; it holds that $\sum_{j \in J_m} \gamma^c_{m,i,j}(k) = 1$.

The main variables referring to the origin links, for each vehicle class $c = 1, \ldots, C$, for each origin link $o = 1, \ldots, O$ and for each time step $k = 0, \ldots, K$, are:

- $d^c_{o,j}(k)$ is the partial origin demand of class c entering origin link o at time instant kT with destination $j \in \bar{J}_o$ [veh^c/h];
- $d^c_o(k)$ is the origin demand of class c entering origin link o at time instant kT [veh^c/h];
- $l^c_{o,j}(k)$ is the partial queue length of class c at origin link o at time instant kT with destination $j \in \bar{J}_o$ [veh^c];
- $l^c_o(k)$ is the queue length of class c at origin link o at time instant kT [veh^c];
- $\gamma^c_{o,j}(k) \in [0, 1]$ is the composition rate, i.e. the portion of flow of class c leaving origin link o at time instant kT having destination $j \in \bar{J}_o$; it holds that $\sum_{j \in \bar{J}_o} \gamma^c_{o,j}(k) = 1$;
- $\theta^c_{o,j}(k) \in [0, 1]$ is the portion of the demand of class c originating in origin link o at time instant kT having destination $j \in \bar{J}_o$; it holds that $\sum_{j \in \bar{J}_o} \theta^c_{o,j}(k) = 1$;
- $q^c_o(k)$ is the traffic flow of class c leaving origin link o during time interval $[kT, (k+1)T)$ [veh^c/h];
- $q_o(k)$ is the total traffic flow leaving origin link o during time interval $[kT, (k+1)T)$ [PCE/h];
- $r^{C,c}_o(k)$ is the *ramp metering control variable*, i.e. the flow of class c, computed by the ramp metering controller, that should enter from origin link o during time interval $[kT, (k+1)T)$ [veh^c/h].

The variables referring to the nodes, for each vehicle class $c = 1, \ldots, C$, for each node $n = 1, \ldots, N$ and for each time step $k = 0, \ldots, K$, are:

- $Q^c_{n,j}(k)$ is the flow of class c entering node n during time interval $[kT, (k+1)T)$ with destination $j \in \bar{\bar{J}}_n$ [veh^c/h];

- $\beta^{c}_{m,n,j}(k) \in [0, 1]$ is the effective splitting rate, i.e. the portion of flow of class c present in node n at time instant kT which chooses link m to reach destination $j \in \bar{\bar{J}}_n$; it holds that $\sum_{\mu \in O_n} \beta^{c}_{\mu,n,j}(k) = 1$;
- $\beta^{\mathrm{C},c}_{m,n,j}(k)$ is the *route guidance control variable*, i.e. the splitting rate defined by a traffic controller, representing the portion of flow of class c present in node n at time instant kT which should choose link m to reach destination $j \in \bar{\bar{J}}_n$;
- $\beta^{\mathrm{N},c}_{m,n,j}(k)$ is the nominal splitting rate, i.e. the portion of flow of class c present in node n at time instant kT which would spontaneously choose link m to reach destination $j \in \bar{\bar{J}}_n$.

The model parameters are: $v^{\mathrm{f},c}_{m,i}$ is the free-flow speed [km/h] in section i of link m for class c, $\rho^{\mathrm{cr}}_{m,i}$ is the critical density [PCE/km/lane] in section i of link m, $\rho^{\max}_{m,i}$ is the jam density [PCE/km/lane] in section i of link m, $c = 1, \ldots, C$, $m = 1, \ldots, M$, $i = 1, \ldots, N_m$, $q^{\max,c}_{o}$ is the capacity of origin link o for class c, $c = 1, \ldots, C$, $o = 1, \ldots, O$, $\varepsilon^{c}_{m,n} \in [0, 1]$ is the compliance rate with the route recommendations for class c, $c = 1, \ldots, C$, $m = 1, \ldots, M$, $N = 1, \ldots, N$, whereas τ^c, υ^c, χ^c, δ^{c}_{on}, ϕ^c are model parameters present in the speed equation and specifically defined for class c, $c = 1, \ldots, C$, and l^c, m^c are parameters of the steady-state speed–density relation, $c = 1, \ldots, C$.

The equations of the multi-class network model are obtained from those of the single-class case, i.e. (4.20)–(4.40), properly extended to consider multiple classes of vehicles. Starting from the freeway links, the dynamic equations for the partial traffic density and the mean speed are

$$\rho^{c}_{m,i,j}(k+1) = \rho^{c}_{m,i,j}(k) + \frac{T}{L_m \lambda_m}\left[\gamma^{c}_{m,i-1,j}(k) q^{c}_{m,i-1}(k) - \gamma^{c}_{m,i,j}(k) q^{c}_{m,i}(k)\right] \tag{4.56}$$

$$\begin{aligned} v^{c}_{m,i}(k+1) &= v^{c}_{m,i}(k) + \frac{T}{\tau^c}\left[V^c\left(\rho_{m,i}(k)\right) - v^{c}_{m,i}(k)\right] \\ &+ \frac{T}{L_m} v^{c}_{m,i}(k)\left[v^{c}_{m,i-1}(k) - v^{c}_{m,i}(k)\right] - \frac{\upsilon^c T\left[\rho_{m,i+1}(k) - \rho_{m,i}(k)\right]}{\tau^c L_m\left[\rho_{m,i}(k) + \chi^c\right]} \end{aligned} \tag{4.57}$$

where $c = 1, \ldots, C, m = 1, \ldots, M, i = 1, \ldots, N_m, j \in J_m, k = 0, \ldots, K-1$, and the following relations hold:

$$\rho^{c}_{m,i}(k) = \sum_{j \in J_m} \rho^{c}_{m,i,j}(k) \tag{4.58}$$

$$\gamma^{c}_{m,i,j}(k) = \frac{\rho^{c}_{m,i,j}(k)}{\rho^{c}_{m,i}(k)} \tag{4.59}$$

$$\rho_{m,i}(k) = \sum_{c=1}^{C} \eta^c \rho^{c}_{m,i}(k) \tag{4.60}$$

The traffic flow in (4.56) and the steady-state speed–density relation in (4.57) are given, respectively, by

$$q_{m,i}^c(k) = \rho_{m,i}^c(k) v_{m,i}^c(k) \lambda_m \tag{4.61}$$

$$V^c\left(\rho_{m,i}(k)\right) = v_{m,i}^{\mathrm{f},c}\left[1 - \left(\frac{\rho_{m,i}(k)}{\rho_{m,i}^{\max}}\right)^{l^c}\right]^{m^c} \tag{4.62}$$

In (4.57), a term can be added to take into account the speed reduction due to merging flows coming from on-ramps. Considering a node in which an origin link o merges, in the first section of link m leaving that node there is a speed reduction given by

$$-\delta_{\mathrm{on}}^c T \frac{v_{m,1}^c(k) q_o(k)}{L_m \lambda_m \left[\rho_{m,1}(k) + \chi^c\right]} \tag{4.63}$$

A further additional term can be added to (4.57), to model the speed reduction due to weaving phenomena in case of lane reductions. By denoting with $\Delta\lambda$ the number of lanes dropped between link m and the following one, the speed reduction in the last section of link m is given by

$$-\phi^c T \Delta\lambda \frac{v_{m,N_m}^c(k)^2 \rho_{m,N_m}(k)}{L_m \lambda_m \rho_m^{\mathrm{cr}}} \tag{4.64}$$

The boundary conditions of (4.57) are the virtual downstream density at the end of the link $\rho_{m,N_m+1}(k)$ and the virtual upstream speed at the beginning of the link $v_{m,0}^c(k)$. If node n (at the end of link m) has more than one leaving link, the virtual downstream density can be computed as in (4.28), i.e.

$$\rho_{m,N_m+1}(k) = \frac{\sum_{\mu \in O_n} \rho_{\mu,1}(k)^2}{\sum_{\mu \in O_n} \rho_{\mu,1}(k)} \tag{4.65}$$

In case node n (at the beginning of link m) has more than one entering link, the virtual upstream speed may be computed as

$$v_{m,0}^c(k) = \frac{\sum_{\mu \in I_n} v_{\mu,N_\mu}^c(k) q_{\mu,N_\mu}^c(k)}{\sum_{\mu \in I_n} q_{\mu,N_\mu}^c(k)} \tag{4.66}$$

Let us now consider the origin links. The dynamic evolution of the partial queue length is given by

$$l_{o,j}^c(k+1) = l_{o,j}^c(k) + T\left[d_{o,j}^c(k) - \gamma_{o,j}^c(k) q_o^c(k)\right] \tag{4.67}$$

where $c = 1, \dots, C$, $o = 1, \dots, O$, $j = 1, \dots, \bar{J}_o$, $k = 0, \dots, K-1$, and the following relations hold:

$$l_o^c(k) = \sum_{j \in \bar{J}_o} l_{o,j}^c(k) \tag{4.68}$$

$$\gamma_{o,j}^c(k) = \frac{l_{o,j}^c(k)}{l_o^c(k)} \tag{4.69}$$

$$d_{o,j}^c(k) = \theta_{o,j}^c(k) d_o^c(k) \tag{4.70}$$

The traffic flow of class c leaving each origin link o, having m as downstream link, is given by

$$q_o^c(k) = \min\left\{ d_o^c(k) + \frac{l_o^c(k)}{T}, q_o^{\max,c}, q_o^{\max,c} \frac{\rho_{m,1}^{\max} - \rho_{m,1}(k)}{\rho_{m,1}^{\max} - \rho_{m,1}^{\mathrm{cr}}} \right\} \tag{4.71}$$

In case the considered origin link o is a controlled on-ramp, the traffic flow of class c leaving origin link o and entering link m is computed as

$$q_o^c(k) = \min\left\{ d_o^c(k) + \frac{l_o^c(k)}{T}, q_o^{\max,c}, r_o^{\mathrm{C},c}(k), q_o^{\max,c} \frac{\rho_{m,1}^{\max} - \rho_{m,1}(k)}{\rho_{m,1}^{\max} - \rho_{m,1}^{\mathrm{cr}}} \right\} \tag{4.72}$$

As for the node model, the total traffic flow entering node n with destination j is computed as

$$Q_{n,j}^c(k) = \sum_{\mu \in I_n} q_{\mu,N_\mu}^c(k) \gamma_{\mu,N_\mu,j}^c(k) + \sum_{o \in \bar{I}_n} q_o^c(k) \gamma_{o,j}^c(k) \tag{4.73}$$

The traffic flow exiting node n and entering the first section of link m is calculated as

$$q_{m,0}^c(k) = \sum_{j \in J_m} \beta_{m,n,j}^c(k) Q_{n,j}^c(k) \tag{4.74}$$

and the relative composition rate is given by

$$\gamma_{m,0,j}^c(k) = \frac{\beta_{m,n,j}^c(k) Q_{n,j}^c(k)}{q_{m,0}^c(k)} \tag{4.75}$$

In presence of route guidance control actions, the splitting rates are computed according to the following relation:

$$\beta_{m,n,j}^c(k) = (1 - \varepsilon_{m,n}^c) \beta_{m,n,j}^{\mathrm{N},c}(k) + \varepsilon_{m,n}^c \beta_{m,n,j}^{\mathrm{C},c}(k) \tag{4.76}$$

References

1. Payne HJ (1971) Models of freeway traffic and control. Math Model Public Syst 28:51–61
2. Whitham GB (1974) Linear and nonlinear waves. Wiley, New York
3. Daganzo CF (1995) Requiem for second-order fluid approximations of traffic flow. Transp Res Part B 29:277–286
4. Aw A, Rascle M (2000) Resurrection of "second order" models of traffic flow. SIAM J Appl Math 60:916–938
5. Zhang HM (2002) A non-equilibrium traffic model devoid of gas-like behavior. Transp Res Part B 36:275–290
6. Garavello M, Piccoli B (2016) Traffic flow on networks. American Institute of Mathematical Sciences
7. Garavello M, Han K, Piccoli B (2006) Models for vehicular traffic on networks. American Institute of Mathematical Sciences
8. Helbing D, Johansson AF (2009) On the controversy around Daganzo's requiem for and Aw-Rascle's resurrection of second-order traffic flow models. Eur Phys J 69:549–562
9. Greenberg JM (2001) Extensions and amplifications of a traffic model of Aw and Rascle. SIAM J Appl Math 62:729–745
10. Rascle M (2002) An improved macroscopic model of traffic flow: derivation and links with the Lighthill-Whitham model. Math Comput Model 35:581–590
11. Lebacque J-P, Mammar S, Haj-Salem H (2007) The Aw-Rascle and Zhang's model: vacuum problems, existence and regularity of the solutions of the Riemann problem. Transp Res Part B 41:710–721
12. Garavello M, Piccoli B (2006) Traffic flow on a road network using the Aw-Rascle model. Commun Partial Differ Equ 31:243–275
13. Herty M, Rascle M (2006) Coupling conditions for a class of second-order models for traffic flow. SIAM J Math Anal 38:595–616
14. Herty M, Moutari S, Rascle M (2006) Optimization criteria for modelling intersections of vehicular traffic flow. Netw Heterog Media 1:275–294
15. Kerner B (1998) Experimental features of self-organization in traffic flow. Phys Rev Lett 81:3797–3800
16. Colombo R (2003) Hyperbolic phase transitions in traffic flow. SIAM J Appl Math 63:708–721
17. Blandin S, Work D, Goatin P, Piccoli B, Bayen A (2011) A general phase transition model for vehicular traffic. SIAM J Appl Math 71:107–127
18. Blandin S, Argote J, Bayen AM, Work DB (2013) Phase transition model of non-stationary traffic flow: definition. properties and solution method. Transp Res Part B 52:31–55
19. Goatin P (2006) The Aw-Rascle vehicular traffic flow model with phase transitions. Math Comput Model 44:287–303
20. Colombo RM, Goatin P, Piccoli B (2010) Road networks with phase transitions. J Hyperbolic Differ Equ 7:85–106
21. Colombo RM, Garavello M (2014) Phase transition model for traffic at a junction. J Math Sci 196:30–36
22. Papageorgiou M, Blosseville J-M, Hadj-Salem H (1989) Macroscopic modelling of traffic flow on the Boulevard Périphérique in Paris. Transp Res Part B 23:29–47
23. Papageorgiou M (1990) Modelling and real-time control of traffic flow on the Southern part of Boulevard Périphérique in Paris: part I: modelling. Transp Res Part A 24:345–359
24. Messmer A, Papageorgiou M (1990) METANET: a macroscopic simulation program for motorway networks. Traffic Eng Control 31:466–470
25. Kotsialos A, Papageorgiou M, Diakaki C, Pavlis Y, Middelham F (2002) Traffic flow modeling of large-scale motorway networks using the macroscopic modeling tool METANET. IEEE Trans Intell Transp Syst 3:282–292
26. Cremer M, May AD (1986) An extended traffic flow model for inner urban freeways. In: Preprints of 5th IFAC/IFIP/IFORS International conference on control in transportation systems, pp 383–388

27. Papageorgiou M, Kotsialos A (2002) Freeway ramp metering: an overview. IEEE Trans Intell Transp Syst 3:271–281
28. Bellemans T, De Schutter B, De Moor B (2006) Model predictive control for ramp metering of motorway traffic: a case study. Control Eng Pract 14:757–767
29. Hegyi A, De Schutter B, Hellendoorn H (2005) Model predictive control for optimal coordination of ramp metering and variable speed limits. Transp Res Part C 13:185–209
30. Hegyi A, De Schutter B, Hellendoorn J (2005) Optimal coordination of variable speed limits to suppress shock waves. IEEE Trans Intell Transp Syst 6:102–112
31. Cremer M (1979) Der Verkehrsfluss auf Schnellstrassen (Traffic flow on freeways), Fachberichte Messen 3, Steuern, Regeln. Springer, Berlin
32. Carlson RC, Papamichail I, Papageorgiou M, Messmer A (2010) Optimal mainstream traffic flow control of large-scale motorway networks. Transp Res Part C 18:193–212
33. Carlson RC, Papamichail I, Papageorgiou M (2011) Local feedback-based mainstream traffic flow control on motorways using variable speed limits. IEEE Trans Intell Transp Syst 12:1261–1276
34. Tang TQ, Huang HJ, Zhao SG, Shang HY (2009) A new dynamic model for heterogeneous traffic flow. Phys Lett A 373:2461–2466
35. Mohan R, Ramadurai G (2017) Heterogeneous traffic flow modelling using second-order macroscopic continuum model. Phys Lett A 381:115–123
36. Deo P, De Schutter B, Hegyi A (2009) Model predictive control for multi-class traffic flows. In: Proceedings of the 12th IFAC symposium on transportation systems, pp 25–30
37. Liu S, De Schutter B, Hellendoorn H (2014) Model predictive traffic control based on a new multi-class METANET model. In: Proceedings of the 19th IFAC world congress, pp 8781–8785
38. Logghe S, Immers LH (2008) Multi-class kinematic wave theory of traffic flow. Transp Res Part B 42:523–541
39. Caligaris C, Sacone S, Siri S (2007) Optimal ramp metering and variable speed signs for multiclass freeway traffic. In: Proceedings of the European control conference, pp 1780–1785
40. Pasquale C, Sacone S, Siri S (2014) Two-class emission traffic control for freeway systems. In: Proceedings of the 19th IFAC world congress, pp 936–941
41. Pasquale C, Papamichail I, Roncoli C, Sacone S, Siri S, Papageorgiou M (2015) Two-class freeway traffic regulation to reduce congestion and emissions via nonlinear optimal control. Transp Res Part C 55:85–99
42. Pasquale C, Sacone S, Siri S, De Schutter B (2017) A multi-class model-based control scheme for reducing congestion and emissions in freeway networks by combining ramp metering and route guidance. Transp Res Part C 80:384–408
43. Special report 209 (1994) Highway capacity manual, 3rd edn. Transportation Research Board, Washington DC
44. Al-Kaisy AF, Hall FL, Reisman ES (2002) Developing passenger car equivalents for heavy vehicles on freeways during queue discharge flow. Transp Res Part A 36:725–742

Chapter 5
Microscopic and Mesoscopic Traffic Models

5.1 Uses and Applications of Traffic Models

Chapters 3 and 4 of this book are focused on *macroscopic* traffic models, which represent the dynamics of traffic flow by means of aggregate variables. The main classifications of macroscopic traffic models distinguish them according to the number of variables whose dynamics is explicitly taken into account, corresponding to first-order, second-order or higher-order models. Macroscopic models, generally allow to represent large road networks with an acceptable computational load. This computational advantage characterising macroscopic models is counterbalanced by the drawback that these models cannot capture some specific traffic phenomena related to the behaviour of individual drivers.

On the opposite side, *microscopic* models describe the dynamic behaviour of each single vehicle in the traffic stream, trying to capture the interactions among vehicles and between vehicles and the road infrastructure. These models can be very detailed and accurate in representing specific features of traffic but, of course, are very demanding from a computational point of view. Another important drawback of microscopic models is that they are often characterised by a very high number of parameters which must be properly calibrated. In case of models including heterogeneity among drivers or vehicles and stochasticity, the number of parameters becomes higher. Section 5.2 is devoted to some of the microscopic models present in literature, but it does not aim to exhaustively cover the wide variety of microscopic traffic models. The interested reader can refer to [1–3] and the references therein for a comprehensive overview on the topic.

A very interesting use of microscopic models is their utilisation inside *traffic simulation tools* (see Sect. 5.2.4). Indeed, the complexity of the traffic stream behaviour and the difficulties in performing experiments with real-world cases make computer simulation an important analysis tool in the traffic engineering field. By making use of different traffic models, generally of microscopic type, one can simulate large-scale real-world situations in great detail [4, 5].

A. Ferrara et al., *Freeway Traffic Modelling and Control*, Advances in Industrial Control, https://doi.org/10.1007/978-3-319-75961-6_5

An intermediate class of traffic models, which bridges the gap between the higher level of detail of microscopic models and the aggregate description of macroscopic models, is constituted by the so-called *mesoscopic* models. These models represent a link between microscopic and macroscopic modelling, where the characteristic aspects of both levels of description are combined. In mesoscopic models, the traffic flow dynamics is described in aggregate terms using probability distribution functions and the dynamics of these distributions is governed by individual drivers' behaviour. In fact, even if mesoscopic models do not distinguish individual vehicles (as it happens instead with microscopic models), they specify individual behaviours in probabilistic terms. In this sense, mesoscopic models provide an intermediate option with their ability to model large road networks with limited coding and calibration effort, while providing a better representation of the traffic dynamics and individual travel behaviour than their macroscopic counterparts. Some mesoscopic traffic models are presented in Sect. 5.3, which, again, does not represent an exhaustive survey of all the mesoscopic models appeared in the literature. The interested reader can refer to [1, 2] and the references therein for a more detailed discussion on mesoscopic models.

Taking into account all the traffic models present in the literature and partly described in this book, i.e. macroscopic, mesoscopic and microscopic models, it can be stated that the variety of modelling options is very wide. Of course, each model is characterised by its own strengths and weaknesses, thus making the choice of the most suitable model to be adopted strictly dependent on the objective of the study under concern and on the scale of the system to be investigated.

Microscopic models are surely more suitable for applications in small size road networks or, better, for specific road sections, especially in the urban context. Moreover, a very common use of microscopic models is for *simulation*, especially in case of offline decisions, such as for long-term planning or road design. In these cases, it is more relevant to have a highly detailed model, possibly stochastic, able to provide accurate predictions of the system dynamics, even if it requires a high computational load, rather than a fast but less accurate simulation. The use of macroscopic models is instead particularly relevant for *model-based estimation and control* purposes, especially when real-time applications are considered and large traffic networks are involved. In addition, if optimal control is applied, not only a small problem to be solved is preferable (i.e. with less variables, as macroscopic models can provide) but also the structure of the problem becomes relevant, and hence linear or linearisable traffic models are aimed for. These aspects will be further discussed in Chap. 7 and Chaps. 8–10, respectively, on traffic state estimation and traffic control, where all the reported approaches are based on macroscopic modelling. It is also worth noting that microscopic models can be used for real-time estimation and control, not as a basis for the method but for validation purposes. There are indeed many research works in which new estimation and control methods are developed and their effectiveness is tested and validated by means of traffic simulators.

It is certainly unquestionable that the new developments in technologies and computing devices will change the possible applications of traffic models. The development of faster computers will probably give a chance to the use of microscopic models for real-time applications, as well as the development of new data sources (e.g. probe vehicles) capturing more detailed aspects of the traffic flow and the individual behaviour of drivers will require the use of more specific traffic models, especially of mesoscopic and microscopic types. Surely, as suggested in [2], the development of multi-class models, as well as the improvement of hybrid models properly combining macroscopic, mesoscopic and microscopic features, seems the most promising future direction for traffic modelling.

5.2 Microscopic Traffic Models

Microscopic traffic models describe the behaviour of each single vehicle in the traffic stream and how it interacts with the other vehicles and with the road infrastructure. Specifically, in microscopic models, the vehicle–driver relation and vehicle–vehicle interactions are represented via differential equations in which the longitudinal (car-following) and/or the lateral (lane-changing) behaviour of individual vehicles can be taken into account. Since microscopic models allow to explicitly represent the dynamics of each single vehicle, it is straightforward to model different typologies of vehicles, e.g. cars and trucks, by properly setting the model parameters to represent the different behaviours of the different classes.

Several microscopic models, considering at different extents the different aspects of individual vehicle dynamics, are present in the literature. Among them, let us consider in this section of the book the following classes of models: car-following models, lane-changing models and cellular automata models.

Car-following models, also known as *follow-the-leader models*, were introduced in the 50s [6–8]. These models represent the position and speed dynamics of each vehicle through continuous-time differential equations, in which it is basically assumed that the speed dynamics of a single vehicle depends on its speed, as well as on the distance from the preceding vehicle and the speed of this latter. In more sophisticated models, the behaviour of a driver depends on a platoon of preceding vehicles instead of on one single leader. As discussed in [1, 9], these models have seen various developments after their first appearance. In a first version proposed by Pipes [7], the distance between the two vehicles (leader and follower) is determined as the safe distance computed on the basis of the vehicle length. Later, in [10], the concepts of perception time, decision time and braking time were introduced, allowing to identify the necessary safety distance to avoid collisions between two vehicles. In other models, stimulus–response concepts were introduced, including terms related to the acceleration [11] and sensitivity factors [12], calculated on the basis of the speed difference between the leader and the follower. Further models including the acceleration dynamics were presented in [13, 14]. Section 5.2.1 reports a brief overview of the main car-following models present in the literature.

Lane-changing models seek to describe the behaviour of drivers when a change of lane occurs, regardless of the reason yielding the lane changing (overtaking of a vehicle, merging to and from secondary roads or freeway on-ramps, need to avoid obstacles and so on). The representation of this phenomenon in a reliable manner is, however, one of the most complex problems that the traffic theoreticians have had to face. The lane-changing behaviour can be schematically subdivided into three steps: the decision on lane changing, the selection of the desired lane and the gap acceptance decision. Most of the modelling efforts focused on the last aspect, i.e. the representation of the gap acceptance. Several lane-changing models can be found in the literature, such as the lane-changing urban driving model described in [15] or the advanced model aiming to capture the merging behaviour in severe jammed traffic conditions proposed in [16]. Some more details on lane-changing models are reported in Sect. 5.2.2.

Another class of microscopic models is represented by *cellular automata models* (see, e.g. [17–19]), where the road topology is described by means of a grid of cells and a discrete-time dynamics is adopted. The dimension of a single cell is generally chosen in such a way that each cell can be occupied by only one vehicle (or it can remain empty), whereas the discretisation in time is carried out considering the reaction time of drivers. The traffic dynamics, given by the movement of vehicles, is represented in terms of the state (free or occupied) of the road cells. The speed is instead defined as the number of cells overtaken by a vehicle in a time step. The dynamic evolution of the speed is defined considering some factors that are the acceleration needed to reach a desired speed, the slowing down in order to decrease the speed according to the distance gap to the preceding vehicle, and a random term accounting for a deceleration which spontaneously decreases the vehicle speed according to a certain probability. Even though cellular automata models are less accurate than car-following ones, they allow to effectively replicate many traffic phenomena with a lower computational burden. An overview of cellular automata models can be found in Sect. 5.2.3.

Microscopic models are often adopted in *traffic simulation tools*, and a review of their application in this field is reported in [4, 5]. Section 5.2.4 reports a description of the most common traffic simulators.

5.2.1 Car-Following Models

Car-following models describe the *longitudinal* interactions of vehicles in a road, i.e. the behaviour according to which a driver follows the preceding vehicle in traffic. The first car-following models appeared in the 50s [7] and, since then, a great number of models of this type were proposed by researchers.

Car-following models differentiate for the considered assumptions, but they present a common notation which considers a pair of vehicles: the preceding vehicle (i.e. the *leader*) is denoted with $n - 1$, whereas the vehicle following the leader (i.e. the *follower*) is denoted with n, as shown in Fig. 5.1. The following notation is used:

- L_{n-1}, L_n are the lengths of vehicles $n - 1$, n, respectively [m];
- $x_{n-1}(t)$, $x_n(t)$ are the positions of vehicles $n - 1$, n, respectively, at time t [m];

Fig. 5.1 The main notation of car-following models

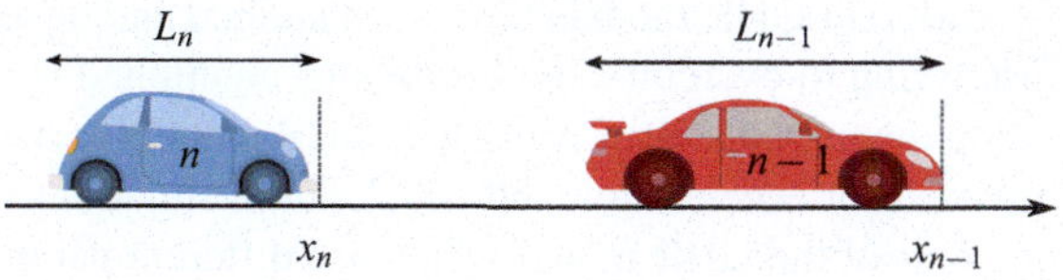

- $v_{n-1}(t)$, $v_n(t)$ are the speeds of vehicles $n-1$, n, respectively, at time t [m/s];
- $a_{n-1}(t)$, $a_n(t)$ are the accelerations of vehicles $n-1$, n, respectively, at time t [m/s^2];
- $\Delta x(t) = x_{n-1}(t) - x_n(t)$ is the space headway between vehicle $n-1$ and n at time t [m];
- $\Delta v(t) = v_{n-1}(t) - v_n(t)$ is the speed difference between vehicle $n-1$ and n at time t [m/s];
- $s_n(t) = \Delta x_n(t) - L_{n-1}$ is the spacing from the front edge of vehicle n to the rear end of vehicle $n-1$ [m];
- T is the reaction time [s].

Several overview papers on car-following models have appeared in the literature, such as the historical survey proposed in [9], the one in [20] specifically focused on driver heterogeneity aspects, and the more recent review especially addressing how human factors are incorporated in car-following models [21]. Each of these papers suggests a different classification of car-following models. In this book, we propose a classification based on four categories: Gazis–Herman–Rothery or stimulus–response models, safety-distance or collision-avoidance models, reference-signal models, and models including human factors.

GHR or Stimulus–Response Models Gazis–Herman–Rothery (GHR) models are probably the most studied models of car-following type. The basic concept of GHR models [12] is the definition of the acceleration of vehicle n at time t as

$$a_n(t) = c\, v_n^m(t) \frac{\Delta v(t-T)}{\Delta x(t-T)^l} \tag{5.1}$$

where c, l and m are the model parameters to be determined.

GHR models are also known as *stimulus–response models*; the stimulus being defined by the speed difference between the preceding vehicle and the follower, and the response being the braking or acceleration of the follower delayed by the reaction time. If GHR models have the great advantage of being simple, they also received a few critiques since they are rather unrealistic to represent some traffic situations. Actually, in free-flow conditions, when the distance headway is very large, the model assumes that drivers keep reacting to speed differences. Moreover, the traffic is considered homogeneous, i.e. all the vehicles are assumed to react in the same way. This is clearly not true in real situations in which heavy vehicles typically behave differently from cars, for instance, slow trucks are not able to adapt their speed to one of the possible leading fast cars.

Different GHR models have been studied and developed during the last decades, also trying to overcome the limitations mentioned above. Among others, it is worth citing the *asymmetrical* version of GHR models in which different parameter values are used for acceleration and deceleration situations (see, e.g. [22]). There are also versions of the GHR model which use different parameter values for congested and non-congested situations (see, e.g. [23]). This allows to model the fact that drivers may have shorter reaction times in congested situations, since they are more alert. A significant amount of work has been devoted to find suitable calibration procedures for the GHR model parameters. The most reliable parameter values, according to [9], are those indicated in [11, 24–26].

An interesting extension of GHR models is based on the use of *fuzzy logic* [27]. In this framework, concepts like "too close" or "too fast" are described using fuzzy sets, and logical rules are introduced to model the corresponding behaviour of drivers. The fuzzy sets may overlap, so that probabilistic density functions must be used to deduce how the driver perceives the considered variable (for instance, given a certain speed of the leader vehicle, the fuzzy model describes whether it is regarded as low, moderate or high by the follower). The first fuzzy version of the GHR model was proposed in [28]. More recently, a fuzzy model has been presented in [29]. A discussion on calibration and validation of car-following models based on fuzzy logic is contained in [30].

Safety-Distance or Collision-Avoidance Models Safety-distance models are also known as collision-avoidance models, since their basic relationship indicates a safety distance between vehicles in order to avoid collisions. This is specified by the so-called *Pipes' rule*, stating that a good rule for following another vehicle at a safe distance is to maintain a distance that is at least the length of a car for every ten miles an hour (i.e. 16.1 km/h) of speed [7]. This rule can be mathematically expressed as follows:

$$D_n(t) = L_n \left[1 + \frac{v(t)}{16.1} \right] \tag{5.2}$$

where $D_n(t)$ is the *prescribed headway* between vehicle $n - 1$ and vehicle n. In alternative, (5.2) can be expressed as

$$D_n(t) = L_n \left[1 + \frac{v(t-T)}{16.1} \right] \tag{5.3}$$

if the reaction time T is taken into account.

Safety distance models differ from GHR models since they assume that drivers react to spacing with respect to the preceding vehicle, rather than to the relative speed. This idea was elaborated in [31], where the proposed model assumes that each vehicle always tries to keep the minimum safety distance from the preceding vehicle, defined as

$$\Delta x(t-T) = \alpha v_{n-1}^2(t-T) + \beta v_n^2(t) + \gamma v_n(t) + d \tag{5.4}$$

where α, β and γ are model parameters, whereas d is the minimum allowed spacing between subsequent vehicles.

Models implementing the same philosophy are those presented in [10, 32, 33]. Yet, the most widely used safety-distance model is the *Gipps* model [34], which is the car-following model implemented in the well-known traffic simulation software Aimsun (see Sect. 5.2.4). The Gipps model assumes that any vehicle tends to travel at the speed which allows to avoid a rear crash if the vehicle performs an emergency braking. Despite presenting several advantages, the Gipps model has the limitation that the following vehicle can only travel exactly at the safe distance with respect to the preceding vehicle, which is clearly unrealistic. More realistic safety-distance car-following models overcome such limitation by better defining the safe distance, for instance, as a function of the relative speed between the leading vehicle and the follower, such as in [35].

Reference-Signal Models This category includes models in which a desired reference signal is explicitly introduced to describe the tendency of any individual driver to adjust his/her behaviour to track that signal. The nature of the reference signal differs from model to model. More specifically, the reference signal can be a prescribed space headway or a desired speed or an adequate time gap.

The first example of model of this class was introduced by *Helly* in [13] and is often known in the literature as the *linear model*. In this model, the acceleration of any vehicle linearly depends on the relative speed and on the difference between the relative distance and the prescribed space headway. The latter is defined by including a term accounting for the follower's acceleration, in contrast with (5.2). This can be expressed mathematically as follows:

$$a_n(t) = C_1 \Delta v(t-T) + C_2 \left[\Delta x(t-T) - D_n(t)\right] \tag{5.5}$$

where the prescribed headway is computed as

$$D_n(t) = \alpha + \beta v_n(t-T) + \gamma a_n(t-T) \tag{5.6}$$

where C_1, C_2, α, β and γ are parameters to be identified on the basis of real data. In particular, Helly observed that C_1 could be considered as dependent on the relative distance between vehicles, whereas C_2 could be made speed dependent. Several works were then devoted to calibrate the Helly model parameters (see, for instance, [36–39]).

Another example of reference-signal models is the so-called *intelligent driver model*, proposed in [40, 41]. In this model, there are two reference signals, the desired speed and the desired space headway, i.e.

$$a_n(t) = a_n^{\max}\left[1 - \left(\frac{v_n(t)}{\tilde{v}_n(t)}\right)^{\beta} - \left(\frac{\tilde{s}_n(t)}{s_n(t)}\right)^{2}\right] \tag{5.7}$$

where $a_n^{\max}$ is the maximum acceleration/deceleration of vehicle n, $\tilde{v}_n(t)$ is the speed reference signal, $\tilde{s}_n(t)$ is the spacing reference signal and β is a model parameter. It is worth noting that, when the spacing between two subsequent vehicles is high, the third term becomes negligible, so that the considered vehicle just follows the speed reference signal. In car-following situations, the spacing reference signal can depend on several factors, such as the speed of vehicle n, the relative speed between vehicles n and $n-1$, the maximum acceleration, the desired time gap and so on.

A further example of reference-signal models is the *optimal speed model*, introduced in [14]. In this model, the reference signal is a speed assumed to be optimal for the considered vehicle, taking into account the distance from the preceding vehicle. Hence, the acceleration of vehicle n can be determined according to the difference between the actual speed and the optimal speed v_n^*, i.e.

$$a_n(t) = \alpha \left[v_n^*(\Delta x_n(t)) - v_n(t) \right] \tag{5.8}$$

where α is a model parameter. Variations of the original optimal speed model were proposed in [42], also to counteract the tendency of the model to produce unrealistic accelerations or decelerations. Further extensions can be found, for instance, in [43–46].

Models Including Human Factors The car-following models described so far are mainly based on physical signals. Nevertheless, as highlighted in [47], the human driving behaviour is not only influenced by physical signals but also by psychological aspects. Moreover, many assumptions of standard car-following models are not always true in real cases, for instance, drivers often adopt strategies that are adequate for the current situation but not optimal, drivers do not continuously react to stimuli, each driver has a different driving style and so on. Based on these considerations, a wide literature has been developed in order to encompass psychophysical aspects, typical of perceptual psychology, into car-following models.

The most famous car-following models which include human factors are the so-called *psychophysical* or *action point* models. The basic idea is that *perception thresholds* characterise the human capability of perceiving spacing and speed differences (see [48, 49] for perception-based experiments to quantify the thresholds). In practice, drivers do not continuously react to speed differences and spacings but only when the current action significantly differs from the action which is regarded as appropriate for the given situation. In other terms, the existence of these perception thresholds makes the acceleration (or, more in general, behavioural changes) occur at asynchronous time instants, named *action points*. The thresholds and time intervals between two subsequent action points are stochastic quantities. Referring in particular to the vehicle acceleration, it is kept constant by the driver until it is significantly different from the acceleration required to maintain the proper spacing with respect to the preceding vehicle. This implies that, in case of large spacing, the following driver tends to act rather independently, i.e. such driver is not influenced by the relative speed, as if this were imperceptible. At small spacings, instead, the driver alertness is higher. The thresholds, and the regimes they define, are typically presented in a relative space–speed diagram for a vehicle pair.

One of the first psychophysical models was introduced by *Wiedermann* [50], a modified version of which has been used in the software tool Vissim (see Sect. 5.2.4). The car-following model implemented in the software tool Paramics (see Sect. 5.2.4) is based instead on the psychophysical model reported in [51]. Other psychophysical models were investigated and can be found in the literature (see, e.g. [51–54]).

Another class of models including human factors are those modelling the driving behaviour related to the *visual angle* subtended by the preceding vehicle. The first car-following model of this type was introduced in [55], where the basic assumption is that drivers approaching a vehicle react to the changes in the apparent size of this vehicle. Then, compared to classical car-following models, the relative spacing and speed are replaced by the visual angle and the angular speed. Different versions of car-following models based on visual angles were developed (see, for instance, [56, 57]).

More sophisticated car-following models were defined by researchers in order to represent aspects related to *risk* and *driving errors*. For instance, driving in risky situations was modelled in [58] as a human decision-making problem, relying on prospect theory [59], and properly defining the subjective probability of being involved in a collision with the preceding vehicle. This model was then extended in [60] to consider response and behaviour of drivers in different surrounding traffic conditions. Further efforts were devoted to include, in car-following models, driving errors and distraction situations, which are the main cause of crashes in real traffic circumstances. For instance, in [61], the Helly model is extended to consider that the time headway is influenced by different aspects, such as visual conditions and driver state, in [62] the intelligent driver model is modified to consider the reactions of the driver to the surrounding traffic environment, and in [63] the Gipps model is extended by considering human perception limitations in processing information and adjusting speed accordingly.

5.2.2 Lane-Changing Models

While car-following models have the main objective of representing the longitudinal interactions among vehicles inside the traffic flow, lane-changing models are instead devoted to describe *lateral* interactions on the road. These two primary modelling tasks have often been treated separately, even if they are two fundamental components of the microscopic traffic flow modelling theory. Although car-following models have been widely studied for many years, lane-changing aspects have received some attention only in recent years [64–66]. This recent interest in lane-changing behaviours has been mainly due to the increasing evidence of their negative impact on traffic safety and traffic congestion.

The impact of lane-changing movements on traffic safety was investigated in some works, such as in [67, 68]. Many studies show that the stress of drivers significantly increases during lane-changing manoeuvres, thus making them more error-prone and dangerous. Moreover, the lane-changing process plays a role in capacity drop phenomena related to bottleneck discharge rate reduction at the onset of congestion [69] and also in the formation and propagation of stop-and-go oscillations [70, 71]. More recently, in [72] it has been shown that lane changing is a primary trigger of oscillations and is responsible for transforming minor and localised oscillations into substantial disturbances.

Research efforts to represent the lane-changing aspects have rapidly increased over the last decade. The main lane-changing models in the literature can be distinguished into two groups: models related to the lane-changing decision-making process (i.e. how a driver reaches the lane-changing decision), and models devoted to quantify the impact of lane-changing behaviours on surrounding vehicles. It can be noted that a comprehensive lane-changing model should take into account both these aspects together with car-following behaviours in order to fully represent the dynamics of vehicles, but a widely recognised modelling tool covering all these aspects is not yet available [66].

The different models developed in the literature differentiate for the way in which they represent the lane-changing decision-making process, but, in any case, they must take into consideration the interactions of the vehicle aiming to change lane with the other vehicles in the surroundings. In particular, as shown in the scheme presented in Fig. 5.2, let us consider the lane changer vehicle, denoted as LC, which is travelling in the lane called *initial lane* and would like to move to the so-called *target lane*. Vehicle LC has to interact, in some way, with the preceding vehicle (i.e. the leader) and the following vehicle (i.e. the follower) in the initial lane, denoted as L^I and F^I, respectively, and with the preceding and following vehicle in the target lane, denoted as L^T and F^T, respectively.

The lane-changing decision-making process is based on several factors, one of which is the so-called *gap acceptance process* which precedes an overtaking manoeuvre. In this process, a driver who wants to overtake a vehicle preceding him estimates both the space he needs and the available space. On the basis of the comparison between required and available space, the driver decides whether to start the lane-changing manoeuvre or not. Several gap acceptance models are present in the liter-

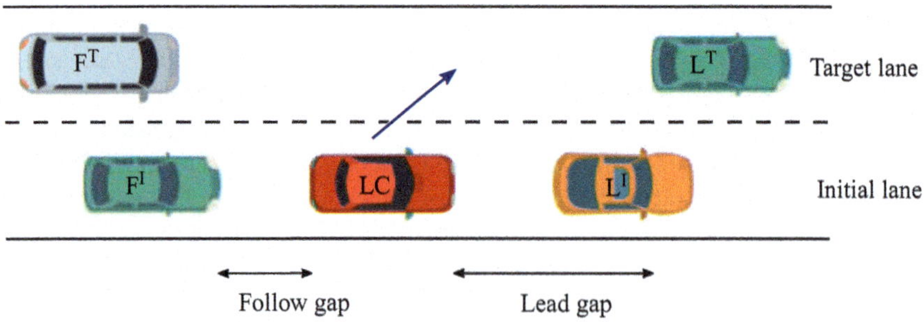

Fig. 5.2 Generic lane-changing process

ature, not only for freeway systems but for any kind of road (and also for pedestrian flows). These models are stochastic and they are based on the definition of a *gap acceptance function* defining the probability that an arbitrary driver accepts an available gap, thus starting the overtaking manoeuvre. A description of gap acceptance models can be found in [16].

A basic model describing a more general lane-changing decision-making process is due to *Gipps* [15], in which various driving situations in an urban street context are considered. In the Gipps model, the driver's behaviour is governed by two basic considerations typically arising in an urban network: the willingness to maintain a desired speed and the desire to be in the correct lane for an intended turning manoeuvre. The drivers' behaviour is considered as deterministic and, thus, a set of deterministic rules to be sequentially evaluated is defined.

One of the first works related to the lane-changing decision process in the freeway context is [73], in which lane-changing movements are classified as either mandatory (when the lane change is necessary due to a lane drop, an accident or the use of an exit junction) or discretionary (when a driver evaluates that in the target lane better driving conditions can be experienced compared to those found in the current lane) and a lane-changing probability is introduced to make the model more realistic. Several variations and extensions were proposed, as, for instance, in [74], in which a novel logic for simplifying and modelling lane-changing decisions is defined in terms of single-lane accelerations.

In [16], the utility theory is applied to model the decision process of lane changing, whereas in [75] Markov processes are used to model mandatory lane-changing actions. Furthermore, several lane-changing decision models based on fuzzy logic have been developed in the last decades [29, 76].

The models discussed so far largely ignore the impact of lane changing on surrounding vehicles. Several studies, such as [69, 77], address the influence between lane-changing and critical traffic phenomena, such as breakdowns, capacity drop and traffic oscillations. Some models for representing the impact of lane changing on surrounding vehicles are reported in [66], where it is also discussed how this aspect still needs to be investigated to define accurate lane-changing models.

5.2.3 Cellular Automata Models

Cellular Automata (CA) models, sometimes also called *Particle Hopping* models, were first proposed in 1948 [78] and then revitalised in the 80s with the work reported in [79]. CA models are basically characterised by four components, i.e. the physical environment, the states of cells, the neighbourhoods of cells and the local transition rules. The physical environment in which CA models are applied for modelling traffic flow is a road segment, which is discretised into cells of the same length, typically equal to the vehicle length, so that any cell can be exactly occupied by a single vehicle. CA models are discrete-time models in which time is discretised and the sample time is generally set equal to 1 s. The speed of a vehicle is then computed as the number of cells that a vehicle hops in one time step (implying that speed is discretised as well). The state of each cell can be either equal to 0 (if the cell is empty) or equal to 1 (if it is occupied).

One of the most famous CA models is the one developed by *Nagel* and *Schreckenberg* [17], which has a stochastic nature. According to this model, the road is discretised into cells (approximately 7.5 m long) and a maximum speed v^{max} is considered. At each time step, the model evolves according to the following predefined rules:

- *acceleration*: if the speed v of a vehicle is lower than v^{max} and if the distance to the vehicle in front is larger than $v + 1$, then the speed is increased by one;
- *deceleration*: if a vehicle in cell i finds the next vehicle in cell $i + j$, with $j \leq v$, then the vehicle decelerates to $j - 1$;
- *randomisation*: the nonzero speed of each vehicle is decreased by one, with probability p;
- *vehicle motion*: each vehicle is advanced by v cells.

The update of the states of cells can be done in different ways, i.e. in the direction of travel, in the opposite direction or even in a random order, without affecting the model behaviour. CA models are very simple and computationally low demanding, and hence large size road networks with a high number of vehicles can be analysed (and simulated) in short computational times, and this is surely a relevant advantage of such models, especially for real-time applications.

Moreover, different traffic Fundamental Diagrams can be established by varying the model parameters, specifically by varying v^{max} and p. Also, CA models describe the spontaneous formation of traffic congestion and stop-and-go waves. As observed in the various survey papers about CA models (see, e.g. the review papers [80–82]), a large number of variations and extensions to the basic CA model have been defined and studied. Let us report in the following a CA model including lane-changing phenomena for a two-class traffic case.

A Two-Class CA Model with Lane Changing The considered model was defined in [83], being based on the model reported in [84]. This model refers to the case in which two classes of vehicles, i.e. cars and trucks, are present in a multi-lane freeway stretch. Specifically, two-lane freeway stretches are taken into account, in which cars can overtake other vehicles by occupying the left lane, with lane-changing rules inspired from [85], while trucks are forced not to overtake other vehicles.

As it is common in CA models, the space is discretised, specifically each lane is subdivided into cells with length equal to 1.5 m. It is assumed that cars have an occupancy of 3 cells, whereas trucks occupy 8 cells. The speed is expressed as the number of cells that one vehicle can go over in one time step, being 1 s the sample time.

The model introduced in [84] presents some important features that makes it more accurate than the original simple model reported in [17]. With reference to Fig. 5.3, considering three vehicles, denoted as n, m and l, the main notation of the model is the following:

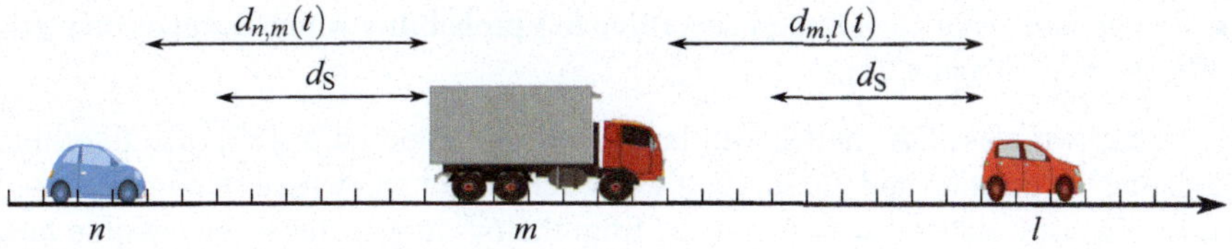

Fig. 5.3 The main notation in the two-class CA model

- d_S is a fixed safety distance between vehicles [number of cells];
- $d_{n,m}(t)$ and $d_{m,l}(t)$ are the number of free cells between vehicles n and m, and between m and l, respectively, at time t [number of cells];
- $v_n(t)$ is the speed of vehicle n at time t, i.e. the number of cells that vehicle n can go over in one time step (analogously $v_m(t)$ and $v_l(t)$ for vehicles m and l) [number of cells];
- $b_n(t) \in \{\text{on, off}\}$ is the state of the brake light of vehicle n at time t (analogously $b_m(t)$ and $b_l(t)$ for vehicles m and l);
- $l_n(t) \in \{\text{straight, right, left}\}$ is the position that vehicle n would like to occupy at time t, which can be obtained by going straight, moving to right or moving to left (analogously $l_m(t)$ and $l_l(t)$ for vehicles m and l);
- $\psi_n \in \{\text{car, truck}\}$ is the typology of vehicle n (analogously ψ_m and ψ_l for vehicles m and l).

The main rules adopted in the model presented in [84] are the following:

- *anticipation*: a generic vehicle n does not only consider the distance from the preceding vehicle m but it also estimates how far this vehicle will move during the time step; this is done by introducing and computing $d^{\text{eff}}_{n,m}(t)$ as

$$d^{\text{eff}}_{n,m}(t) = d_{n,m}(t) + \max\left\{v^{\min}_m(t) - d_S, 0\right\} \tag{5.9}$$

  where $v^{\min}_m(t)$ is given by

$$v^{\min}_m(t) = \min\left\{d_{m,l}(t), v_m(t)\right\} - 1 \tag{5.10}$$

- *brake lights*: again considering a generic vehicle n, a time interval $\tau^S_n(t)$ is referred to the interaction with the brake light of the vehicle in front; specifically, vehicle n reacts to the state $b_m(t)$ of the brake light if $\tau^H_{n,m}(t) < \tau^S_n(t)$, where quantities $\tau^H_{n,m}(t)$ and $\tau^S_n(t)$ are defined follows:

$$\tau^H_{n,m}(t) = \frac{d_{n,m}(t)}{v_m(t)} \tag{5.11}$$

$$\tau^S_n(t) = \min\left\{v_n(t), \nu\right\} \tag{5.12}$$

  where ν is a model parameter;

- *slow-to-start*: vehicle n brakes according to a probability which depends on $v_n(t)$, $b_m(t)$, $\tau^{\mathrm{H}}_{n,m}(t)$ and $\tau^{\mathrm{S}}_n(t)$.

In the two-class CA model with lane changing proposed in [83], the algorithm updating the position and the speed of every vehicle for every time step is composed of four phases: definition of entrances from the on-ramps, check for possible lane changes, application of vehicle motion and definition of exits through the off-ramps. Each of these four phases is detailed in the following.

1. *Entrances from on-ramps*. The presence of vehicles at the on-ramps is modelled by means of queues, where vehicles wait to access the mainstream. The queue can contain up to $q^{\max}$ vehicles, and the number of vehicles accessing the queue is generated at each time step according to a given probability p_{in} depending on the vehicle class. Moreover, the number of vehicles which enter the mainstream depends on the space available in the mainstream (this number is reduced if the freeway is congested) and on a maximum value of κ vehicles (where κ is a given parameter related to the on-ramp capacity).
2. *Lane change*. As in [84], two different rules are adopted to define the lane-changing process, from the right lane to the left one and vice versa. Moreover, it is imposed that trucks cannot move to the left lane; hence, these two-lane change rules are applied only to cars. Let us consider these two different rules separately.

   - Rule for moving *from right to left*: let us consider vehicle n in the right lane and let us identify the preceding vehicle m in the same lane, the preceding vehicle s in the left lane and the vehicle r before vehicle s in the left lane (see Fig. 5.4); the variable $l_n(t)$ is first set as follows:

$$l_n(t) = \text{straight} \tag{5.13}$$

   Then, it is checked if the lane change is possible for vehicle n, i.e.

$$\begin{aligned}&\text{If} \quad (b_n(t) = \text{off}) \wedge \big(d_{n,m}(t) < v_n(t)\big) \wedge \Big(d^{\text{eff}}_{n,s}(t) \geq v_n(t)\Big) \wedge \big(d_{r,n}(t) \geq v_r(t)\big)\\ &\text{then} \quad l_n(t) = \text{left}\end{aligned} \tag{5.14}$$

   If, by applying (5.14), it results $l_n(t)$ = left, then vehicle n moves to the left lane.

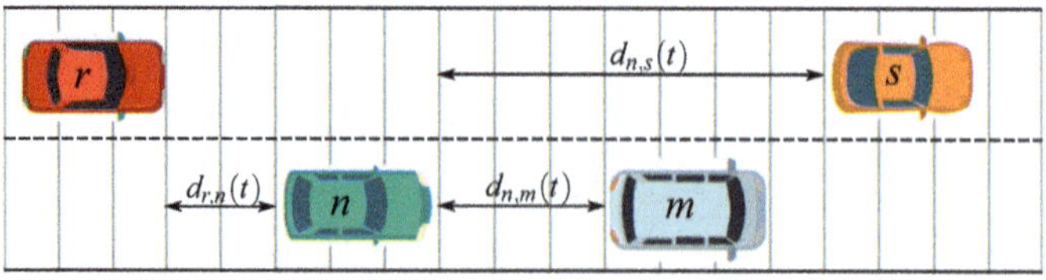

Fig. 5.4 Lane change from right to left in the two-class CA model

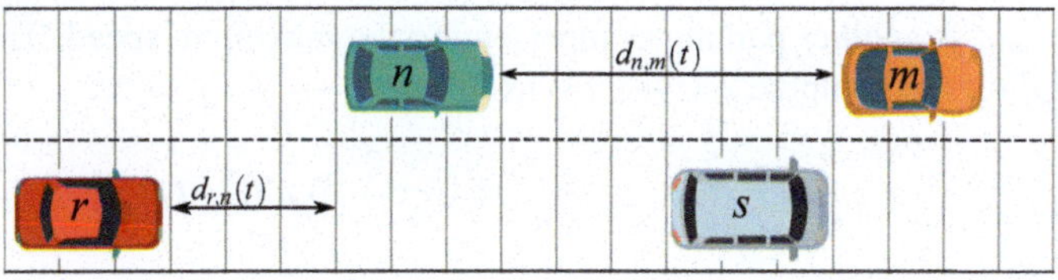

Fig. 5.5 Lane change from left to right in the two-class CA model

- Rule for moving *from left to right*: let us consider vehicle n in the left lane and let us identify the preceding vehicle m in the left lane, the preceding vehicle s in the right lane and vehicle r before vehicle s in the right lane (see Fig. 5.5); the variable $l_n(t)$ is initially fixed as

$$l_n(t) = \text{straight} \tag{5.15}$$

Then, the possibility of lane change is checked for vehicle n, i.e.

$$\begin{aligned} &\text{If} \quad (b_n(t) = \text{off}) \wedge \left(\tau^{\text{H}}_{n,s}(t) > \xi\right) \wedge \left(\tau^{\text{H}}_{n,m} > \upsilon \vee v_n(t) > d_{n,m}(t)\right) \\ &\qquad \wedge \left(d_{r,n}(t) > v_r(t)\right) \\ &\text{then} \quad l_n(t) = \text{right} \end{aligned} \tag{5.16}$$

where ξ and υ are other parameters. Once (5.16) has been applied, if $l_n(t) =$ right, then vehicle n moves to the right lane.

Summarising, for the two classes of vehicles, the lane changing rule is the following:

$$\begin{aligned} &\text{If} \quad \psi_n = \text{truck} \\ &\text{then} \quad l_n(t) = \text{straight} \qquad \text{(lane change not allowed)} \\ &\text{else} \quad \text{conditions (5.13)–(5.16) hold} \qquad \text{(lane change allowed)} \end{aligned} \tag{5.17}$$

3. *Vehicle motion*. The vehicle motion phase is the core of the algorithm and is executed for every vehicle in each lane at each time step. Vehicle motion is based on a set of rules in order to obtain the speed $v_n(t+1)$ of vehicle n through some consecutive steps, in which the intermediate values $v_n(t+1/3)$ and $v_n(t+2/3)$ are computed. More specifically, let us consider vehicle n and the next vehicle in front m, and let us set $b_n(t+1) = \text{off}$. According to the *acceleration* phase, $v_n(t+1/3)$ is computed as

$$v_n(t+1/3) = \begin{cases} v_n(t) & \text{if } (b_n(t) = \text{on}) \vee \left(b_m(t) = \text{on} \right. \\ & \qquad \left. \wedge\ \tau^{\text{H}}_{n,m}(t) < \tau^{S}_{n}(t)\right) \\ \min\{v_n(t)+1, v_{\max}\} & \text{otherwise} \end{cases} \tag{5.18}$$

where $v_{\max}$ is another parameter representing the maximum speed. The *braking* phase allows to compute $v_n(t + 2/3)$ as follows:

$$v_n(t + 2/3) = \min \left\{ v_n(t + 1/3), d_{n,m}^{\text{eff}}(t) \right\} \tag{5.19}$$

and the following rule is applied:

$$\begin{aligned} &\text{If} \quad v_n(t + 2/3) < v_n(t) \\ &\text{then} \quad b_n(t + 1) = \text{on} \end{aligned} \tag{5.20}$$

According to the *randomisation* phase, the value of $v_n(t + 1)$ is obtained as

$$v_n(t + 1) = \begin{cases} \max \{v_n(t + 2/3) - 1, 0\} & \text{with probability } p \\ v_n(t + 2/3) & \text{otherwise} \end{cases} \tag{5.21}$$

and the following rule is applied:

$$\begin{aligned} &\text{If} \quad p = p_0 \ \wedge \ v_n(t + 1) < v_n(t + 2/3) \\ &\text{then} \quad b_n(t + 1) = \text{on} \end{aligned} \tag{5.22}$$

where p_0 is a parameter. Finally, according to the *move* rule, the position of each vehicle is updated according to the speed just determined, i.e.

$$x_n(t + 1) = x_n(t) + v_n(t + 1) \tag{5.23}$$

4. *Exits from off-ramps*. The number of vehicles exiting a freeway stretch is defined by means of a probability p_{out} depending on the vehicle class. Note that more advanced approaches should consider the assignment of the final destination to every vehicle. Moreover, it would be possible to model the off-ramp as a finite-capacity buffer, so that, when the buffer is full, a queue grows backwards in the freeway stretch creating a spillback phenomenon.

5.2.4 *Traffic Simulation Tools*

Traffic simulation tools are software systems with a large variety of applications, both in the urban and in the freeway context. These tools generally implement different types of traffic models, such as microscopic and mesoscopic models, and provide a visual framework useful for experimental studies, also in case of large-scale traffic systems.

In the following, an analysis of the characteristics of the main traffic simulators available on the market is reported, without presuming to provide in this book a complete list of all the traffic simulation tools present worldwide. The interested

Table 5.1 Traffic simulation tools

Name	Developer	Type of license
Paramics	Quadstone	Commercial
Aimsun	Aimsun	Commercial
PTV Vissim	PTV	Commercial
TSIS-CORSIM	McTrans	Commercial
MATSim	Open Community	Open Source
MITSIMLab	Massachusetts Institute of Technology	Open Source
SUMO	DLR	Open Source

reader can find more details in the books [4, 5], which are specifically dedicated to the topic.

Some of the traffic simulators that are presently used by traffic experts and research centres working on modelling, planning and control of road traffic systems are listed in Table 5.1. Among the commercial traffic simulators, it is worth citing Paramics, Aimsun, PTV Vissim and TSIS-CORSIM. *Paramics*, developed by Quadstone, is a microscopic traffic simulation software used by researchers, engineers and planners worldwide, and it provides solutions for both freeway and urban networks, including public transport, pedestrian modelling and ITS applications. *Aimsun* is an integrated transport modelling software which has grown from being a micro-simulator to becoming a fully integrated application with features of travel demand modelling, macroscopic functionalities and mesoscopic–microscopic hybrid simulation allowing to represent the traffic behaviour in a very detailed way, while preserving computational efficiency. *PTV Vissim* is a microscopic multimodal traffic flow simulation software package developed by PTV. It is conceived for motorised private transport, goods transport, rail and road public transport, pedestrians and cyclists, and allows to make a detailed analysis and planning of urban and extra-urban road infrastructure. *TSIS-CORSIM* is a microscopic traffic simulation software package for urban signalised traffic systems, freeway traffic systems or combined urban-freeway systems. It is based on microscopic traffic models to represent the movements of individual vehicles, including the influences of geometric conditions, drivers' behaviours, presence of traffic control implementations and so on.

Besides these commercial software tools, many open-source traffic simulators have been developed by open communities or research groups worldwide. Among them, it is worth mentioning MATSim, MITSIMLab and SUMO. *MATSim* is an agent-based micro-simulator, in which every part of the traffic system is represented as an agent specified by a dynamic behaviour, and the evolution of the entire system is given by interactions among the various agents. The intermodal simulation is supported as well and advanced users can extend the source code, written in Java, to create customised releases adapted to their own purposes. *MITSIMLab* is an open-source application, written in C++, developed at the MIT Intelligent Transportation Systems Program. This platform includes MITSIM, i.e. the traffic simulation tool, implementing microscopic traffic models, and TMS, i.e. the traffic management simulator, which models the implementation of traffic control strategies, such as ramp metering, mainline control, route guidance and so on. *SUMO* is a free and open traffic simulation suite developed in C++, basically devoted to urban mobility, including intermodal traffic composed of road vehicles, public transport and pedestrians.

Each traffic simulation tool has its own characteristics and it is sometimes difficult to find the best tool to be used for the simulation of a given traffic case. Some works in the literature deal with the comparison among the characteristics and the performance of different software tools for traffic simulations. These comparisons, and the conclusions drawn in these works, are of course dependant on the considered test case and on the software version that has been adopted. For instance, [86] reports a comparison among three traffic simulation software programs, that are CORSIM, Vissim and Paramics, referring to a test case of an intersection between the U.S. Highway 50 and the Missouri Flat Road interchange near Placerville, California, U.S. In this study, the application to the test case showed for instance that Paramics and Vissim are characterised by a larger number of parameters compared to CORSIM, allowing to more accurate simulations but making the set-up phase more difficult.

Another comparison among traffic simulation tools was done and reported in [87], where Aimsun, Paramics and Vissim are analysed with specific attention to the effectiveness of car-following models. This comparative analysis was carried out considering a real car-following experiment, set in Germany, in which instrumented vehicles were used to record the speeds and relative distances on a one-lane road. The same setting was implemented with the three traffic simulators and the simulated results were compared with field data. The results show that the lowest errors are obtained with the Gipps-based models implemented in AIMSUN, while higher errors are obtained with the psychophysical models used in Paramics and Vissim.

Another comparison related with the car-following rules was discussed in [88], where the simulators Aimsun, Paramics, Vissim and MITSIM were compared considering the same test case. According to this study, the number of parameters present in Vissim and Paramics is very high, whereas MITSIM and Aimsun are characterised by fewer parameters, and, also, in Aimsun the parameters have a more intuitive meaning. In this study, some specific microscopic aspects are analysed in detail and the way how they can be represented with the four traffic simulators is described. For instance, referring to the reaction time of drivers, in [88] it is observed that AIMSUN uses a driver reaction time equal to the simulation time step, which is equal for all drivers, MITSIM assigns possibly different individual reaction times to every vehicle, while Vissim and Paramics do not model reaction times explicitly.

The results obtained from the described comparisons highlight how each simulator has strengths and weaknesses; the choice is subject to specific user needs and a trade-off between different features and performance. Despite the commercial simulators offer the most comprehensive options with programming frameworks that are carefully designed and optimised, guaranteeing support to the users, the open-source simulators have the strength that the user can use the source code and properly modify it. This aspect is relevant for two main reasons: the former is the possibility for the users to create an ad-hoc version of the software that meets their precise needs and the latter lies in the contribution that individual users can give to the developers' community.

5.3 Mesoscopic Traffic Models

The class of mesoscopic traffic models represents an intermediate approach between macroscopic traffic models, relying on the dynamics of aggregate variables, and microscopic traffic models, representing instead the dynamics of each vehicle in the traffic flow. Mesoscopic models describe the traffic flow dynamics in an aggregate way but represent the individual behaviour of drivers using probability distribution functions. In the literature, different mesoscopic modelling approaches are present [1]. Among them, three main classes can be identified related to *headway distribution models*, *cluster models* and *gas-kinetic models*. Sections 5.3.1–5.3.3 describe, respectively, these three types of mesoscopic models.

5.3.1 Headway Distribution Models

In headway distribution models, attention is posed on the statistical properties of *time headways*. Starting from an empirical observation of the distribution of time headways (or, alternatively, of vehicle spacings) and assuming that they are independent and identical distributed random variables, headway distribution models are based on the definition of suitable probability density functions for such distributions.

In a first set of works dealing with headway distribution models (see, for example, [89–91]), *stationary* distribution models were addressed. These models have shown to effectively fit empirical data in free-flow traffic conditions but they are not completely adequate in congested situations. *Mixed* headway distribution models tackle this drawback by distinguishing between free-driving vehicles and following vehicles, with the headways of the two categories characterised by different probability density distributions (see, e.g. [92]).

The characteristic of some stationary distribution models of being mainly suitable for free-flow conditions is often motivated by the fact that they support an incomplete representation of the interactions among vehicles, which are typically weak and negligible in free-flow conditions and consistent, instead, in congested traffic cases. More recently, *dynamic* headway distribution models have been developed to improve the way in which the dynamic role of traffic is considered. To this end, in [93] different vehicle types in the different phases of traffic are explicitly modelled, whereas random matrix theory is used in [94] to predict headway distributions in a model in which traffic is represented as a set of strongly linked particles under fluctuations. A further work on the topic is, for instance, [95], in which a variance-driven adaptation mechanism is defined, according to which drivers increase their safety time gaps when the local traffic dynamics is unstable or largely varying.

5.3.2 Cluster Models

Cluster models represent the dynamics of traffic flow by describing the formation of *clusters of vehicles*, i.e. groups of vehicles which share a specific property. Clusters usually emerge because of restricted lane-changing possibilities or due to prevailing weather or ambient conditions. Different aspects of clusters can be considered, such as their size (the number of vehicles in a cluster) and their speed. Generally, the size of a cluster is dynamic, i.e. clusters can grow and decay. Clusters are typically considered as homogeneous, in a sense that the conditions of vehicles inside a cluster, e.g. their headways or the speed differences, are not explicitly taken into account (see, for example, [96, 97]).

In particular, cluster models deal with the rules of cluster formation, the conditions under which clusters can appear and their characteristics. The basic idea is to find a physically motivated assumption for the transition rates of the attachment and detachment of individual vehicles to a cluster consistent with the empirical observations in real traffic.

Cluster models are first referred to the simplified case in which only one cluster is present in the traffic system [97], and then extended to a multi-cluster case [98]. In the case in which a *single cluster* is considered, the cluster is specified by its size n, which is the number of aggregated vehicles. Its internal parameters, namely the headway distance and, consequently, the speed of vehicles in the cluster, are treated as fixed values independent of the cluster size n. As depicted in Fig. 5.6, a cluster grows when free vehicles join it at its upstream boundary, and it becomes instead shorter when vehicles located near its downstream boundary accelerate to leave it.

The processes yielding changes in the cluster size are described as random processes in which the probability function $P(n, t)$ for the cluster to have size n at time t is defined. This function evolves, thanks to the so-called *one-step master equation* expressed as follows:

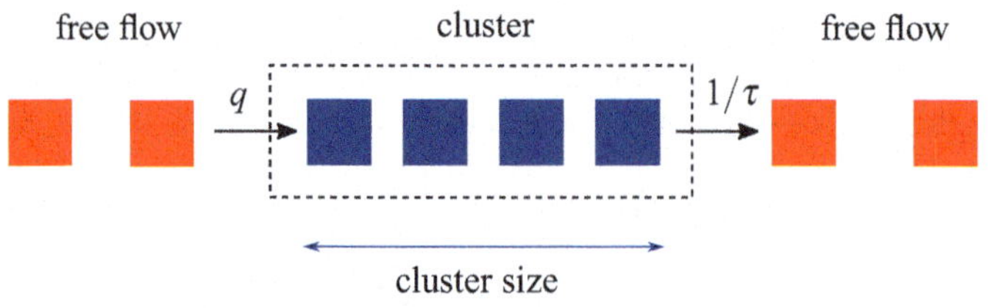

Fig. 5.6 Sketch of a single cluster

$$\frac{\partial P(n,t)}{\partial t} = w_+(n-1)P(n-1,t) + w_-(n+1)P(n+1,t) \\ -[w_+(n)P(n,t) + w_-(n)P(n,t)] \tag{5.24}$$

where $w_+(n)$ and $w_-(n)$ are the attachment rate and the rate of vehicles leaving the cluster when it has size n, respectively. These rates can be considered as constant values and expressed as $w_+ = q$, $w_- = 1/\tau$, where q is the traffic flow before the cluster and τ is the characteristic time needed to the first vehicle in the cluster to leave it and to go out from its downstream boundary at a distance approximately equal to the headway distance in the free-flow state.

On the basis of the balance equation (5.24) and the Fokker–Planck approximation to calculate mean first passage times or escape rates, it is possible to determine the dynamics of the traffic pattern formation and, specifically, the time in which the traffic conditions vary from free-flow to congested, including the influence of the parameters affecting the discharge and adhesion rates.

The single-cluster case can then be extended to consider the presence of *several clusters* of different sizes [98]. It is in this case necessary to model the dynamics of all the sizes of the clusters and to make the transition rates of the attachment and detachment of individual vehicles to a cluster consistent with the empirical observations in real traffic. To this end, the analogy with first-order phase transitions and nucleation phenomena in physical systems (like supersaturated vapour) is exploited.

In order to make a comparison with real measurements, the results are represented with a Fundamental Diagram of traffic flow (i.e. the steady-state flow–density relation), and, then, compared with empirical data. It is also possible to analyse different traffic conditions (free-flow, congested mode and heavy viscous traffic) and to include on-ramp effects.

5.3.3 Gas-Kinetic Models

Among mesoscopic approaches, the most known models are gas-kinetic models, in which an analogy between the dynamics of gases and the dynamics of traffic flows is exploited. In these models, some concepts of statistical physics are introduced, such as the *reduced phase-space density*, which is related to the expected number of vehicles present in an infinitesimal region, travelling with a speed defined on the basis of a probability distribution function. Such a concept can be seen as the mesoscopic version of the macroscopic traffic density. Moreover, the distribution function of the speed is affected by three processes: the process of convection, the process of acceleration towards the desired speed and the process of deceleration due to the interaction among vehicles.

An initial proposal of these models was presented by *Prigogine* and *Herman* in [99, 100]. These works introduce the concept of reduced phase-space density $\tilde{\rho}(x, v, t)$. Specifically, the reduced phase-space density $\tilde{\rho}(x, v, t)$ can be used to compute the expected number of vehicles present at time t in the infinitesimal region between position x and $x + \mathrm{d}x$, with $\mathrm{d}x \to 0$, moving with a speed between v and $v + \mathrm{d}v$, with $\mathrm{d}v \to 0$. This expected number of vehicles can be obtained as $\tilde{\rho}(x, v, t)\mathrm{d}x\,\mathrm{d}v$.

The first relation encountered in gas-kinetic traffic flow models is the following partial differential equation:

$$\frac{\partial \tilde{\rho}(x, v, t)}{\partial t} + v\frac{\partial \tilde{\rho}(x, v, t)}{\partial x} = \left(\frac{\partial \tilde{\rho}(x, v, t)}{\partial t}\right)_{\text{acc}} + \left(\frac{\partial \tilde{\rho}(x, v, t)}{\partial t}\right)_{\text{int}} \tag{5.25}$$

where

- the second term of the left-hand side is the so-called *convection* term describing the propagation of the phase-space density with speed v;
- the first term of the right-hand side is the *acceleration/relaxation* term modelling the fact that vehicles tend to reach an equilibrium or desired speed;
- the second term of the right-hand side represents the *interactions* with surrounding vehicles; in this term the probability of overtaking is explicitly considered.

According to [100], the acceleration term depends on the desired speed distribution, denoted as $V_0(x, v, t)$, and can be written with the following expression:

$$\left(\frac{\partial \tilde{\rho}(x, v, t)}{\partial t}\right)_{\text{acc}} = -\frac{\partial}{\partial v}\left(\tilde{\rho}(x, v, t)\frac{V_0(x, v, t) - v}{\tau}\right) \tag{5.26}$$

where τ denotes the acceleration time. This expression represents a collective relaxation towards an equilibrium speed dependent on the traffic composition, thus assuming that there is not a correlation between the speeds of slowing-down vehicles and the speeds of impeding vehicles.

For the interaction term in (5.25), the model by Prigogine and Herman is based on a set of assumptions, including the so-called *vehicular chaos assumption*, which are listed below:

- the length of vehicles can be neglected;
- the interactions affect at most two vehicles;
- if a fast vehicle moving with speed v reaches a vehicle moving with speed $w < v$, the fast vehicle either overtakes or reduces its speed to w and:
  - the speed w of the slow vehicle is not affected by the interaction;
  - the fast vehicle slows down immediately and overtakes immediately;
  - the speed of the fast vehicle after overtaking remains equal to v;
  - the overtaking event is associated with a probability π, while the slowing-down event is associated with probability $1 - \pi$.

To model the interactions between pairs of vehicles, the Prigogine–Herman models consider couples of vehicles located in the infinitesimal positions $[x, x + \mathrm{d}x)$ and $[x', x' + \mathrm{d}x']$, driving with speeds $[v + \mathrm{d}v)$ and $[v' + \mathrm{d}v')$, respectively, and introduces a two-vehicle distribution function $\tilde{\phi}(x, v, x', v', t)$. This function has the following meaning: $\tilde{\phi}(x, v, x', v', t)\mathrm{d}x\,\mathrm{d}v\,\mathrm{d}x'\,\mathrm{d}v'$ is the expected number of vehicle pairs located at the given infinitesimal areas and with the defined speeds. It can be noted that the previous assumptions and, specifically, the vehicle-chaos assumption, imply the following:

$$\tilde{\phi}(x, v, x', v', t) = \tilde{\rho}(x, v, t)\tilde{\rho}(x', v', t) \tag{5.27}$$

Then, the interaction is modelled with the so-called *collision equation* given by

$$\left(\frac{\partial \tilde{\rho}(x, v, t)}{\partial t}\right)_{\text{int}} = (1 - \pi) \int (w - v)\tilde{\phi}(x, v, x, w, t)\text{d}w \tag{5.28}$$

which, by exploiting (5.27), becomes

$$\left(\frac{\partial \tilde{\rho}(x, v, t)}{\partial t}\right)_{\text{int}} = (1 - \pi)\tilde{\rho}(x, v, t) \int (w - v)\tilde{\rho}(x, w, t)\text{d}w \tag{5.29}$$

This model has received some critiques regarding both the acceleration/relaxation term and the interaction term. Specifically, the acceleration/relaxation term has been criticised referring to the fact that the speeds of slowing-down vehicles and the speeds of impeding vehicles cannot be considered as uncorrelated quantities, meaning that individual relaxation terms in place of the collective one should be more suitable to be adopted. A way of overcoming this assumption was proposed in [101], where a quadratic *Boltzmann term* is used to represent slowing-down and speeding-up interactions. Suitable models for driver reaction and vehicular correlation are used to determine the adopted Boltzmann term.

Other approaches modelling the acceleration term in different ways have been proposed, by taking into account the individual desired speed v_0 and a class-specific acceleration time τ_0. Let us consider in particular the two following extreme cases:

1. all the vehicles can accelerate towards v_0 with an acceleration time equal to τ_0 (see, e.g. [102]);
2. only vehicles in free-flow conditions can accelerate towards v_0 with τ_0 as acceleration time. Vehicles which are constrained (possibly gathered in platoons) do not accelerate at all (see, e.g. [103]).

If the former assumption holds, it is

$$V_0(x, v, t) = v_0 \qquad \tau = \tau_0 \tag{5.30}$$

and these terms must be substituted in (5.26). In case, instead, the latter assumption is considered, the expected fraction θ of platooning vehicles is defined and the following relation holds:

$$V_0(x, v, t) = \theta v + (1 - \theta)v_0 \qquad \tau = \tau_0 \tag{5.31}$$

again to be inserted in (5.26).

As aforementioned, the Prigogine–Herman model received some critiques regarding the relaxation term, since it appears that the relaxation of the distribution function is a property of the road, it does not describe the behaviour of drivers, and it corresponds to discontinuous speed changes. Again in [102], it is discussed that also the collision term (as the acceleration term) proposed in the gas-kinetic model by Prigogine and Herman is only valid when vehicles are not platooning. By considering a scenario in which a free-flowing vehicle encounters a platoon, two cases are analysed in [102]:

1. the free-flowing vehicle overtakes the whole queue of vehicles constituting the platoon;
2. the free-flowing vehicle overtakes each single vehicle in the platoon as if it were alone.

In [102], it is shown that the Prigogine–Herman model is represented by the second case, while the real cases stand between these two extreme situations. Moreover, in [102], a new model is proposed, often known as the *Paveri–Fontana* model, which considers a phase-space density explicitly dependent on the individual desired speed v_0, i.e. $\rho(x, v, v_0, t)$, being

$$\tilde{\rho}(x, v, t) = \int \rho(x, v, v_0, t) \mathrm{d}v_0 \tag{5.32}$$

Moreover, the interaction term is expressed as

$$\begin{aligned} \left(\frac{\partial \tilde{\rho}(x, v, t)}{\partial t}\right)_{\text{int}} &= -\rho(x, v, v_0, t) \int_0^v (1-\pi)(v-\omega)\tilde{\rho}(x, \omega, t)\mathrm{d}\omega \\ &+ \tilde{\rho}(x, v, t) \int_v^{+\infty} (1-\pi)(\omega - v)\rho(x, \omega, v_0, t)\mathrm{d}\omega \end{aligned} \tag{5.33}$$

The overall Paveri–Fontana model can then be written in the following form:

$$\begin{aligned} &\frac{\partial \tilde{\rho}(x, v, t)}{\partial t} + v\frac{\partial \tilde{\rho}(x, v, t)}{\partial x} + \frac{\partial}{\partial v}\left(\tilde{\rho}(x, v, t)\frac{v_0 - v}{\tau_0}\right) \\ &= -\rho(x, v, v_0, t) \int_0^v (1-\pi)(v-\omega)\tilde{\rho}(x, \omega, t)\mathrm{d}\omega \\ &+ \tilde{\rho}(x, v, t) \int_v^{+\infty} (1-\pi)(\omega - v)\rho(x, \omega, v_0, t)\mathrm{d}\omega \end{aligned} \tag{5.34}$$

The complete Paveri–Fontana equation has not been solved in an analytical way, but it is numerically solved in some special cases. However, it is used as the starting point to construct macroscopic and mesoscopic models based on the gas-kinetic theory.

Another issue which is raised with reference to the basic gas-kinetic models is that the assumption that there are some drivers desiring to drive at any speed, no matter how small, seems somewhat unrealistic. The work in [104] addresses the case in which this assumption does not hold, by showing that at high densities it happens that a two-parameter family of solutions exist and, thus, continuously distributed mean speeds can be identified for each density value. This result also gives reason to the well-known scattering of observed data related to the relationship between speed and density for high density values.

An extension of the Paveri–Fontana model was proposed in [105], in which a multi-lane case is considered. Lane changing is explicitly modelled in the following way: indicating with j the lane index, a multi-lane phase-space density $\rho_j(x, v, v_0, t)$ is defined and the following expression holds:

$$\begin{aligned}\frac{\partial \rho_j(x, v, v_0, t)}{\partial t} + v\frac{\partial \rho_j(x, v, v_0, t)}{\partial x} &= \left(\frac{\partial \rho_j(x, v, v_0, t)}{\partial t}\right)_{\text{acc}} \\ &+ \left(\frac{\partial \rho_j(x, v, v_0, t)}{\partial t}\right)_{\text{int}} + \left(\frac{\partial \rho_j(x, v, v_0, t)}{\partial t}\right)_{\text{vc}} \\ + \left(\frac{\partial \rho_j(x, v, v_0, t)}{\partial t}\right)_{\text{lc}} &+ \nu_j^+(x, v, v_0, t) - \nu_j^-(x, v, v_0, t) \end{aligned} \tag{5.35}$$

where $\nu_j^+(x, v, v_0, t)$ and $\nu_j^-(x, v, v_0, t)$ are the rates of vehicles entering and leaving the road at place x, respectively. These rates are different from zero only for merging lanes at entrances and exits. As for the acceleration term in (5.35), it is assumed that vehicles are split into a set of vehicles that can move freely and a set of impeded vehicles that have to move slower than desired, since they are queued behind other vehicles. As in [103], a proportion of freely moving vehicles is, then, defined and the acceleration term is only related to the acceleration of these vehicles.

Moreover, the interaction term in (5.35) is similar to the one used in Paveri–Fontana model, whereas four terms have been added to that previous model. The first of these terms is a speed diffusion term expressed as

$$\left(\frac{\partial \rho_j(x, v, v_0, t)}{\partial t}\right)_{\text{vc}} \tag{5.36}$$

modelling individual fluctuations of the speed due to imperfect driving, while the second is a lane-changing term given by

$$\left(\frac{\partial \rho_j(x, v, v_0, t)}{\partial t}\right)_{\text{lc}} \tag{5.37}$$

representing the changes in the phase-space density of a lane due to vehicles moving to and from the lane itself. Finally, the third and forth terms are the rates of vehicles entering and exiting the road through merging lanes.

A further extension of the basic gas-kinetic models refers to the explicit representation of different vehicle classes belonging to a set $\mathscr{U}$. In [106], a multi-class phase-space density $\rho_u(x, v, v_0, t)$ is introduced, with the index $u \in \mathscr{U}$ related to the vehicle class. In this case, a relation analogous to (5.25) is considered for each vehicle class with

$$\left(\frac{\partial \rho_u(x, v, v_0, t)}{\partial t}\right)_{\text{acc}} = -\frac{\partial}{\partial v}\left(\rho_u(x, v, v_0, t)\frac{v_0 - v}{\tau_u}\right) \tag{5.38}$$

where τ_u is the acceleration time of vehicles of class u. Moreover, the interaction term is defined by separately considering the interactions of vehicles of class u with vehicles of the same class and with vehicles of other classes. To this end, the two terms $I_{u,s}(x, t)$ and $R_{u,s}(x, t)$ are introduced and expressed respectively as

$$I_{u,s}(x, t) = \int_0^v (v - w)\rho_u(x, v, v_0, t)\rho_s(x, w, w_0, t)\mathrm{d}w\,\mathrm{d}w_0 \tag{5.39}$$

$$R_{u,s}(x, t) = \int_v^{+\infty} (w - v)\rho_u(x, w, v_0, t)\rho_s(x, v, w_0, t)\mathrm{d}w\,\mathrm{d}w_0 \tag{5.40}$$

The interaction term is given by

$$\left(\frac{\partial \rho_u(x, v, v_0, t)}{\partial t}\right)_{\text{int}} = -(1 - \pi_u)\sum_s I_{u,s}(x, t) - R_{u,s}(x, t) \tag{5.41}$$

where π_u is the probability associated with an overtaking event for vehicles of class u. It can be noted that the presence of different classes of users results in an asymmetric slowing-down process for fast vehicles, i.e. fast vehicles are influenced by slow vehicles more frequently than vice versa.

In [107], a generic traffic model including multi-lane and multi-class aspects together with the presence of platoons is described. This model gathers all the features of existing gas-kinetic approaches for representing the traffic behaviour. Specifically, a phase-state density $\rho_{u,j,c}(x, v, v_0, t)$ is defined, depending on the vehicle class u, on the road lane j and on the possible belonging of vehicles to a platoon ($c = 2$) or not ($c = 1$). Several drawbacks of previous gas-kinetic models are tackled, since the model describes separately free-flowing and platooning vehicles instead of considering vehicles as independent moving entities. This overcomes the limitations due to the vehicular chaos assumption. Also, the acceleration term is determined in the model by the platoon leader, as it happens in real cases.

These mesoscopic principles present in gas-kinetic models have been also exploited to extend macroscopic models. For instance, in [108, 109], a macroscopic traffic model based on gas-kinetic logics is introduced for the case of multiple classes of vehicles. Analogously, gas-kinetic traffic flow modelling is the basis for a macroscopic model considering adaptive cruise control policies in [110]. Specifically, in [110], two approaches are considered, the former is adapted from [111], while the latter is a novel one and is based on the introduction of a new relaxation term which satisfies the time/space-gap principle of adaptive cruise control systems. The kinetic theory is also used in [112] to derive a new mathematical model of vehicular traffic, in which the assumption on the continuously distributed spatial position and speed of the vehicles is relaxed, consequently resulting in a discretisation of position and speed of the vehicles.

References

1. Hoogendoorn SP, Bovy PHL (2001) State-of-the-art of vehicular traffic flow modelling. Proc Inst Mech Eng Part I, J Syst Control Eng 215:283–303
2. van Wageningen-Kessels F, van Lint H, Vuik K, Hoogendoorn SP (2015) Genealogy of traffic flow models. EURO J Transp Logist 4:445–473
3. Treiber M, Kesting A (2013) Traffic flow dynamics: data, models and simulation. Springer-Verlag, Berlin Heidelberg
4. Kitamura R, Kuwahara M (2005) Simulation approaches in transportation analysis. Springer, US
5. Barceló J (2010) Fundamentals of traffic simulation. Springer, New York
6. Reuschel A (1950) Vehicle movements in a platoon. Oesterreichisches Ingenieur-Archir 4:193–215
7. Pipes LA (1953) An operational analysis of traffic dynamics. J Appl Phys 24:274–281
8. Gazis DC, Herman R, Potts RB (1959) Car-following theory of steady-state traffic flow. Oper Res 7:499–505
9. Brackstone M, McDonald M (1999) Car-following: a historical review. Transp Res Part F 2:181–196
10. Leutzbach W (1988) Introduction to the theory of traffic flow. Springer, Berlin
11. Chandler RE, Herman R, Montroll EW (1958) Traffic dynamics: studies in car following. Oper Res 6:165–184
12. Gazis DC, Herman R, Rothery RW (1961) Nonlinear follow the leader models of traffic flow. Oper Res 9:545–567
13. Helly W (1959) Simulation of bottlenecks in single-lane traffic flow. In: Proceedings of the symposium on theory of traffic flow, pp 207–238
14. Bando M, Hasebe K, Nakayama A, Shibata A, Sugiyama Y (1995) Dynamical model of traffic congestion and numerical simulation. Phys Rev E 51:1035–1042
15. Gipps PG (1986) A model for the structure of lane-changing decisions. Transp Res Part B 20:403–414
16. Ahmed KI (1999) Modeling drivers' acceleration and lane changing behavior. PhD thesis, Massachusetts Institute of Technology
17. Nagel K, Schreckenberg M (1992) A cellular automaton model for freeway traffic. J Phys I 2:2221–2229
18. Rickert M, Nagel K, Schreckenberg M, Latour A (1996) Two lane traffic simulations using cellular automata. Phys A 231:534–550
19. Nagel K (1998) From particle hopping models to traffic flow theory. Transp Res Rec 1644:1–9
20. Ossen S, Hoogendoorn SP (2011) Heterogeneity in car-following behavior: theory and empirics. Transp Res Part C 19:182–195
21. Saifuzzaman M, Zheng Z (2014) Incorporating human-factors in car-following models: a review of recent developments and research needs. Transp Res Part C 48:379–403
22. Koutsopoulos HN, Farah H (2012) Latent class model for car following behavior. Transp Res B 46:563–578
23. Ceder A, May AD (1976) Further evaluation of single and two regime traffic flow models. Transp Res Rec 567:1–30
24. Herman R, Potts RB (1959) Single lane traffic theory and experiment. In: Proceedings of the symposium on theory of traffic flow, pp 147–157

25. Treiterer J, Myers JA (1974) The hysteresis phenomenon in traffic flow. In: Proceedings of the sixth international symposium on transportation and traffic theory, pp 3–38
26. Ozaki H (1993) Reaction and anticipation in the car following behaviour. In: Proceedings of the 13th international symposium on traffic and transportation theory, pp 349–366
27. Zadeh LA (1975) Fuzzy logic and approximate reasoning. Synthese 30:407–428
28. Kikuchi C, Chakroborty P (1992) Car following model based on a fuzzy inference system. Transp Res Rec 1365:82–91
29. Wu J, Brackstone M, McDonald M (2005) Fuzzy sets and systems for a motorway microscopic simulation model. Fuzzy Sets Syst 116:65–76
30. Ross TJ (2010) Fuzzy logic with engineering applications. Wiley, United Kingdom
31. Kometani E, Sasaki T (1959) Dynamic behaviour of traffic with a non-linear spacing-speed relationship. In: Proceedings of the symposium on theory of traffic flow, pp 105–119
32. Newell GF (1961) Nonlinear effects in the dynamics of car following. Oper Res 9:209–229
33. Dijker T, Bovy PHL, Vermijs RGMM (1998) Car-following under congested conditions: empirical findings. Transp Res Rec 1644:20–28
34. Gipps PG (1981) A behavioural car-following model for computer simulation. Transp Res Part B 15:105–111
35. Yang D, Zhu LL, Yu D (2014) An enhanced safe distance car-following model. J Shanghai Jiaotong Univ 19:115–120
36. Hanken A, Rockwell TH (1967) A model of car following derived empirically by piece-wise regression analysis. In: Proceedings of the 3rd international symposium on the theory of traffic flow, pp 40–41
37. Bekey GA, Burnham GO, Seo J (1977) Control theoretic models of human drivers in car following. Hum Factors 19:399–413
38. Aron M (1988) Car following in an urban network: simulation and experiments. In: Proceedings of the 16th PTRC meeting, pp 27–39
39. Xing J (1995) A parameter identification of a car following model. In: Proceedings of the second world congress on ATT, pp 739–1745
40. Treiber M, Hennecke A, Helbing D (2000) Congested traffic states in empirical observations and microscopic simulations. Phys Rev E 62:1805–1824
41. Treiber M, Helbing D (2003) Memory effects in microscopic traffic models and wide scattering in flow-density data. Phys Rev E 68:046119
42. Helbing D, Tilch B (1998) Generalized force model of traffic dynamics. Phys Rev E 58:133–138
43. Lenz H, Wagner C, Sollacher R (1999) Multi-anticipative car-following model. Eur Phys J B 7:331–335
44. Jiang R, Wu Q, Zhu Z (2001) Full velocity difference model for a car-following theory. Phys Rev E 64:017101
45. Gong H, Liu H, Wang BH (2008) An asymmetric full velocity difference car-following model. Phys A 387:2595–2602
46. Peng G, Sun D (2010) A dynamical model of car-following with the consideration of the multiple information of preceding cars. Phys Lett A 374:1694–1698
47. Boer ER (1999) Car following from the driver's perspective. Transp Res Part F 2:201–206
48. Evans L, Rothery R (1973) Experimental measurement of perceptual thresholds in car following. Highway Res Rec 64:13–29
49. Evans L, Rothery R (1977) Perceptual thresholds in car following—a recent comparison. Transp Sci 11:60–72
50. Wiedemann R (1974) Simulation des Strassenverkehrsflusses. Schriftenreihe des Instituts fr Verkehrswesen der Universitt Karlsruhe, Germany
51. Fritzsche HT (1994) A model for traffic simulation. Traffic Eng Control 317–321
52. Leutzbach W, Wiedemann R (1986) Development and applications of traffic simulation models at the Karlsruhe Institut fur Verkehrwesen. Traffic Eng Control 27:270–278
53. Burnham GO, Bekey GA (1976) A heuristic finite state model of the human driver in a car following situation. IEEE Trans Syst Man Cybern 6:554–562

54. Wiedemann R, Reiter U (1992) Microscopic traffic simulation: the simulation system MISSION, background and actual state. CEC Project ICARUS, Final Report, vol 2, Appendix A
55. Michaels R (1963) Perceptual factors in car following. In: Proceedings of the 2nd international symposium on the theory of road traffic flow, pp 44–59
56. Andersen GJ, Sauer CW (2007) Optical information for car following: the driving by visual angle (DVA) model. Hum Factors: J Hum Factors Ergonomics Soc 49:878–896
57. Jin S, Wang DH, Huang ZY, Tao PF (2011) Visual angle model for car-following theory. Phys A 390:1931–1940
58. Hamdar SH, Mahmassani HS (2008) From existing accident-free car-following models to colliding vehicles: exploration and assessment. Transp Res Rec 2088:45–56
59. Kahneman D, Tversky A (1979) Prospect theory: an analysis of decision under risk. Econometrica 47:263–291
60. Talebpour A, Mahmassani HS, Hamdar SH (2011) Multiregime sequential risk-taking model of car-following behavior. Transp Res Rec 2260:60–66
61. Van Winsum W (1999) The human element in car following models. Transp Res Part F 2:207–211
62. Treiber M, Kesting A, Helbing D (2006) Delays, inaccuracies and anticipation in microscopic traffic models. Physica A 360:71–88
63. Bevrani K, Chung E (2012) A safety adapted car following model for traffic safety studies. In: Stanton NA (ed) Advances in human aspects of road and rail transportation. CRC Press, USA, pp 550–559
64. Toledo T (2007) Driving behaviors: models and research directions. Transp Rev 27:65–84
65. Moridpour S, Sarvi M, Rose G (2010) Lane changing models: a critical review. Transp Lett 2:157–173
66. Zheng Z (2014) Recent developments and research needs in modeling lane changing. Transp Res Part B 60:16–32
67. Pande A, Abdel-Aty M (2006) Assessment of freeway traffic parameters leading to lane-change related collisions. Accid Anal Prev 38:936–948
68. Zheng Z, Ahn S, Monsere CM (2010) Impact of traffic oscillations on freeway crash occurrences. Accid Anal Prev 42:626–636
69. Cassidy M, Rudjanakanoknad J (2005) Increasing the capacity of an isolated merge by metering its on-ramp. Transp Res Part B 39:896–913
70. Kerner B, Rehborn H (1996) Experimental features and characteristics of traffic jams. Phys Rev E 53:1297–1300
71. Ahn S, Cassidy M (2007) Freeway traffic oscillations and vehicle lane-change maneuvers. In: Proceedings of the 17th international symposium on transportation and traffic theory, pp 691–710
72. Zheng Z, Ahn S, Chen D, Laval J (2011) Freeway traffic oscillations: microscopic analysis of formations and propagations using wavelet transform. Transp Res Part B 45:1378–1388
73. Yang Q, Koutsopoulos HN (1996) A microscopic traffic simulator for evaluation of dynamic traffic management systems. Transp Res Part C 4:113–129
74. Kesting A, Treiber M, Helbing D (1999) General lane-changing model MOBIL for car-following models. Transp Res Rec 2007:86–94
75. Sheu JB, Ritchie SG (2001) Stochastic modelling and real-time prediction of vehicular lane-changing behavior. Transp Res Part B 35:695–716
76. Moridpour S, Rose G, Sarvi M (2009) Modelling the heavy vehicle drivers? Lane changing decision under heavy traffic conditions. J Road Transp Res 18:49–57
77. Laval JA, Daganzo CF (2006) Lane-changing in traffic streams. Transp Res Part B 40:251–264
78. von Neumann J (1948) The general and logical theory of automata. In: Jeffress LA (ed) Cerebral mechanisms in behavior. Wiley, NewYork, pp 1–41
79. Wolfram S (1983) Statistical mechanics of cellular automata. Rev Mod Phys 55:601–644
80. Chowdhury D, Santen L, Schadschneider A (2000) Statistical physics of vehicular traffic and some related systems. Phys Rep 329:199–329

81. Knospe W, Santen L, Schadschneider A, Schreckenberg M (2004) An empirical test for cellular automaton models of traffic flow. Phys Rev E 70:016115
82. Maerivoet S, De Moor B (2005) Cellular automata models of road traffic. Phys Rep 419:1–64
83. Caligaris C, Sacone S, Siri S (2009) Model predictive control for multiclass freeway traffic. In: Proceedings of the European control conference, pp 1764–1769
84. Hafstein SF, Chrobok R, Pottmeier A, Schreckenberg M, Mazur FC (2004) A high-resolution cellular automata traffic simulation model with application in a freeway traffic information system. Comput-Aided Civil Infrastruct Eng 19:338–350
85. Nagel K, Wolf DE, Wagner P, Simon P (1998) Two-lane traffic rules for cellular automata: a systematic approach. Phys Rev E 58:1425–1437
86. Choa F, Milam RT, AICP, Stanek D (2004) CORSIM, PARAMICS, and VISSIM: what the manuals never told you. In: Proceedings of the Ninth TRB conference on the application of transportation planning methods
87. Panwai S, Dia H (2005) Comparative evaluation of microscopic car-following behavior. IEEE Trans Intell Transp Syst 6:314–325
88. Olstam JJ, Tapani A (2004) Comparison of car-following models. Report of the Swedish National Road and Transport Research Institute
89. Buckley DJ (1968) A semi-poisson model of traffic flow. Transp Sci 2:107–133
90. Wasielewski P (1974) An integral equation for the semi-poisson headway distribution model. Transp Sci 8:237–247
91. Branston D (1976) Models of single lane time headway distributions. Transp Sci 10:125–148
92. Cowan RJ (1975) Useful headway models. Transp Res 9:371–375
93. Hoogendoorn SP, Bovy PHL (1998) A new estimation technique for vehicle-type specific headway distribution. Transp Res Rec 1646:18–28
94. Krbalek M, Seba P, Wagner P (2001) Headways in traffic flow: remarks from a physical perspective. Phys. Rev. E 64:066119
95. Treiber M, Kesting A, Helbing D (2006) Understanding widely scattered traffic flows, the capacity drop, and platoons as effects of variance-driven time gaps. Phys Rev E 74:016123
96. Herrmann M, Kerner BS (1998) Local cluster effect in different traffic flow models. Phys A 55:163–188
97. Mahnke R, Kühne R (2007) Probabilistic description of traffic breakdown. In: Schadschneider A, Poschel T, Kühne R, Schreckenberg M, Wolf DE (eds) Traffic and granular flow. Springer, New York, pp 527–536
98. Mahnke R, Kaupužs J, Lubashevsky I (2005) Probabilistic description of traffic flow. Phys Rep 408:1–130
99. Prigogine I (1961) A Boltzmann-like approach to the statistical theory of traffic flow. In: Theory of traffic flow. Elsevier, Amsterdam, pp 158–164
100. Prigogine I, Herman R (1971) Kinetic theory of vehicular traffic. American Elsevier, New York
101. Nelson P (1995) A kinetic theory of vehicular traffic and its associated bimodal equilibrium solutions. Transp Theory Stat Phys 24:383–409
102. Paveri-Fontana SL (1975) On Boltzmann-Like treatments for traffic flow: a critical review of the basic model and an alternative proposal for dilute traffic analysis. Transp Res Part B 9:225–235
103. Helbing D (1997) Verkehrsdynamik. Springer, Berlin
104. Nelson P, Sopasakis A (1998) The Prigogine-Herman kinetic model predicts widely scattered traffic flow data at high concentrations. Transp Res Part B 32:589–604
105. Helbing D (1997) Modeling multi-lane traffic flow with queuing effects. Phys A 242:175–194
106. Hoogendoorn SP, Bovy PHL (2000) Modelling multiple user-class traffic flow. Transp Res Part B 34:123–146
107. Hoogendoorn SP, Bovy PHL (2001) Generic gas-kinetic traffic systems modeling with applications to vehicular traffic flow. Transp Res Part B 35:317–336
108. Hoogendoorn SP, Bovy PHL (2000) Continuum modeling of multiclass traffic flow. Transp Res Part B 34:123–146

109. Hoogendoorn SP, Bovy PHL (2001) Platoon-based multiclass modeling of multilane traffic flow. Netw Spatial Econ 1:137–166
110. Delis AI, Nikolos IK, Papageorgiou M (2015) Macroscopic traffic flow modeling with adaptive cruise control: development and numerical solution. Comput Math Appl 70:1921–1947
111. Ngoduy D (2013) Instability of cooperative adaptive cruise control traffic flow: a macroscopic approach. Commun Nonlinear Sci Numer Simul 18:2838–2851
112. Fermo L, Tosin A (2013) A fully-discrete-state kinetic theory approach to modeling vehicular traffic. SIAM J Appl Math 73:1533–1556

Chapter 6
Emission Models for Freeway Traffic Systems

6.1 Features and Applications

Transportation and its impact on the society and the environment are themes that have been and are being debated in any scientific, professional, politic and social contexts. Transportation yields substantial socio-economic benefits for the society, but at the same time it brings negative effects on the environment. On the one hand, transportation activities support increasing mobility demands for passengers and freight and, on the other hand, they are associated with growing levels of environmental externalities.

Among the different modes of transport, road transportation presents several advantages, since it is cost effective, fast for short-term operations, flexible (services, routes and timings can be adjusted and changed without incurring in too high costs), and typically comfortable for passengers. However, one of the main negative effects due to the predominance of the road system compared with other modes of transport is the environmental impact, which is mainly linked to the introduction of chemical pollutants, heat or greenhouse gases in the environment. *Air pollution* from automotive sources is one of the major causes of air pollution in the environment and this is even more serious for urban areas located near major roadways (see Fig. 6.1). The European Union is not insensitive to these issues, and therefore, in the past decades, severe legislations on pollutants produced by single vehicles and on their concentrations in the environmental system have been introduced. Most of the other countries all over the world have defined specific rules and legislations to deal with pollution issues as well.

The environmental issues must then be considered also in the definition of freeway traffic control schemes, which are the main topic of the present book. Besides taking care of traffic congestion phenomena, which have been central in the definition of any traffic control scheme in the last decades, it is more and more important to also tackle the environmental impact of road arterials and to explicitly consider the reduction of air pollution among the different requisites of the traffic control schemes.

A. Ferrara et al., *Freeway Traffic Modelling and Control*, Advances in Industrial Control, https://doi.org/10.1007/978-3-319-75961-6_6

Fig. 6.1 Van Brienenoord bridge on the A16 freeway, the Netherlands (courtesy of Rijkswaterstaat, Photo: Essencia Communication/Rob de Voogd)

In view of this, it is essential to design models allowing to estimate the environmental damage produced by traffic streams or to predict the impact due to the introduction of new infrastructures. These models must provide a reliable knowledge about the sources and causes of pollution, about the technological and behavioural parameters and about the potentials of different control strategies. Moreover, these models need to be suitably integrated with the models representing the traffic behaviour in order to constitute a solid model basis to be used in control schemes. Among the several environmental impacts of road traffic, this chapter is mainly devoted to consider air pollution and to analyse some existing and innovative models for traffic emissions. It can be noted that also models for evaluating fuel consumption could be found in the literature, but they will be not analysed in detail in this book, being less interesting for traffic control purposes.

The vehicle emission models are conceived in order to evaluate the impact of traffic flows on air quality. Indeed, these models require as inputs the traffic data from adequate traffic flow models or from real measurements (e.g. traffic flow, traffic composition, vehicle speed and vehicle acceleration). Such models generally estimate the air pollution produced in a specific traffic scenario, providing the emission level for each type of pollutant, related to a reference time or space unit. The outputs produced by emission (or consumption) models may be distinguished in two categories, as done in [1]. The two categories are the following:

- *local emission factors*, which describe the produced emissions (or the consumed fuel) per space unit;
- *instantaneous emission factors*, which describe the produced emissions (or the consumed fuel) per time unit.

In general, both local and instantaneous emission factors depend on several aspects, such as the ones listed in the following:

- *mechanical characteristics of vehicles*: the presence of pollutant elements in the exhaust gases is mainly due to anomalies during the combustion process, whereas the introduction of CO_2 and heat is an inevitable consequence of this phase. Other factors that determine the production of pollutants depend on the air–fuel ratio, the ignition timing, the compression ratio of the engine and the geometric characteristics of the cylinders. Finally, the presence of devices that enable the reduction of pollutant emissions must be considered, i.e. the adoption of filters and traps, exhaust gas recirculation, advanced systems of valves control and many others;
- *fuel characteristics*: the type and quantity of emitted pollutants largely depend on the type of fuel used (such as gasoline, diesel, liquefied petroleum gas, etc.) and by its quality;
- *vehicle operating conditions*: driving style and operating conditions have a significant influence on the emission rates and fuel consumption (see for instance the study reported in [2]). In particular, the greatest contribution depends on the engine temperature and on the driving phases, i.e. acceleration, deceleration and cruise mode. Specifically, it is experimentally analysed that, during the acceleration phases, the quantity of polluting substances emitted is greater than the one produced during the other phases of motion. Another relevant aspect is represented by road morphology: the presence of slopes or intersections may negatively affect the level of produced emissions.

The emission factors obtained from traffic emission models may be used either for offline evaluations, or for applications in real-time monitoring tools. With reference to the first purpose, in case no effective emission measurements are available, emission models are applicable to generate emission inventories. An example of these approaches is illustrated in [3], where different models have been adopted and compared in order to propose a new inventory approach based on mean speed distributions.

Additional offline procedures require the adoption of emission models in order to quantify the environmental impact caused by the modification of the transport offer. In this regard, in [4] the benefits of traffic light coordination on the reduction of pollutant emissions are evaluated, in [5] the effect of roundabout operations on the environment is illustrated, whereas in [6] a study to assess the air quality near traffic intersections is conducted. An interesting similarity with the Braess Paradox is shown in [7]. Indeed, in this latter work, in analogy with [8, 9], it is found that the improvement of the network capacity can lead to an increase in the emissions generated by traffic.

Moreover, the emission models may be adopted to verify the effectiveness of traffic control measures in the abatement of the environmental damage. This was done, for instance in [10], which illustrates the advantages, in terms of emission reduction, produced through the implementation of traffic control systems during the 2008 Olympic Games in Beijing, or in [11–13], where the effect of variable speed limits on the level of traffic emissions is discussed. With reference to real-time applications, the methods that include the use of emission models within control schemes are discussed in detail in Chap. 10: some examples are the works [14–17] in the freeway context and [18] in the urban context.

Finally, besides the evaluation of emission factors, the concentrations of harmful substances in the environment may be quantified through dispersion emission models. These models, on the basis of the emission source, the morphological and meteorological characteristics of the site, produce an indication on the concentration of pollutants in the environmental system. In the literature, many models of pollutant dispersion have been proposed, some of these are reported in [19–21]. In this chapter, only traffic emission models and their adoption in traffic flow models are analysed, whereas dispersion models are not treated, since all the aspects related to the dispersion of pollutants are out of the scope of the book.

6.2 Classification

The existing traffic emission models aimed at estimating pollutant emission levels can be classified according to the complexity of the necessary input information and the aggregation level of the variables that describe them. Indeed, in the different models used for evaluating emissions, the model variables and their relations may describe this phenomenon at different levels of detail. Analogously to traffic flow models, emission models can be classified in

- *macroscopic models*, based on aggregate variables that allow to compute, in a simplified way, the overall pollutant emissions on road portions;
- *microscopic models*, where emission factors are associated with single vehicles and are computed starting from an accurate description of the physical processes underlying the phenomenon;
- *mesoscopic models*, that represent an intermediate description level between microscopic and macroscopic models.

In the following, an outline of the main models that belong to each of these categories and that can be useful for freeway traffic control is reported. Then, two models are described in detail, i.e. the COPERT model (Sect. 6.3) and the VERSIT+ model (Sect. 6.4), which have already been applied in traffic control schemes, as deeply discussed in Chap. 10.

6.2.1 Macroscopic Emission Models

Macroscopic emission models provide an approximate representation of the real emission processes, adopting aggregate parameters and variables. Different from microscopic models, they are not based on a faithful representation of the physical phenomena that generate the emission rates, but they take into account aggregate and average information. Thanks to this feature, macroscopic models allow to analyse the entire transport system without requiring the high computational effort needed by microscopic representations, and this surely represents a relevant advantage of such models. In light of these considerations, macroscopic emission models are the most suitable for being adopted in freeway traffic control schemes, also because these latter are normally based on traffic models which are macroscopic as well.

Besides the so-called *area-wide models* that are mainly used for planning purposes (and are, thus, outside the scope of this book), macroscopic models are further classified in average-speed emission models, traffic-situation emission models and traffic-variable emission models.

Average-speed emission models allow to make a rather realistic estimation with the lowest computational effort and, therefore, they are quite suitable to be embedded in on-line traffic control schemes. These models provide the average values of the emission factors of each harmful substance, for different categories of vehicles, as a function of the average speed in a certain road link. The output produced by these models is a local emission factor, namely the mass of pollutant emitted per space unit and per vehicle. Generally, these models are formulated so that the average speeds implicitly consider the various phases of motion, thereby increasing the accuracy of the model. An example is the COPERT model proposed in [22–24] and discussed in detail in Sect. 6.3. Another model of this category is MOBILE [25], which is, instead, insensitive to changes in the driving cycle and requires very detailed information about the type of vehicles, the used fuel and the environmental conditions. Other average-speed emission models are the Elemental model [26, 27], developed in urban contexts, which expresses the consumption of fuel through a linear function of the average trip time for a unit distance, and the Watson model [28], where the variation of speed during the trip is partially taken into account.

Another class of macroscopic models is represented by *traffic-situation emission models*, which express the relations between emission factors (and fuel consumption) and specific traffic conditions. More specifically, instead of the average-speed trajectory, this kind of models receives in input several sets of driving patterns. Each driving pattern reproduces the behaviour of different driving conditions (e.g. free-flow, congested, stop-and-go) and traffic scenarios (freeway, rural road, arterial road, urban road). Examples of traffic-situation emission models are, for instance, the HBEFA model [29], where the traffic emissions are related to different types of vehicles, to traffic situations and to the adopted fuel, and the ARTEMIS model, proposed in [30], in which the effect of different driving conditions is considered through a set of sub-models.

The third class of macroscopic models is represented by *traffic-variable emission models*. The emission factors generated by these models are dependent on the average dynamic traffic variables (speed, density, flow and queue length) and on the characteristics of the transport infrastructures. In some cases, suitable correction factors are introduced to consider the variance of the traffic variables. An example of such models is reported in [31].

6.2.2 Microscopic Emission Models

Within the wide range of models for estimating traffic emissions, microscopic models surely provide the most accurate evaluation of vehicular pollutants. Indeed, compared with macroscopic models, these models are based on a more precise knowledge of the dynamics of individual vehicles (e.g. instantaneous speed and acceleration), on the road geometry and on environmental features (such as temperature and air humidity). Microscopic emission models are especially used in the assessments of local pollution conditions, relying on their disaggregation level and the high number of required input data. Examples of such applications are shown in [4, 5, 18].

In the literature, several microscopic emission models exist, which may be classified in speed-profile emission models and modal emission models. Readers interested in a more comprehensive description are referred to [1]. In this section, only the main features of these models are addressed by referring to the most recent literature.

Speed-profile emission models use as input data the speed trajectories of a single vehicle with high temporal resolution. These trajectories are not directly used by the model, but grouped into some speed-profile factors identified for specific driving cycles. These speed-profile factors, suitably completed with additional information (such as classes of vehicles, environmental factors, road information), allow to generate the instantaneous or local emission factors for several pollutants. Among these types of models, it is worth recalling the MEASURE model [32, 33].

Differently from speed-profile emission models, which evaluate the substances emitted from vehicular traffic through some aggregation factors of the single vehicle speed profile, the *modal emission models* directly adopt the instantaneous information obtained from microscopic flow models or from traffic detectors. Modal emission models may also be distinguished in three categories:

- *emission map models*: they are presented in the form of matrices in which, for each kind of emission types and vehicle categories, one dimension represents the range of the possible speeds while the other indicates the possible areas of specific power or acceleration. Hence, the instantaneous emissions are assigned to each cell of the matrix representing a combination of the vehicle speed and acceleration (or power) observed at a specific time instant. Several limitations characterise these models, since these maps may be sparse and sensitive to the driving cycle used to generate them, as well as they may be not flexible to changes in the boundary conditions.

Further information about properties and possible applications of emission maps are discussed in [34];

- *regression-based models*: they generally make use of instantaneous speed and acceleration relations obtained from linear regressions. These models, on the one hand, allow to overcome the inaccuracies of emission map models, but, on the other hand, the absence of an accurate physical relation can lead to unrealistic results. In the literature, several approaches concerning statistical models are described. Some examples can be found in the modelling framework proposed in [35], the POLY model [36], the CMEM model [37], the VT-micro model [38, 39] and the VERSIT+ model [40]. This latter model will be described in detail in Sect. 6.4;
- *load-based models*: they rely on a careful analysis of the physical and chemical processes that give rise to pollutant emissions, where the main variable is represented by the rate of fuel consumption. Although these models are very effective in the description of emission and consumption phenomena, they require a high number of input parameters that makes them more suitable for punctual applications than for the analysis of traffic flows. An example of load-based models can be found in [41], whereas a detailed description of this model is reported in [42].

6.2.3 Mesoscopic Emission Models

Mesoscopic emission models represent an intermediate description level between microscopic and macroscopic models. In fact, in analogy with traffic flow models, mesoscopic emission models have a higher aggregation level of variables than the microscopic ones, while they are more detailed compared with the macroscopic models. In contrast with macroscopic models, that are based on average variables (i.e. the average speed), mesoscopic models can carry out a more accurate estimation without reaching the high level of detail of microscopic models.

One example is the approach proposed in [43], where fuel consumption is computed by decomposing the driving cycle in its primary components, i.e. idling acceleration, cruise and deceleration phases. This model is similar to the one used in the TRANSYT-7F simulation tool [44]. Another type of mesoscopic model is the mesoscopic version of the VT-model presented in [45], where the consumption rates of fuel are estimated as functions of the average speed, the number of stops and the average length of each stop.

6.3 The COPERT Model for Freeway Traffic Systems

In the definition of freeway traffic control schemes, a suitable choice is the adoption of aggregate models computing emissions as dependent on the main traffic variables. Actually, in this way, the control scheme can include both the traffic flow model describing the system dynamic evolution and the model for the computation of the emissions. Consequently, the control scheme can take into account explicitly, not only the reduction of traffic congestion phenomena, but also the emissions produced by vehicles in the freeway system.

Average-speed emission models have been used in some control approaches (see for instance [15, 46, 47]) to compute the impact of freeway traffic control on air quality. One of the major advantages in adopting average-speed emission models for control purposes is that they require a low computational effort.

A straightforward and widely known average-speed emission model is represented by the so-called COPERT model. COPERT has been introduced by the European Environment Agency in order to realise national and regional emission inventories for the CORINAIR project. The CORINAIR project was developed for the first time in 1985 [48] and successively updated in 1990 [49]. The aim of this project was to produce an extensive inventory of anthropogenic emissions (not only those generated by the road sector), by dividing them into different categories. The COPERT model was initially proposed in [50] and then implemented in the tool COPERT II [22]. Subsequent modifications to the model were made in COPERT III [23], in COPERT 4 [24] and in COPERT 5, which represents the most recent version of the model.

The most updated versions of the model cover a wide variety of pollutant types, in particular the following ones are examined:

- chemical compounds as carbon monoxide (CO), hydrocarbons (HC) (both methane (CH_4) and non-methane volatile organic compounds (NMVOC)), nitrogen oxides NO_x, sulphur oxides (SO_x) and particulate matter (PM);
- greenhouse gases such as nitrous oxide (N_2O), methane (CH_4), sulphur (SF_6) and carbon dioxide (CO_2);
- toxic substances as dioxins, furans, heavy metals, and carcinogenic species as persistent organic pollutants (POPs) and polycyclic aromatic hydrocarbons (PAHs).

However, if for substances such as CO, NO_x, NMVOC and PM, the model produces a rather accurate estimation, for the emissions of CO_2, SO_x, N_2O, CH_4, the evaluations produced by the model are derived from estimates of fuel consumptions and are, thus, less precise. Furthermore, in order to meet the technological advances required by the European Union for the reduction of emissions of harmful substances, COPERT computes the emission factors of a considerable range of vehicles, distinguishing them on the basis of the emission control technologies installed on board of vehicles.

In the COPERT model, the total emissions are computed as the sum of three components, i.e. hot emissions produced during the stabilised engine operation, cold emissions produced during the warming-up phase following the cold starts of the vehicle, and evaporative emissions associated with the evaporative phenomenon. In order to evaluate the impact of freeway traffic, in this chapter and in the remaining chapters of the book only *hot emissions* will be considered. The interested reader may refer to [24, 51] for the calculation of the other types of emissions.

In the following, the COPERT model is firstly introduced for different types of vehicles (Sect. 6.3.1) and, then, in Sect. 6.3.2 its adoption in a multi-class traffic flow model is discussed.

6.3.1 The COPERT Model

In the COPERT model, the emission factors concerning hot emissions are exclusively dependent on the average speed of vehicles, through rather simple relations. COPERT covers a broad range of vehicles, only some of which are illustrated in this section, with specific attention to the relations used in traffic control problems and recalled in Chap. 10.

An important distinction is between passenger vehicles and heavy-duty vehicles, that can be associated with cars and trucks, respectively, in case a two-class traffic flow model is adopted (see Sect. 6.3.2).

Let us start from *passenger cars* and let us consider J different legislation emission categories; for instance, in [24], $J = 4$ categories are considered, from Euro 1 to Euro 4, whereas, in [51], they have been extended to $J = 6$ categories, adding EURO 5 and EURO 6. By relying for instance on the COPERT model proposed in [24], for a gasoline passenger car of legislation emission category j, $j = 1, \ldots, 4$, the local emission factor for each single vehicle, related to the hot emissions of a given type of pollutant, is function of the mean speed v_j^{car}, with v_j^{car} between 10 and 130 [km/h]. Specifically, the emission factor of a single car $\Xi_j^{\text{car}}(v_j^{\text{car}})$ [g/km] is obtained as

$$\Xi_j^{\text{car}}(v_j^{\text{car}}) = \frac{a_j^{\text{car}} + e_j^{\text{car}} v_j^{\text{car}} + f_j^{\text{car}} (v_j^{\text{car}})^2}{1 + b_j^{\text{car}} v_j^{\text{car}} + d_j^{\text{car}} (v_j^{\text{car}})^2} \tag{6.1}$$

where a_j^{car}, b_j^{car}, d_j^{car}, e_j^{car} and f_j^{car}, $j = 1, \ldots, 4$, are parameters assuming specific values according to the considered type of pollutant [24].

Figure 6.2 shows the curves of CO emissions depending on the traffic mean speed for cars for the four legislation emission categories, according to (6.1) and with the values of the parameters defined in [24]. It is worth highlighting that such profiles strongly change according to the legislation emission category; Euro 1 and Euro 2 cars present the lowest emission factors for intermediate values of the average speed, whereas Euro 3 and Euro 4 cars show increasing curves, i.e. the lowest emission factors correspond to the lowest speeds.

Analogously, for *trucks, buses and coaches*, the emission factor computed by COPERT depends on the average speed of vehicles, again divided in classes, from Euro 1 to Euro 6. Several relations may be used to calculate the emission factor of a generic vehicle of type h, with h indicating a specific class of heavy vehicles, with given loading conditions and specific characteristics of the slope of the road, and referring to a specific legislation emission category j, $j = 1, \ldots, J$ [24, 51].

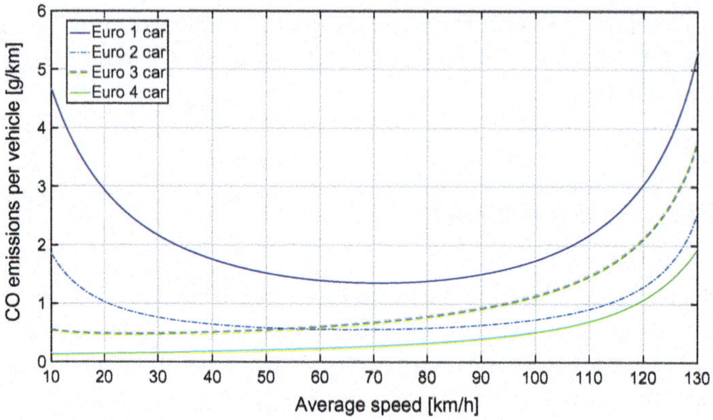

Fig. 6.2 CO emissions for passenger cars in COPERT model

Specifically, the emission factor of a single vehicle $\Xi_j^h(v_j^h)$ [g/km] depends on its main speed v_j^h. The most common relations are

$$\Xi_j^h(v_j^h) = a_j^h + b_j^h v_j^h + \frac{(c_j^h - b_j^h)(1 - \exp(-d_j^h v_j^h))}{d_j^h} \tag{6.2}$$

$$\Xi_j^h(v_j^h) = e_j^h + a_j^h \exp(-b_j^h v_j^h) + c_j^h \exp(-d_j^h v_j^h) \tag{6.3}$$

$$\Xi_j^h(v_j^h) = \frac{1}{c_j^h (v_j^h)^2 + b_j^h v_j^h + a_j^h} \tag{6.4}$$

$$\Xi_j^h(v_j^h) = \frac{1}{a_j^h + b_j^h (v_j^h)^{c_j^h}} \tag{6.5}$$

$$\Xi_j^h(v_j^h) = \frac{1}{a_j^h + b_j^h v_j^h} \tag{6.6}$$

$$\Xi_j^h(v_j^h) = a_j^h - b_j^h \exp\left(-c_j^h (v_j^h)^{d_j^h}\right) \tag{6.7}$$

$$\Xi_j^h(v_j^h) = a_j^h + \frac{b_j^h}{1 + \exp\left(-c_j^h + d_j^h \ln(v_j^h) + e_j^h v_j^h\right)} \tag{6.8}$$

$$\Xi_j^h(v_j^h) = c_j^h + a_j^h \exp(-b_j^h v_j^h) \tag{6.9}$$

$$\Xi_j^h(v_j^h) = c_j^h + a_j^h \exp(b_j^h v_j^h) \tag{6.10}$$

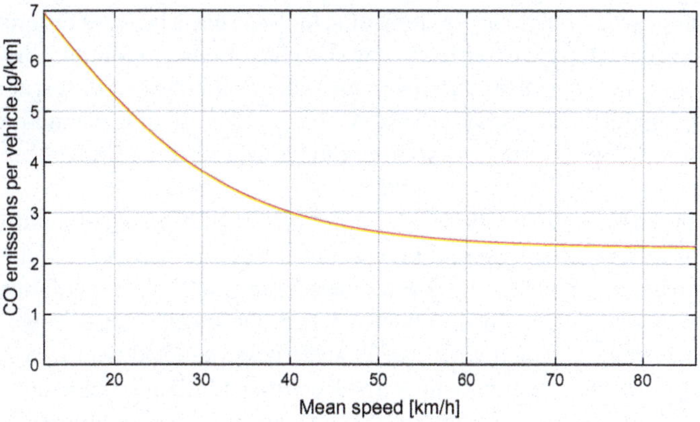

Fig. 6.3 CO emissions for Euro 1 trucks in COPERT model

$$\Xi_j^h(v_j^h) = \exp\left(a_j^h + \frac{b_j^h}{v_j^h}\right) + c_j^h \ln(v_j^h) \tag{6.11}$$

where $a_j^h, b_j^h, c_j^h, d_j^h, d_j^h$ are the model parameters, $j = 1, \ldots, J$.

Let us consider now a specific case, i.e. half-loaded trucks of Euro 1 class (i.e. $j = 1$) for roads with no slope. The COPERT model [24] computes the emission factor of a single vehicle of this type, denoted ad $\Xi_1^{\text{truck}}(v_1^{\text{truck}})$ [g/km], as function of the mean speed v_1^{truck}, with v_1^{truck} between 12 and 86 [km/h], as

$$\Xi_1^{\text{truck}}(v_1^{\text{truck}}) = a_1^{\text{truck}} + \frac{b_1^{\text{truck}}}{1 + \exp\left(-c_1^{\text{truck}} + d_1^{\text{truck}} \ln(v^{\text{truck}}) + e_1^{\text{truck}} v_1^{\text{truck}}\right)} \tag{6.12}$$

which is derived from (6.8).

Figure 6.3 shows the profile of CO emissions for half-loaded trucks of Euro 1 class in case of roads with no slope, as functions of the mean speed of the vehicle, according to (6.12), with the values of the parameters defined in [24]. It is worth noting that the curve of the emission factor shows a decreasing profile with the average speed, i.e. the highest emissions are produced with low speeds (this behaviour is the opposite compared to the one of the Euro 3 and Euro 4 cars shown in Fig. 6.2).

6.3.2 *Use of COPERT in a Traffic Flow Model*

The COPERT model can be used associated with a traffic flow model in order to compute the emissions in a given freeway stretch or in a given freeway network. Generally speaking, if an emission model is adopted, it is useful to consider a multi-class traffic flow model instead of a single-class traffic model representing the entire flow as composed of only one typology of vehicles. This is because different types of vehicles are characterised by different emission factors and, to compute the emissions in an accurate way, it is useful to distinguish the traffic flow in different classes, at least by distinguishing cars and trucks. Moreover, the use of a second-order model seems more appropriate than a first-order model, since in the former the mean speed is explicitly modelled with its own dynamics.

Referring in particular to the second-order METANET model of multi-class type, the case of a freeway stretch and the case of a freeway network should be distinguished, as discussed in Sect. 4.3. Let us consider the multi-class second-order model for a freeway stretch described in Sect. 4.3.1, in the specific case of two classes of vehicles, i.e. cars (indicated with $c = 1$) and trucks (denoted with $c = 2$). Considering a given freeway stretch with N road sections (L_i being the length of section $i = 1, \dots, N$) and a time horizon of K time steps, it is first of all possible to compute the number of vehicles of class c present in the mainstream, in a given section $i = 1, \dots, N$ and at a given time step $k = 0, \dots, K$. This number of vehicles is denoted as $n_i^{M,c}(k)$ and depends on the density $\rho_i^c(k)$, i.e.

$$n_i^{M,c}(k) = L_i \rho_i^c(k) \tag{6.13}$$

Then, it is straightforward to compute the mainstream emissions $E_i^M(k)$ [g/km] in a given section i and at a given time step k, by summing the emission factors of each single vehicle over the number of vehicles present in the road section at that time step, i.e.

$$E_i^M(k) = n_i^{M,1}(k) \sum_{j=1}^{J} \gamma_j^1 \Xi_j^1 \left(v_j^1(k)\right) + n_i^{M,2}(k) \Xi_1^2 \left(v_1^2(k)\right) \tag{6.14}$$

where $\Xi_j^1(v_j^1(k))$ and $\Xi_1^2(v_1^2(k))$ are computed, respectively, as in (6.1) and (6.12), while γ_j^1, $j = 1, \dots, J$ represent the composition rates of cars of legislation emission j. Obviously, these composition rates must be such that $\sum_{j=1}^{J} \gamma_j^1 = 1$. Note that, by adopting (6.12) in (6.14), it is assumed that all the trucks are of the same type, but it is quite easy to extend (6.14) to consider multiple types of trucks, with emission factors defined according to (6.2)–(6.11).

Analogously, it is possible to compute the on-ramp emissions starting from the number of vehicles in the on-ramp $n_i^{R,c}(k)$ given by

$$n_i^{R,c}(k) = l_i^c(k) \tag{6.15}$$

The on-ramp emissions $E_i^R(k)$ [g/km] in section i at time step k are obtained by summing the emission factors over the number of vehicles involved, i.e.

$$E_i^R(k) = n_i^{R,1}(k) \sum_{j=1}^{J} \gamma_j^1 \alpha_j^1 + n_i^{R,2}(k) \alpha_1^2 \tag{6.16}$$

where α_j^1, $j = 1, \ldots, J$, are constant emission factors obtained from (6.1) in case of minimum average speed equal to 10 km/h. Analogously, α_1^2 is obtained from (6.12) with the speed equal to 12 km/h.

Note that the application of COPERT to a multi-class second-order model for a freeway network (as the one described in Sect. 4.3.2) is very similar to the case of the freeway stretch, with a slightly different notation.

6.4 The VERSIT+ Model for Freeway Traffic Systems

Average-speed models, such as the COPERT model, are surely efficient from a computational point of view, since they are based on speed measurements (or estimations) only. It is, however, evident that a detailed evaluation of pollutant emissions should also depend on accelerations.

VERSIT+ is a statistical emission model that belongs to the class of regression-based models and computes the emission factor of each single vehicle on the basis of its speed and acceleration. It allows to compute many types of pollutant emissions, such as HC, CO, NO_x, PM_{10} and CO_2, for a wide range of vehicles and for several traffic conditions. The study was proposed for the first time in [52], significantly reflecting the Dutch fleet composition, and was based on over 20.000 measurements (performed both on cold and hot engines) and on more than 3.200 vehicles, for a period longer than 20 years. The chosen population size and the duration of the experimentation allowed to obtain a significant sample in terms of traffic scenarios, vehicle technologies, levels of maintenance and types of fuels. In the original version of the model, the computation of the emission factor was exclusively dependent on the average speed. An improved version of the VERSIT+ model was proposed in [40] in order to achieve a more accurate estimation.

The adoption of VERSIT+ in a traffic model allows to obtain an accurate estimate of the emissions, without a too high computational load, thanks to the limited number of parameters and the rather simple formulation of the model. This is the reason why the use of VERSIT+ in a traffic model is suitable for implementation in online control schemes.

In order to adopt VERSIT+ in a macroscopic traffic flow model, it is necessary to estimate the average acceleration of vehicles starting from the average speed provided by the model. In Sect. 6.4.1 the VERSIT+ model is described, while its use in a traffic model is discussed in Sect. 6.4.2, where a procedure to compute the average accelerations of vehicles is reported.

6.4.1 The VERSIT+ Model

As aforementioned, the VERSIT+ model [40] includes, in the computation, not only the average speed, but also the average acceleration. In particular, the emission factor of a generic vehicle of type h produced by the model depends on two terms. The former is a combination of the acceleration a^h [m/s^2] and the speed v^h [km/h], included in the model through the dynamic variable w^h defined as

$$w^h = a^h + 0.014v^h \tag{6.17}$$

The latter term is the speed value v^h [km/h], which is divided in four categories corresponding to different driving conditions: idling conditions when $v^h < 5$ and $a^h < 0.5$, urban driving with $v^h \leq 50$, rural driving with $50 < v^h \leq 80$ and freeway driving with $v^h > 80$.

Specifically, the emission factor Ξ^h for each vehicle of type h [kg/s] is given by

$$\Xi^h = \begin{cases} u^{h,0} & \text{if } v^h < 5 \text{ and } a^h < 0.5 \\ u^{h,1} + u^{h,2}w_+^h + u^{h,3}(w^h - 1)_+ & \text{if } v^h \leq 50 \\ u^{h,4} + u^{h,5}w_+^h + u^{h,6}(w^h - 1)_+ & \text{if } 50 < v^h \leq 80 \\ u^{h,7} + u^{h,8}(w^h - 0.5)_+ + u^{h,9}(w^h - 1.5)_+ & \text{if } v^h > 80 \end{cases} \tag{6.18}$$

where $u^{h,j}$, with $j = 0, \ldots, 9$, are coefficients of the emission model, whereas the function $(x)_+$ imposes the non-negativity of the variable x, i.e. $(x)_+ = 0$ if $x < 0$ and $(x)_+ = x$ otherwise.

6.4.2 Use of VERSIT+ in a Traffic Flow Model

Analogously to COPERT (see Sect. 6.3.2), also the VERSIT+ model can be associated with a traffic flow model to compute the emissions in a freeway stretch or network. In particular, a multi-class traffic model is surely more suitable than a single-class model, since it is able to distinguish different classes of vehicles that can present quite different emission factors. If the adoption of COPERT in a traffic flow model is rather straightforward, the application of VERSIT+ requires the computation of accelerations, that are not directly provided by the traffic model.

The extension of VERSIT+ to be used in a macroscopic traffic model was introduced in [14, 53], where two types of acceleration have been identified, i.e. the segmental acceleration considering the speed variation within a road section, and the cross-segmental acceleration, which concerns the speed variation of vehicles moving from one road section to the next one, between two consecutive time steps. In [54, 55], such accelerations were extended to the multi-class case, while in [56] the model was extended to add the computation of the emissions at the on-ramps.

Let us refer to the second-order METANET model of multi-class type representing a freeway stretch, described in Sect. 4.3.1, in which the time horizon is partitioned into time intervals with sample time T, with K the number of time steps, and the freeway stretch is divided into N road sections, with L_i being the length of road section $i = 1, \ldots, N$. In order to apply VERSIT+, it is first of all necessary to provide a methodology for evaluating the average accelerations of vehicles for each road section, for each class of vehicles and for every simulation time step. This methodology is different for vehicles travelling in the mainstream and for vehicles moving, instead, at the on-ramps. These two aspects are separately described in the following.

Mainstream Emissions In order to evaluate the emissions due to vehicles travelling in the mainstream, the average acceleration and the number of vehicles involved have to be computed for each road section $i = 1, \ldots, N$, for each class of vehicles $c = 1, \ldots, C$, and for every time step $k = 0, \ldots, K$.

Specifically, two types of acceleration are considered in the freeway links, i.e.

- the *segmental acceleration* $a_i^{\text{seg},c}(k)$ represents the speed variation of vehicles of class c within section i between time step k and time step $k+1$; the number of vehicles subject to this acceleration is denoted as $n_i^{\text{seg},c}(k)$;
- the *cross-segmental acceleration* $a_{i,i+1}^{\text{cross},c}(k)$ is the speed variation of vehicles of class c moving from section i to section $i+1$ between time step k and $k+1$; the number of vehicles involved is indicated with $n_{i,i+1}^{\text{cross},c}(k)$.

The two types of acceleration are computed on the basis of the mean speed $v_i^c(k)$, respectively, as follows:

$$a_i^{\text{seg},c}(k) = \frac{v_i^c(k+1) - v_i^c(k)}{T} \tag{6.19}$$

$$a_{i,i+1}^{\text{cross},c}(k) = \frac{v_{i+1}^c(k+1) - v_i^c(k)}{T} \tag{6.20}$$

Moreover, the number of vehicles subject to segmental and cross-segmental accelerations is obtained depending on the traffic density $\rho_i^c(k)$ and the traffic flow $q_i^c(k)$, i.e.

$$n_i^{\text{seg},c}(k) = L_i \rho_i^c(k) - T q_i^c(k) \tag{6.21}$$

$$n_{i,i+1}^{\text{cross},c}(k) = T q_i^c(k) \tag{6.22}$$

By taking into account the computation of the average accelerations and the number of vehicles involved, it is possible to evaluate the mainstream emissions associated with each road section and each time step. More specifically, by taking into account (6.18), the emission factor due to the segmental acceleration, referred to section i and time step k, can be computed as follows:

$$\Xi_i^{\text{seg,c}}(k) = \begin{cases} u^{c,0} & \text{if } v_i^c(k) < 5 \text{ and} \\ & a_i^{\text{seg,c}}(k) < 0.5 \\ u^{c,1} + u^{c,2} w_i^{\text{seg,c}}(k)_+ + u^{c,3}(w_i^{\text{seg,c}}(k) - 1)_+ & \text{if } v_i^c(k) \leq 50 \\ u^{c,4} + u^{c,5} w_i^{\text{seg,c}}(k)_+ + u^{c,6}(w_i^{\text{seg,c}}(k) - 1)_+ & \text{if } 50 < v_i^c(k) \leq 80 \\ u^{c,7} + u^{c,8}(w_i^{\text{seg,c}}(k) - 0.5)_+ + u^{c,9}(w_i^{\text{seg,c}}(k) - 1.5)_+ & \text{if } v_i^c(k) > 80 \end{cases} \tag{6.23}$$

where the dynamic variable $w_i^{\text{seg,c}}(k)$ is computed according to (6.17), i.e.

$$w_i^{\text{seg,c}}(k) = a_i^{\text{seg,c}}(k) + 0.014 v_i^c(k) \tag{6.24}$$

Analogously, the emission factor due to the cross-segmental acceleration, referred to section i and time step k, is given by

$$\Xi_{i,i+1}^{\text{cross,c}}(k) = \begin{cases} u^{c,0} & \text{if } v_i^c(k) < 5 \text{ and} \\ & a_{i,i+1}^{\text{cross,c}}(k) < 0.5 \\ u^{c,1} + u^{c,2} w_{i,i+1}^{\text{cross,c}}(k)_+ + u^{c,3}(w_{i,i+1}^{\text{cross,c}}(k) - 1)_+ & \text{if } v_i^c(k) \leq 50 \\ u^{c,4} + u^{c,5} w_{i,i+1}^{\text{cross,c}}(k)_+ + u^{c,6}(w_{i,i+1}^{\text{cross,c}}(k) - 1)_+ & \text{if } 50 < v_i^c(k) \leq 80 \\ u^{c,7} + u^{c,8}(w_{i,i+1}^{\text{cross,c}}(k) - 0.5)_+ + u^{c,9}(w_{i,i+1}^{\text{cross,c}}(k) - 1.5)_+ & \text{if } v_i^c(k) > 80 \end{cases} \tag{6.25}$$

where $w_{i,i+1}^{\text{cross,c}}(k)$ is computed as

$$w_{i,i+1}^{\text{cross,c}}(k) = a_{i,i+1}^{\text{cross,c}}(k) + 0.014 v_i^c(k) \tag{6.26}$$

According to (6.23) and (6.25), it is possible to compute the mainstream emissions $E_i^M(k)$ [kg/s] in a given section i and a given time step k, summing the emission factors of each single vehicle over the number of vehicles present in the section at that time step, i.e.

$$E_i^M(k) = \sum_{c=1}^{C} \left[n_i^{\text{seg,c}}(k) \Xi_i^{\text{seg,c}}(k) + n_{i,i+1}^{\text{cross,c}}(k) \Xi_{i,i+1}^{\text{cross,c}}(k) \right] \tag{6.27}$$

On-ramp Emissions When dealing with freeways, it is important to evaluate the emissions of vehicles not only in the mainstream, but also at the on-ramps, in order to correctly take into account the emission phenomena along the overall system. In fact, the operating conditions of vehicles queuing at the on-ramps are quite important and the associated emissions should be included in the total calculation of traffic emissions.

Referring to a generic on-ramp of road section i, four groups of vehicles are introduced and four types of acceleration are considered:

- the acceleration $a_i^{\mathrm{a,c}}(k)$ of *arriving vehicles*, i.e. vehicles of class c arriving at the on-ramp of section i at time step k and waiting in the queue at $k+1$; the number of arriving vehicles is denoted as $n_i^{\mathrm{a,c}}(k)$;
- the acceleration $a_i^{\mathrm{w,c}}(k)$ of *waiting vehicles*, i.e. vehicles of class c moving within the queue of the on-ramp of section i between time step k and $k+1$; let $n_i^{\mathrm{w,c}}(k)$ indicate the number of waiting vehicles;
- the acceleration $a_i^{\mathrm{ls,c}}(k)$ of *leaving vehicles with stop*, i.e. vehicles of class c being in the queue of the on-ramp of section i at time step k and exiting the on-ramp at $k+1$; let $n_i^{\mathrm{ls,c}}(k)$ indicate the number of leaving vehicles with stop;
- the acceleration $a_i^{\mathrm{lns,c}}(k)$ of *leaving vehicles without stop*, i.e. vehicles of class c arriving at the on-ramp of section i at time step k and exiting the on-ramp at $k+1$ without any intermediate stop in the queue; let $n_i^{\mathrm{lns,c}}(k)$ indicate the number of leaving vehicles without stop.

Analogously to the mainstream emissions, it is necessary to compute the mean accelerations and the number of vehicles involved, for each of these four groups of vehicles.

The acceleration of arriving vehicles is given by

$$a_i^{\mathrm{a,c}}(k) = \frac{v_i^{\mathrm{idl,c}}(k+1) - v_i^{\mathrm{on,c}}(k)}{T} \tag{6.28}$$

where $v_i^{\mathrm{on,c}}(k)$ is the speed of vehicles arriving at the on-ramp and $v_i^{\mathrm{idl,c}}(k)$ is the speed of vehicles moving within the queue of the on-ramp.

The acceleration of waiting vehicles is computed as

$$a_i^{\mathrm{w,c}}(k) = \frac{v_i^{\mathrm{idl,c}}(k+1) - v_i^{\mathrm{idl,c}}(k)}{T} \tag{6.29}$$

The acceleration of leaving vehicles with stop is obtained as

$$a_i^{\mathrm{ls,c}}(k) = \frac{v_i^{c}(k+1) - v_i^{\mathrm{idl,c}}(k)}{T} \tag{6.30}$$

while the acceleration of leaving vehicles without stop is given by

$$a_i^{\mathrm{lns,c}}(k) = \frac{v_i^{c}(k+1) - v_i^{\mathrm{on,c}}(k)}{T} \tag{6.31}$$

The number of vehicles that belong to each group is computed depending on the value of the flow $r_i^c(k)$ leaving the on-ramp and entering the mainstream. In particular, two cases may be distinguished:

1. if $0 \le r_i^c(k) \le \frac{l_i^c(k)}{T}$, corresponding to the case in which the vehicles entering the mainstream are fewer than the vehicles in the queue (see Fig. 6.4), the number of vehicles of the four groups is given by

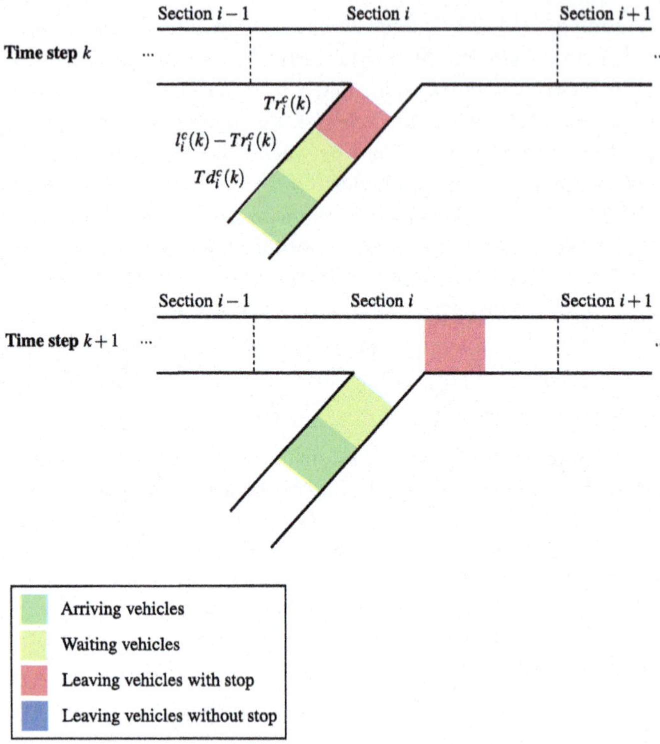

Fig. 6.4 Groups of on-ramp vehicles if $0 \leq r_i^c(k) \leq \frac{l_i^c(k)}{T}$

$$n_i^{\mathrm{a,c}}(k) = Td_i^c(k) \tag{6.32}$$

$$n_i^{\mathrm{w,c}}(k) = l_i^c(k) - Tr_i^c(k) \tag{6.33}$$

$$n_i^{\mathrm{ls,c}}(k) = Tr_i^c(k) \tag{6.34}$$

$$n_i^{\mathrm{lns,c}}(k) = 0 \tag{6.35}$$

2. if $\frac{l_i^c(k)}{T} < r_i^c(k) \leq d_i^c(k) + \frac{l_i^c}{T}(k)$, corresponding to the case in which the vehicles entering the mainstream are more than the vehicles in the queue (see Fig. 6.5), the number of vehicles is obtained as

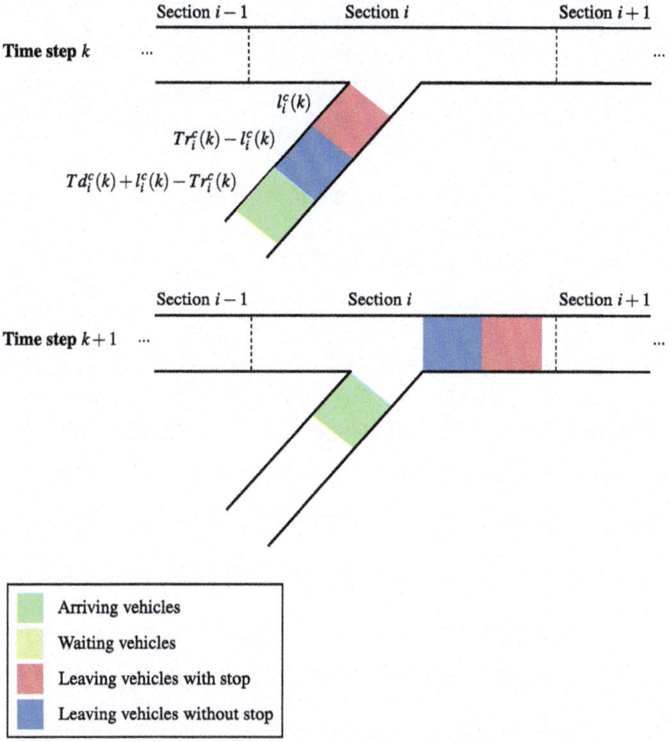

Fig. 6.5 Groups of on-ramp vehicles if $\frac{l_i^c(k)}{T} < r_i^c(k) \le d_i^c(k) + \frac{l_i^c}{T}(k)$

$$n_i^{\mathrm{a,c}}(k) = Td_i^c(k) + l_i^c(k) - Tr_i^c(k) \tag{6.36}$$

$$n_i^{\mathrm{w,c}}(k) = 0 \tag{6.37}$$

$$n_i^{\mathrm{ls,c}}(k) = l_i^c(k) \tag{6.38}$$

$$n_i^{\mathrm{lns,c}}(k) = Tr_i^c(k) - l_i^c(k) \tag{6.39}$$

The emission factors for the four groups of vehicles at the on-ramps can be computed analogously to the mainstream case. For notational purposes, let us define the speed values related to the four groups of vehicles $y \in Y = \{\mathrm{a, w, lns, ls}\}$, in the on-ramp of section i at time step k, as follows:

$$v_i^{y,c}(k) = \begin{cases} v_i^{\text{on},c}(k) & \text{if } y = \text{a, lns} \\ v_i^{\text{idl},c}(k) & \text{if } y = \text{w, ls} \end{cases} \tag{6.40}$$

The emission factors related to the generic on-ramp group $y \in Y$ in the on-ramp of section i at time step k are computed as

$$\Xi_i^{y,c}(k) = \begin{cases} u^{c,0} & \text{if } v_i^{y,c}(k) < 5 \text{ and} \\ & \quad a_i^{y,c}(k) < 0.5 \\ u^{c,1} + u^{c,2} w_i^{y,c}(k)_+ + u^{c,3}(w_i^{y,c}(k) - 1)_+ & \text{if } v_i^{y,c}(k) \leq 50 \\ u^{c,4} + u^{c,5} w_i^{y,c}(k)_+ + u^{c,6}(w_i^{y,c}(k) - 1)_+ & \text{if } 50 < v_i^{y,c}(k) \leq 80 \\ u^{c,7} + u^{c,8}(w_i^{y,c}(k) - 0.5)_+ + u^{c,9}(w_i^{y,c}(k) - 1.5)_+ & \text{if } v_i^{y,c}(k) > 80 \end{cases} \tag{6.41}$$

where the dynamic variable $w_i^{y,c}(k)$ is calculated as

$$w_i^{y,c}(k) = a_i^{y,c}(k) + 0.014 v_i^{y,c}(k) \tag{6.42}$$

By taking into account (6.41), the on-ramp emissions $E_i^R(k)$ [kg/s] in section i at time step k are obtained by summing the emission factors over the number of vehicles, i.e.

$$E_i^R(k) = \sum_{c=1}^{C} \sum_{y \in Y} n_i^{y,c}(k) \Xi_i^{y,c}(k) \tag{6.43}$$

Note that the application of VERSIT+ to the multi-class second-order model for a freeway network (described in Sect. 4.3.2) is rather similar to the one of a freeway link, but a slightly different notation should be adopted, specifically at the boundary between two adjacent links, i.e. when vehicles move between the last section of a link and the first section of the downstream link. Further details can be found in [16, 57].

References

1. Treiber M, Kesting A (2013) Traffic flow dynamics: data, models and simulation. Springer, Berlin
2. Van Mierlo J, Maggetto G, Van de Burgwal E, Gense R (2004) Driving style and traffic measures-influence on vehicle emissions and fuel consumption. P I Mech Eng D-J Aut 218:43–50
3. Smit R, Poelman M, Schrijver J (2008) Improved road traffic emission inventories by adding mean speed distributions. Atmos Environ 42:916–926
4. De Coensel B, Can A, Degraeuwe B, De Vlieger I, Botteldooren D (2012) Effects of traffic signal coordination on noise and air pollutant emissions. Environ Model Softw 35:74–83
5. Coelho MC, Farias TL, Rouphail NM (2006) Effect of roundabout operations on pollutant emissions. Transp Res Part D 11:333–343

6. Pandian S, Gokhale S, Ghoshal AK (2009) Evaluating effects of traffic and vehicle characteristics on vehicular emissions near traffic intersections. Transp Res Part D 14:180–196
7. Nagurney A (2000) Congested urban transportation networks and emission paradoxes. Transp Res Part D 5:145–151
8. Braess PDD (1968) Über ein Paradoxon aus der Verkehrsplanung. Unternehmensforschung 12:258–268
9. Frank M (1981) The braess paradox. Math Program 20:283–302
10. Zhou Y, Wu Y, Yang L, Fu L, He K, Wang S, Hao J, Chen J, Li C (2010) The impact of transportation control measures on emission reductions during the 2008 Olympic games in Beijing, China. Atmos Environ 44:285–293
11. Coelho MC, Farias TL, Rouphail NM (2005) Impact of speed control traffic signals on pollutant emissions. Transp Res Part D 10:323–340
12. Panis LI, Broekx S, Liu R (2006) Modelling instantaneous traffic emission and the influence of traffic speed limits. Sci Total Environ 371:270–285
13. Yang H, Wang X, Yin Y (2012) The impact of speed limits on traffic equilibrium and system performance in networks. Transp Res Part B 46:1295–1307
14. Zegeye S, De Schutter B, Hellendoorn J, Breunesse E, Hegyi A (2013) Integrated macroscopic traffic flow. emission, and fuel consumption model for control purposes. Transp Res Part C 31:158–171
15. Pasquale C, Papamichail I, Roncoli C, Sacone S, Siri S, Papageorgiou M (2015) Two-class freeway traffic regulation to reduce congestion and emissions via nonlinear optimal control. Transp Res Part C 55:85–99
16. Pasquale C, Sacone S, Siri S, De Schutter B (2017) A multi-class model-based control scheme for reducing congestion and emissions in freeway networks by combining ramp metering and route guidance. Transp Res Part C 80:384–408
17. Liu S, Hellendoorn H, De Schutter B (2017) Model predictive control for freeway networks based on multi-class traffic flow and emission models. IEEE Trans Intell Transp Syst 18:306–320
18. Lin S, De Schutter B, Xi Y, Hellendoorn H (2013) Integrated urban traffic control for the reduction of travel delays and emissions. IEEE Trans Intell Transp Syst 14:1609–1619
19. Holmes NS, Morawska L (2006) A review of dispersion modelling and its application to the dispersion of particles: an overview of different dispersion models available. Atmos Environ 40:5902–5928
20. Venkatram A, Fitz D, Bumiller K, Du S, Boeck M, Ganguly C (1999) Using a dispersion model to estimate emission rates of particulate matter from paved roads. Atmos Environ 33:1093–1102
21. Venkatram A, Isakov V, Thoma E, Baldauf R (2007) Analysis of air quality data near roadways using a dispersion model. Atmos Environ 41:9481–9497
22. Ntziachristos L, Samaras Z (1997) COPERT II - computer programme to calculate emissions from road transport, User's Manual. European environmental agency, European topic center on air emissions
23. Ntziachristos L, Samaras Z (2000) Speed-dependent representative emission factors for catalyst passenger cars and influencing parameters. Atmos Environ 34:4611–4619
24. Ntziachristos L, Kouridis C. (2007) Road transport emission chapter of the EMEP/CORINAIR emission inventory guidebook (European environment agency Technical Report)
25. User's guide to MOBILE 6.1 and MOBILE 6.2 (U.S. Environmental Protection Agency, 2009)
26. Chang MF, Herman R (1981) Trip time versus stop time and fuel consumption characteristics in cities. Transp Sci 15:183–209
27. Evans L, Herman R (1978) Automobile fuel economy on fixed urban driving schedules. Transp Sci 12:137–152
28. Watson HC, Milkins EE, Marshall GA (1979) A simplified method for quantifying fuel consumption of vehicles in urban traffic. In: Proceedings of the 4th automotive engineering conference
29. Hausberger S, Rexeis M, Zallinger M, Luz R (2009) Emission factors from the model PHEM for the HBEFA version 3 (Report Nr. I-20/2009 Haus-Em 33)

30. Joumard R, Andre JM, Rapone M, Zallinger M, Kljun N, Andre M, Samaras Z, Roujol S, Laurikko J, Weilenmann M, Markewitz K, Geivanidis S, Ajtay D, Paturel L (2007) Emission factor modelling and database for light vehicles (Artemis deliverable. Report LTE 0523:3)
31. Negrenti E (1999) The 'corrected average speed' approach in ENEA's TEE model: an innovative solution for the evaluation of the energetic and environmental impacts of urban transport policies. Sci Total Environ 235:411–413
32. Bachman W, Sarasua W, Guensler R (1996) Geographic information system framework for modeling mobile-source emissions. Transp Res Rec 1551:123–132
33. Bachman W, Sarasua W, Hallmark S, Guensler R (2000) Modeling regional mobile source emissions in a geographic information system framework. Transp Res Part C 8:205–229
34. Sturm PJ (1998) Instantaneous emission data and their use in estimating passenger car emissions (Technical Report MEET Deliverable 6. Technischen Universitat Graz, Austria)
35. Rakha H, Ahn K (2004) Integration modeling framework for estimating mobile source emissions. J Transp Eng 130
36. Qi Y, Teng H, Yu L (2004) Microscale emission models incorporating acceleration and deceleration. J Transp Eng 130:348–359
37. An F, Barth M, Norbeck J, Ross M (1997) Development of comprehensive modal emissions model: operating under hot-stabilized conditions. Transp Res Rec 1587:52–62
38. Ahn K, Trani AA, Rakha H, Van Aerde M (1999) Microscopic fuel consumption and emission models. In Proceedings of the 78th annual meeting of the transportation research board
39. Ahn K, Rakha H, Trani AA, Van Aerde M (2002) Estimating vehicle fuel consumption and emissions based on instantaneous speed and acceleration levels. J Transp Eng 128:182–190
40. Ligterink NE, De Lange R, Schoen E (2009) Refined vehicle and driving-behavior dependencies in the VERSIT+ emission model. In Proceedings of the ETAPP symposium, pp 177–186
41. Cappiello A, Chabini I, Nam EK, Lue A, Abou ZM (2002) A statistical model of vehicle emissions and fuel consumption. In Proceedings of the IEEE 5th international conference on intelligent transportation systems, pp 801–809
42. Barth M, An F, Younglove T, Levine C, Scora G, Ross M, Wenzel T (2000) Development of a comprehensive modal emissions model. National cooperative highway research program, Transportation research board of the national academies
43. Akcelik R (1985) An interpretation of the parameters in the simple average travel speed model of fuel consumption. Australian road research 15
44. Wallace CE, Courage KG, Reaves DP, Schoene GW, Euler GW (1984) TRANSYT-7F user's manual
45. Yue H (2008) Mesoscopic Fuel Consumption and Emission Modeling, PhD Thesis, Virginia Polytechnic Institute
46. Pasquale C, Sacone S, Siri S (2013) Multi-class local ramp metering to reduce traffic emissions in freeway systems. In Proceedings of the IFAC workshop on advances in control and automation theory for transportation applications, pp 43–48
47. Pasquale C, Sacone S, Siri S (2014) Ramp metering control for two vehicle classes to reduce traffic emissions in freeway systems. In Proceedings of the european control conference, pp 2588–2593
48. Eggleston S, Gori*β*en N, Joumard R, Rijkeboer RC, Samaras Z, Zierock KH (1989) CORINAIR working group on emissions factors for calculating 1985 emissions from road traffic. In: Methodology and emission factors (Final Report Contract EUR 12260 EN, 1989), vol 1
49. Eggleston S, Gaudioso D, Gori*β*en N, Joumard R, Rijkeboer RC, Samaras Z, Zierock KH (1993) CORINAIR working group on emissions factors for calculating 1990 emissions from road traffic. In: Methodology and emission factors (Final report, Document of the European commission, 1993), vol 1
50. Ahlvik P, Eggleston S, Gori*β*en N, Hassel D, Hickman AJ, Joumard R, Ntziachristos L, Rijkeboer R, Samaras Z, Zierock KH (1997) COPERT II methodology and emission factors (Technical Report for the European environment agency, 1997)
51. EMEP/EEA air pollutant emission inventory guidebook 2016: technical guidance to prepare national emission inventories (EEA Report, No 21/2016, European environment agency, 2016)

52. Smit R, Smokers R, Schoen E (2005) VERSIT+LD: development of a new emission factor model for passenger cars linking real-world emissions to driving cycle characteristics. In Proceedings of the 14th symposium on transport and air pollution, pp 77–186
53. Zegeye SK, De Schutter B, Hellendoorn H, Breunesse EA, Hegyi A (2012) A predictive traffic controller for sustainable mobility using parameterized control policies. IEEE Trans Intell Transp Syst 13:1420–1429
54. Liu S, De Schutter B, Hellendoorn H (2013) Multi-class traffic flow and emission control for freeway networks. In Proceedings of the 16th international IEEE conference on intelligent transportation systems, pp 2223–2228
55. Liu S, De Schutter B, Hellendoorn H (2014) Integrated traffic flow and emission control based on FASTLANE and the multi-class VT-macro model. In Proceedings of the 2014 European control conference, pp 2908–2913
56. Pasquale C, Liu S, Siri S, Sacone S, De Schutter B (2015) A new emission model including on-ramps for two-class freeway traffic control. In Proceedings of the 18th IEEE international conference on intelligent transportation systems, pp 1143–1149
57. Pasquale C, Sacone S, Siri S, De Schutter B (2016) A multi-class ramp metering and routing control scheme to reduce congestion and traffic emissions in freeway networks. In Proceedings of the 14th IFAC symposium on control in transportation systems, vol 49, pp 329–334

Chapter 7
State Estimation in Freeway Traffic Systems

7.1 Overview of Freeway Traffic State Estimation Techniques

A common feature of feedback control schemes is the necessity of acquiring measurements of the controlled system state. In case of freeways, this normally implies the necessity of measuring traffic density, flow and speed on all the road segments of the considered freeway system. Traditional *stationary sensors* (e.g. inductive loop detectors, ultrasonic sensors, radar sensors and cameras) collect traffic data of the vehicles passing the location in which they are installed. The most common traffic sensors are inductive loop detectors (Fig. 7.1), which are able to count the number of vehicles passing at the sensor location and to measure the traffic occupancy. Speed information can also be measured or derived from such sensors. It is generally instead quite difficult to measure directly traffic density.

In order to design efficient traffic management and control tools, it is necessary to know the values of the traffic variables in real time and with a high spatio-temporal resolution. Since some of the traffic variables cannot be measured or are measured only in specific locations and, in any case, are subject to sensor inaccuracies and failures, the only possibility is to reconstruct such variables with appropriate *estimation techniques*. Among them, the estimation of traffic density is particularly relevant for traffic systems, since this quantity cannot be measured directly and, at the same time, it represents a very relevant information for the design of traffic control tools.

Recently, a new type of traffic data has become very common and surely will be more and more widespread in the near future. These are *mobile data* collected by mobile sensors such as vehicles provided with Global Positioning System (GPS) technology. Since these intelligent and connected vehicles have the possibility, not only to measure traffic information, but also to communicate it in real time, they represent a relevant source of traffic data that will be more and more common in the near future (see e.g. [1] describing a field experiment to obtain traffic data from GPS-enabled mobile phones). This will require the development and enhancement of traffic estimation techniques to account for the disaggregated and inhomogeneous nature

A. Ferrara et al., *Freeway Traffic Modelling and Control*, Advances in Industrial Control, https://doi.org/10.1007/978-3-319-75961-6_7

Fig. 7.1 Inductive loop detectors installed on a freeway on-ramp in San Diego, U.S. (courtesy of Michael Ballard)

of these data. For a more detailed discussion on future trends of traffic estimation methods, the reader is referred to Sect. 7.5.

Different estimation techniques, capable of providing the adopted traffic controllers with estimates of the information they need, have appeared in the literature in the past decades. Several works are related to this topic, each of them exploiting different methodologies and strategies. A complete and well-developed survey of the major contributions to traffic state estimation on freeway systems can be found in [2], which lists a large number of papers dealing with classical and unconventional approaches, such as the very recent ones also incorporating mobile data.

In [2], a useful distinction is made among three different approaches to produce traffic state estimations:

- *model-driven approaches*, which rely on suitably accurate models describing the traffic dynamics. The models have to be calibrated taking into account sets of data collected on the freeway system which are sufficiently informative, i.e. able to cover the different working conditions of the system itself;
- *data-driven approaches*, which mainly rely on large sets of historical data and their statistical processing;
- *streaming-data-driven approaches*, which only use real-time data (they can be for instance GPS probe vehicle data, data coming from the cellular phones of the drivers or data exchanged among vehicles which are connected with each other and/or with the infrastructure).

The present chapter does not aim to exhaustively cover the topic of traffic state estimation. In fact, in view of the scope of the book, traffic state estimation is addressed according to the perspective of control engineers. These latter design observer-based control schemes when relevant measurements are missing and some information about the dynamics of the process can be translated into a simple model of the process to control. For this reason, in this chapter, we will focus on approaches which, according to the classification given in [2], can be regarded as belonging to the model-driven category. These methods rely on a sufficiently accurate dynamic model of the freeway traffic system and typically utilise *observers* to produce the traffic state estimates.

7.1.1 Model-Driven State Estimation for Freeway Traffic

One of the most common techniques adopted for traffic state estimation is the *extended Kalman filter*, which has been applied in traffic surveillance systems since the early 70s [3, 4]. The extended Kalman filter is used for non-linear system models, that, for traffic systems, surely applies for the case in which the METANET model is adopted (more details on METANET can be found in Sect. 4.2). Such a model is considered for instance in [5], where a general approach for the real-time estimation of the whole traffic state in freeway stretches is proposed and an interesting review of the literature on traffic state estimation is presented. The extension of this work to the case of freeway networks can be found in [6], where the software tool RENAISSANCE is presented and described. The estimation framework proposed in [7] allows not only the real-time estimation of the traffic variables, but also the estimation of some model parameters, such as the free-flow speed, the critical density, and the capacity.

An extended Kalman filter is also adopted in [8] but applied to the CTM (see Sect. 3.3 for more details on the CTM). Another work on traffic state estimation relying on the CTM was presented in [9], in which also the prediction of the state of a road network is addressed on the basis of speed and flow measurements on some links. Different sources of uncertainty are taken into account, associated with the adopted model, the demand and the measurements. Many other works in the literature use Kalman filter techniques, either in the original formulation, or in the extended, unscented and ensemble version (see for instance [10, 11] and the references therein). Other filtering techniques were exploited to provide reliable freeway traffic estimates, also including particle filters [12, 13] or the adaptive smoothing filter in its original or extended version (see for instance [14, 15]).

Other research works on traffic state estimation adopt *switched observers*, which generally use switched models to represent the system dynamics. For instance, the works [16, 17] rely on the Switching-Mode Model (SMM) proposed in [18], which is a piecewise-affine version of the CTM (see more details on the SMM in Sect. 3.3.6). The Luenberger-like observer based on the SMM described in [16] is the basis of a traffic density estimation framework applied in the Grenoble Traffic Lab, which is a platform for online collection of traffic data coming from a wireless sensor network installed in the South Ring of Grenoble, France [19].

The Luenberger-like observer proposed in [20] is based on a similar piecewise-affine traffic model, which is a switched version of the Asymmetric Cell Transmission Model (ACTM). This model, already introduced in Sect. 3.3.6, is accurately described in Sect. 7.2.1 for estimation purposes. The Luenberger-like observer proposed in [20] is then described in detail in Sect. 7.3.

Given the distributed nature of freeway traffic systems, *distributed observers* appear to be particularly suitable to deal with large freeway networks. In such systems, sensors are placed only in specific locations, and the number of sensors is typically very low compared to the number of cells or links in which the traffic system is subdivided for control purposes. In large-scale freeway networks, centralised estimation techniques are not advisable, since they are not able to provide reliable estimates in acceptable computational times. According to the considered framework, distributed observers can provide an estimation of the entire freeway system state or a part of it, allowing to take local decisions based on a total or partial knowledge of the entire state. The distributed schemes are also differentiated depending on the type of possible communication among observers, varying from the all-to-all communication topology to distributed schemes in which each observer can communicate only with a subset of other observers.

A paper dealing with distributed state estimation is [21], where a distributed local Kalman consensus filter is adopted for traffic density estimation in large-scale freeway traffic systems. The traffic model adopted in that estimation framework is the SMM, while the idea for distributed estimation is to partition the state into local overlapping subsets, so that each agent estimates a single subset of the state but it communicates with neighbouring agents to exchange messages on measurements and state estimates. This work was extended in [22], where the proposed filter is modified to be scalable both in the sense of computation and as for communication. A similar distributed estimation scheme is described in [23], where, again, the SMM is used to model the traffic dynamics, but the considered distributed consensus-based switched observers are able to estimate the state of whole freeway stretch (see Sect. 7.4.1 for further details). A similar switched model, taking also into account the on-ramps and off-ramps, is adopted in [24], where distributed consensus-based switched observers are applied in a traffic system controlled via ramp metering (see Sect. 7.4.2 for a more detailed description of this estimation scheme).

7.2 A Modelling Framework for Traffic State Estimation

This section proposes a modelling framework that can be used for traffic state estimation. The prediction model is based on the ACTM, which is a modified version of the CTM described in Sect. 3.3. In the following, the adopted model is described for the two cases with and without ramps, and the relative reachability and observability

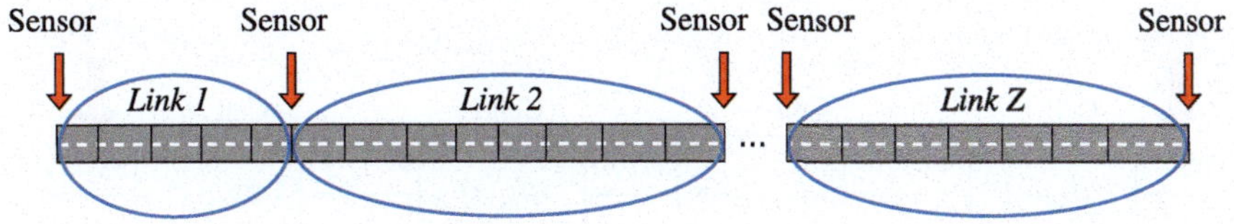

Fig. 7.2 Subdivision of the freeway into links

properties are analysed hereinafter, since they are fundamental for designing efficient controllers and observers for freeway traffic systems.

7.2.1 *The Traffic Model for Freeway Links*

In order to design a suitable state estimation framework, the freeway is supposed to be divided into a set of links, each one further subdivided into a limited number of cells. The *links* are defined according to the position of the sensors, which are assumed to provide traffic flow measurements. Specifically, the sensors are located at the link boundaries, as shown in Fig. 7.2. Let Z denote the number of links, N_l the number of cells composing link l, and $L_{i,l}$ [km] the length of cell i of link l, $l = 1, \dots, Z$, $i = 1, \dots, N_l$.

The considered model based on the ACTM is a discrete-time model in which T denotes the sample time [h] and K the number of time steps. Let us distinguish two cases, the simple case in which no ramps are present in the freeway links and the case in which ramps are taken into account.

Links Without Ramps For each link $l = 1, \dots, Z$, for each cell $i = 1, \dots, N_l$, and for each time step $k = 0, \dots, K$, let us define the following quantities:

- $\rho_{i,l}(k)$ is the traffic density of cell i at time kT [veh/km];
- $\phi_{i,l}(k)$ is the interface flow entering cell i from cell $i - 1$ during time interval $[kT, (k + 1)T)$ [veh/h].

The model includes the following parameters for each link l: $v_{i,l}$ is the free-flow speed of cell i [km/h], $w_{i,l}$ is the congestion wave speed of cell i [km/h], $\rho_{i,l}^{\max}$ is the jam density of cell i [veh/km].

Figure 7.3 shows a sketch of the division of a generic link l into cells and the relative notation. Note that the upstream and downstream flows measured by sensors are denoted, respectively, with $\phi_l^u(k)$ and $\phi_l^d(k)$. According to the adopted notation, it yields that $\phi_{1,l}(k) = \phi_l^u(k)$ and $\phi_{N_l+1,l}(k) = \phi_l^d(k)$.

By assuming that the mainstream capacity of the cells is sufficiently high to be neglected, the system dynamics is given by the following equation

$$\rho_{i,l}(k+1) = \rho_{i,l}(k) + \frac{T}{L_{i,l}}\left[\phi_{i,l}(k) - \phi_{i+1,l}(k)\right] \tag{7.1}$$

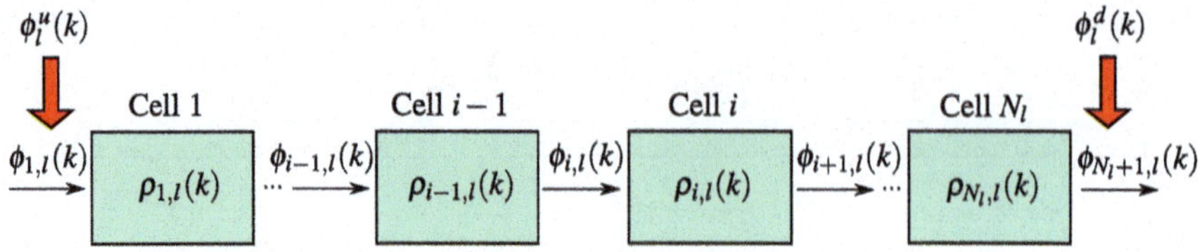

Fig. 7.3 Subdivision of link l into cells (case without ramps)

where the mainstream flow is given by

$$\phi_{i,l}(k) = \min\left\{v_{i-1,l}\rho_{i-1,l}(k), w_{i,l}\left(\rho_{i,l}^{\max} - \rho_{i,l}(k)\right)\right\} \tag{7.2}$$

By taking into account (7.1) and (7.2), a freeway link can be seen as a piecewise-affine system [25], in which each link l is characterised by a number of modes M_l equal to $2^{(N_l+1)}$. Some works (see e.g. [16, 18]) make the assumption that only one congestion wave exists, appearing at the end of the link and propagating upstream: in this case, the number of modes M_l is reduced to $2(N_l + 1)$.

The piecewise-affine system switches among different sets of linear difference equations. In particular, the model has the following state-space representation

$$\begin{cases} \underline{\rho}_l(k+1) = A_l^{s_l(k)}\underline{\rho}_l(k) + D_l^{s_l(k)}\underline{\phi}_l(k) + E_l^{s_l(k)}\underline{\rho}_l^{\max} \\ s_l(k) = \sigma_l(\underline{\rho}_l(k), \underline{\phi}_l(k)) \\ y_l(k) = C_l^{s_l(k)}\underline{\rho}_l(k) + F_l^{s_l(k)}\underline{\rho}_l^{\max} \end{cases} \tag{7.3}$$

where $\underline{\rho}_l(k) = [\rho_{1,l}(k), \ldots, \rho_{N_l,l}(k)]^{\mathrm{T}}$ is the state vector of the traffic densities, $\underline{\phi}_l(k) = [\phi_l^u(k), \phi_l^d(k)]^{\mathrm{T}}$ is the exogenous input vector, $\underline{\rho}_l^{\max} = [\rho_{1,l}^{\max}, \ldots, \rho_{N_l,l}^{\max}]^{\mathrm{T}}$ is a vector of parameters, with $A_l^{s_l(k)}$, $C_l^{s_l(k)}$, $D_l^{s_l(k)}$, $E_l^{s_l(k)}$, $F_l^{s_l(k)}$ matrices of suitable dimensions. The mode selector $\sigma_l(\cdot, \cdot)$ computes the current mode of the system $s_l(k) \in \mathscr{S}_l = \{1, \ldots, M_l\}$, on the basis of the state and exogenous input vectors.

Let us now consider the entire freeway system subdivided into Z links, such that each link l, $l = 1, \ldots, Z$, is modelled according to (7.3). Also, the whole freeway system can be seen as a piecewise-affine system switching among $M = \prod_{l=1}^{Z} M_l$ modes, with the following state-space representation

$$\begin{cases} \underline{x}(k+1) = A^{s(k)}\underline{x}(k) + D^{s(k)}\underline{d}(k) + E^{s(k)}\underline{\pi} \\ s(k) = \sigma\left(\underline{x}(k), \underline{d}(k)\right) \\ \underline{y}(k) = C^{s(k)}\underline{x}(k) + F^{s(k)}\underline{\pi} \end{cases} \tag{7.4}$$

where $\underline{x}(k)$ is the state vector gathering the traffic densities of all the cells in the freeway, $\underline{d}(k)$ is the exogenous input vector, $\underline{\pi}$ is a vector gathering all the parameters of each link, $\underline{y}(k)$ is the vector of the system outputs, while $A^{s(k)}$, $C^{s(k)}$, $D^{s(k)}$, $E^{s(k)}$, $F^{s(k)}$ are matrices of suitable dimensions, in which the values vary depending on the overall freeway mode $s(k)$. The mode selector $\sigma(\cdot, \cdot)$ computes the current mode

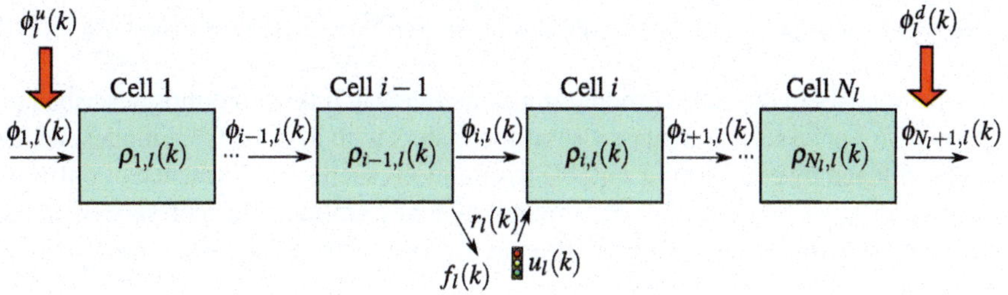

Fig. 7.4 Subdivision of link l into cells (case with ramps and $\delta_{i,l} = 1$)

of the system $s(k) \in \mathscr{S} = \{1, \ldots, M\}$, on the basis of the state vector and the exogenous input vector.

Links with Ramps An extension of the previous model includes the presence of on-ramps and off-ramps in the freeway links. Let us specifically focus on the case of one on-ramp and one off-ramp in each link. The off-ramp and on-ramp are located at the interface between two cells; let $\delta_{i,l}, i = 1, \ldots, N_l+1$, indicate the location of the ramps in the link, in particular $\delta_{i,l} = 1$ means that the off-ramp and the on-ramp are present between cell $i-1$ and cell i, while $\delta_{i,l} = 0$ has the opposite meaning. Since each link has only one off-ramp and one on-ramp, it is $\sum_{i=1}^{N_l+1} \delta_{i,l} = 1, l = 1, \ldots, Z$. Moreover, it is assumed that $\delta_{1,l} = 0$ and $\delta_{N_l+1,l} = 0$.

In case of ramps, the following quantities are added to the previous model, for each link $l = 1, \ldots, Z$, and for each time step $k = 0, \ldots, K$ (see Fig. 7.4):

- $r_l(k)$ is the flow entering the mainstream from the on-ramp during time interval $[kT, (k+1)T)$ [veh/h];
- $u_l(k)$ is the ramp metering control variable, i.e. the flow determined by the ramp metering controller to enter the mainstream from the on-ramp during time interval $[kT, (k+1)T)$ [veh/h];
- $f_l(k)$ is the flow exiting through the off-ramp during time interval $[kT, (k+1)T)$ [veh/h].

In case of ramps, the model includes an additional parameter for each link l, i.e. $\beta_l \in [0, 1]$, which represents the off-ramp split ratio.

By assuming that the mainstream capacity and the off-ramp capacity are sufficiently high to be neglected, and by considering that $r_l(k) = u_l(k)$, the system dynamics is given by the following equation

$$\rho_{i,l}(k+1) = \rho_{i,l}(k) + \frac{T}{L_{i,l}} \left[\phi_{i,l}(k) + \delta_{i,l} u_l(k) - \phi_{i+1,l}(k) - \delta_{i+1,l} f_l(k) \right] \tag{7.5}$$

where the flow exiting the off-ramp and the mainstream flow are given, respectively, by

$$f_l(k) = \frac{\beta_l}{1-\beta_l} \sum_{i=1}^{N_l} \delta_{i,l} \phi_{i,l}(k) \tag{7.6}$$

$$\phi_{i,l}(k) = \min\left\{(1-\delta_{i,l}\beta_l)v_{i-1,l}\rho_{i-1,l}(k),\, w_{i,l}(\rho_{i,l}^{\max} - \rho_{i,l}(k)) - \delta_{i,l}u_l(k)\right\} \quad (7.7)$$

Analogously to the case without ramps, the freeway link with ramps described by (7.5)–(7.7) can be seen as a piecewise-affine system with $M_l = 2^{(N_l+1)}$ modes, which can be reduced by considering suitable assumptions such as the presence of only one congestion front. Specifically, the model switches among different sets of linear difference equations, and can be written in the following state-space representation

$$\begin{cases} \underline{\rho}_l(k+1) = A_l^{s_l(k)}\underline{\rho}_l(k) + B_l^{s_l(k)}u_l(k) + D_l^{s_l(k)}\underline{\phi}_l(k) + E_l^{s_l(k)}\underline{\rho}_l^{\max} \\ s_l(k) = \sigma_l(\underline{\rho}_l(k), u_l(k), \underline{\phi}_l(k)) \\ y_l(k) = C_l^{s_l(k)}\underline{\rho}_l(k) + F_l^{s_l(k)}\underline{\rho}_l^{\max} \end{cases} \quad (7.8)$$

where, compared with (7.3), the additional notation includes the control input $u_l(k)$, the matrix $B_l^{s_l(k)}$, and the mode selector $\sigma_l(\cdot,\cdot,\cdot)$ which here computes the current mode of the system $s_l(k) \in \mathcal{S}_l$ on the basis of the state vector, the control vector and the exogenous input vector.

Considering now the entire freeway system subdivided into Z links, such that each link l, $l = 1, \ldots, Z$, is modelled as (7.8), the freeway system can be seen as a piecewise-affine system switching among $M = \prod_{l=1}^{Z} M_l$ modes, with the following state-space representation

$$\begin{cases} \underline{x}(k+1) = A^{s(k)}\underline{x}(k) + B^{s(k)}\underline{u}(k) + D^{s(k)}\underline{d}(k) + E^{s(k)}\underline{\pi} \\ s(k) = \sigma\left(\underline{x}(k), \underline{u}(k), \underline{d}(k)\right) \\ \underline{y}(k) = C^{s(k)}\underline{x}(k) + F^{s(k)}\underline{\pi} \end{cases} \quad (7.9)$$

where $\underline{u}(k)$ is the vector gathering all the control inputs and $B^{s(k)}$ is a matrix of suitable dimensions, while the other quantities have the same meaning as in (7.4).

7.2.2 Reachability and Observability Analysis

The structural properties of dynamic systems are related to the input-to-state and state-to-output interactions. Among them, *reachability* and *observability* play an important role in the design of estimation and control schemes (see, for instance, [26] for an introduction to these concepts).

More precisely, in the definition of traffic state estimation schemes, only the observability analysis is needed and, in fact, only the results referring to this property will be used in the rest of this chapter. For the sake of completeness, we report here also the reachability analysis which is useful for the design of freeway traffic control schemes adopting the model described in Sect. 7.2.1. In particular, the reachability and observability analysis is developed considering model (7.5)–(7.7), since it is more general than model (7.1)–(7.2), and the observability properties of the two models are analogous.

Table 7.1 Modes of the 4-cells case

$s_l(k)$	Interfaces	$A_l^{s_l(k)}$	$B_l^{s_l(k)^{\mathrm{T}}}$	$C_l^{s_l(k)}$
1	UUUUU	A_a	$[0\ 0\ T/L_{3,l}\ 0]$	$[0\ 0\ 0\ v_{4,l}]$
2	UUUUC	A_a	$[0\ 0\ T/L_{3,l}\ 0]$	$[0\ 0\ 0\ 0]$
3	CUUUU	A_a	$[0\ 0\ T/L_{3,l}\ 0]$	$[-w_{1,l}\ 0\ 0\ v_{4,l}]$
4	CUUUC	A_a	$[0\ 0\ T/L_{3,l}\ 0]$	$[-w_{1,l}\ 0\ 0\ 0]$
5	UUUCU	A_b	$[0\ 0\ T/L_{3,l}\ 0]$	$[0\ 0\ 0\ v_{4,l}]$
6	UUUCC	A_b	$[0\ 0\ T/L_{3,l}\ 0]$	$[0\ 0\ 0\ 0]$
7	CUUCU	A_b	$[0\ 0\ T/L_{3,l}\ 0]$	$[-w_{1,l}\ 0\ 0\ v_{4,l}]$
8	CUUCC	A_b	$[0\ 0\ T/L_{3,l}\ 0]$	$[-w_{1,l}\ 0\ 0\ 0]$
9	UUCUU	A_c	$[0\ T/(L_{2,l}(1-\beta_l))\ 0\ 0]$	$[0\ 0\ 0\ v_{4,l}]$
10	UUCUC	A_c	$[0\ T/(L_{2,l}(1-\beta_l))\ 0\ 0]$	$[0\ 0\ 0\ 0]$
11	CUCUU	A_c	$[0\ T/(L_{2,l}(1-\beta_l))\ 0\ 0]$	$[-w_{1,l}\ 0\ 0\ v_{4,l}]$
12	CUCUC	A_c	$[0\ T/(L_{2,l}(1-\beta_l))\ 0\ 0]$	$[-w_{1,l}\ 0\ 0\ 0]$
13	UCUUU	A_d	$[0\ 0\ T/L_{3,l}\ 0]$	$[0\ 0\ 0\ v_{4,l}]$
14	UCUUC	A_d	$[0\ 0\ T/L_{3,l}\ 0]$	$[0\ 0\ 0\ 0]$
15	CCUUU	A_d	$[0\ 0\ T/L_{3,l}\ 0]$	$[-w_{1,l}\ 0\ 0\ v_{4,l}]$
16	CCUUC	A_d	$[0\ 0\ T/L_{3,l}\ 0]$	$[-w_{1,l}\ 0\ 0\ 0]$
17	UUCCU	A_e	$[0\ T/(L_{2,l}(1-\beta_l))\ 0\ 0]$	$[0\ 0\ 0\ v_{4,l}]$
18	UUCCC	A_e	$[0\ T/(L_{2,l}(1-\beta_l))\ 0\ 0]$	$[0\ 0\ 0\ 0]$
19	CUCCU	A_e	$[0\ T/(L_{2,l}(1-\beta_l))\ 0\ 0]$	$[-w_{1,l}\ 0\ 0\ v_{4,l}]$
20	CUCCC	A_e	$[0\ T/(L_{2,l}(1-\beta_l))\ 0\ 0]$	$[-w_{1,l}\ 0\ 0\ 0]$
21	UCCUU	A_f	$[0\ T/(L_{2,l}(1-\beta_l))\ 0\ 0]$	$[0\ 0\ 0\ v_{4,l}]$
22	UCCUC	A_f	$[0\ T/(L_{2,l}(1-\beta_l))\ 0\ 0]$	$[0\ 0\ 0\ 0]$
23	CCCUU	A_f	$[0\ T/(L_{2,l}(1-\beta_l))\ 0\ 0]$	$[-w_{1,l}\ 0\ 0\ v_{4,l}]$
24	CCCUC	A_f	$[0\ T/(L_{2,l}(1-\beta_l))\ 0\ 0]$	$[-w_{1,l}\ 0\ 0\ 0]$
25	UCUCU	A_g	$[0\ 0\ T/L_{3,l}\ 0]$	$[0\ 0\ 0\ v_{4,l}]$
26	UCUCC	A_g	$[0\ 0\ T/L_{3,l}\ 0]$	$[0\ 0\ 0\ 0]$
27	CCUCU	A_g	$[0\ 0\ T/L_{3,l}\ 0]$	$[-w_{1,l}\ 0\ 0\ v_{4,l}]$
28	CCUCC	A_g	$[0\ 0\ T/L_{3,l}\ 0]$	$[-w_{1,l}\ 0\ 0\ 0]$
29	UCCCU	A_h	$[0\ T/(L_{2,l}(1-\beta_l))\ 0\ 0]$	$[0\ 0\ 0\ v_{4,l}]$
30	UCCCC	A_h	$[0\ T/(L_{2,l}(1-\beta_l))\ 0\ 0]$	$[0\ 0\ 0\ 0]$
31	CCCCU	A_h	$[0\ T/(L_{2,l}(1-\beta_l))\ 0\ 0]$	$[-w_{1,l}\ 0\ 0\ v_{4,l}]$
32	CCCCC	A_h	$[0\ T/(L_{2,l}(1-\beta_l))\ 0\ 0]$	$[-w_{1,l}\ 0\ 0\ 0]$

The analysis of the structural properties of the freeway model (7.5)–(7.7) is realised by standard techniques that are addressed for an example case of a link l with $N_l = 4$ cells having the off-ramp and on-ramp located between the second and the third cell. In this case, there are $M_l = 2^5 = 32$ possible operating modes, according to the two possible terms in the minimum function in (7.7), for the computation of $\phi_{i,l}(k)$, $i = 1, \ldots, 5$. Let us denote with U (*uncongested*) the case in which $\phi_{i,l}(k)$ in (7.7) is equal to the first term, and with C (*congested*) the case in which it is equal to the second term.

Table 7.1 reports the 32 modes, with the corresponding $A_l^{s_l(k)}$, $B_l^{s_l(k)}$ and $C_l^{s_l(k)}$ matrices. In particular, the 8 possible types of $A_l^{s_l(k)}$ are specified in the following:

$$A_a = \begin{bmatrix} 1-\frac{T}{L_{1,l}}v_{1,l} & 0 & 0 & 0 \\ \frac{T}{L_{2,l}}v_{1,l} & 1-\frac{T}{L_{2,l}}v_{2,l} & 0 & 0 \\ 0 & \frac{T}{L_{3,l}}(1-\beta_l)v_{2,l} & 1-\frac{T}{L_{3,l}}v_{3,l} & 0 \\ 0 & 0 & \frac{T}{L_{4,l}}v_{3,l} & 1 \end{bmatrix} \tag{7.10}$$

$$A_b = \begin{bmatrix} 1-\frac{T}{L_{1,l}}v_{1,l} & 0 & 0 & 0 \\ \frac{T}{L_{2,l}}v_{1,l} & 1-\frac{T}{L_{2,l}}v_{2,l} & 0 & 0 \\ 0 & \frac{T}{L_{3,l}}(1-\beta_l)v_{2,l} & 1 & \frac{T}{L_{3,l}}w_{4,l} \\ 0 & 0 & 0 & 1-\frac{T}{L_{4,l}}w_{4,l} \end{bmatrix} \tag{7.11}$$

$$A_c = \begin{bmatrix} 1-\frac{T}{L_{1,l}}v_{1,l} & 0 & 0 & 0 \\ \frac{T}{L_{2,l}}v_{1,l} & 1 & \frac{1}{1-\beta_l}\frac{T}{L_{2,l}}w_{3,l} & 0 \\ 0 & 0 & 1-\frac{T}{L_{3,l}}(w_{3,l}+v_{3,l}) & 0 \\ 0 & 0 & \frac{T}{L_{4,l}}v_{3,l} & 1 \end{bmatrix} \tag{7.12}$$

$$A_d = \begin{bmatrix} 1 & \frac{T}{L_{1,l}}w_{2,l} & 0 & 0 \\ 0 & 1-\frac{T}{L_{2,l}}(w_{2,l}+v_{2,l}) & 0 & 0 \\ 0 & \frac{T}{L_{3,l}}(1-\beta_l)v_{2,l} & 1-\frac{T}{L_{3,l}}v_{3,l} & 0 \\ 0 & 0 & \frac{T}{L_{4,l}}v_{3,l} & 1 \end{bmatrix} \tag{7.13}$$

$$A_e = \begin{bmatrix} 1-\frac{T}{L_{1,l}}v_{1,l} & 0 & 0 & 0 \\ \frac{T}{L_{2,l}}v_{1,l} & 1 & \frac{1}{1-\beta_l}\frac{T}{L_{2,l}}w_{3,l} & 0 \\ 0 & 0 & 1-\frac{T}{L_{3,l}}w_{3,l} & \frac{T}{L_{3,l}}w_{4,l} \\ 0 & 0 & 0 & 1-\frac{T}{L_{4,l}}w_{4,l} \end{bmatrix} \tag{7.14}$$

$$A_f = \begin{bmatrix} 1 & \frac{T}{L_{1,l}}w_{2,l} & 0 & 0 \\ 0 & 1-\frac{T}{L_{2,l}}w_{2,l} & \frac{1}{1-\beta_l}\frac{T}{L_{2,l}}w_{3,l} & 0 \\ 0 & 0 & 1-\frac{T}{L_{3,l}}(w_{3,l}+v_{3,l}) & 0 \\ 0 & 0 & \frac{T}{L_{4,l}}v_{3,l} & 1 \end{bmatrix} \tag{7.15}$$

$$A_g = \begin{bmatrix} 1 & \frac{T}{L_{1,l}}w_{2,l} & 0 & 0 \\ 0 & 1-\frac{T}{L_{2,l}}(w_{2,l}+v_{2,l}) & 0 & 0 \\ 0 & \frac{T}{L_{3,l}}(1-\beta_l)v_{2,l} & 1 & \frac{T}{L_{3,l}}w_{4,l} \\ 0 & 0 & 0 & 1-\frac{T}{L_{4,l}}w_{4,l} \end{bmatrix} \tag{7.16}$$

$$A_h = \begin{bmatrix} 1 & \frac{T}{L_{1,l}} w_{2,l} & 0 & 0 \\ 0 & 1 - \frac{T}{L_{2,l}} w_{2,l} & \frac{1}{1-\beta_l} \frac{T}{L_{2,l}} w_{3,l} & 0 \\ 0 & 0 & 1 - \frac{T}{L_{3,l}} w_{3,l} & \frac{T}{L_{3,l}} w_{4,l} \\ 0 & 0 & 0 & 1 - \frac{T}{L_{4,l}} w_{4,l} \end{bmatrix} \tag{7.17}$$

The reachability and observability results for the considered 4-cell link are summarised in Table 7.2. It is worth noting that, in all the existing modes, there are some traffic densities which belong to the reachable part of the system. This means that, in any mode, it is always worth regulating the system since the control action can always influence the traffic behaviour. On the contrary, by analysing the observability properties, it can be seen that there are some modes of the system in which no state components are observable.

7.3 Luenberger-Like Observers for Traffic State Estimation

State observers are dynamic systems which provide an estimate of the internal state of a system relying on measurements of the system output. The state estimation based on *Luenberger observers* [27] is a widely used methodology to deal with systems which admit a linear model or whenever there exists a linearising transformation of the non-linear system dynamics [28]. In case of freeway traffic, the simplest models that can be chosen are switched models, i.e. models that are linear when a specific mode is featured, as described in Sect. 7.2. Therefore, classical Luenberger observers cannot be adopted.

Nonetheless, in order to exploit the effectiveness of the Luenberger observers, *Luenberger-like observers* for systems represented through switched models are described in this section. In particular, the traffic state estimation of freeway links with ramps is addressed, being the case without ramps simpler (note that the obtained results are analogous). In particular, one observer is associated with each road link and it provides an estimate of the traffic densities in the corresponding link (see Fig. 7.5). The information necessary to each observer is given by the control action and the measurements of the traffic flows at the link boundaries.

Referring to the freeway link dynamics given by (7.8), the Luenberger-like observer associated with link l and capable of estimating the unknown densities of link l can be expressed in the form

$$\begin{aligned} \hat{\underline{\rho}}_l(k+1) = {} & A_l^{s_l(k)} \hat{\underline{\rho}}_l(k) + B_l^{s_l(k)} u_l(k) + D_l^{s_l(k)} \underline{\phi}_l(k) + E_l^{s_l(k)} \underline{\rho}_l^{\max} \\ & + L_l^{s_l(k)} \left[C_l^{s_l(k)} \hat{\underline{\rho}}_l(k) + F_l^{s_l(k)} \underline{\rho}_l^{\max} - y_l(k) \right] \end{aligned} \tag{7.18}$$

where $\hat{\underline{\rho}}_l(k)$ represents the estimation of the densities $\underline{\rho}_l(k)$ of link l and $L_l^{s_l(k)}$ is a gain matrix depending on the operation mode $s_l(k)$. The estimation problem corresponds

Table 7.2 Reachability and observability in the 4-cells case

		Reachability				Observability			
$s_l(k)$	Interfaces	$\rho_{1,l}$	$\rho_{2,l}$	$\rho_{3,l}$	$\rho_{4,l}$	$\rho_{1,l}$	$\rho_{2,l}$	$\rho_{3,l}$	$\rho_{4,l}$
1	UUUUU	No	No	Yes	Yes	Yes	Yes	Yes	yes
2	UUUUC	No	No	Yes	Yes	No	No	No	No
3	CUUUU	No	No	Yes	yes	Yes	Yes	Yes	Yes
4	CUUUC	No	No	Yes	Yes	Yes	No	No	No
5	UUUCU	No	No	Yes	No	No	No	No	Yes
6	UUUCC	No	No	Yes	No	No	No	No	No
7	CUUCU	No	No	Yes	No	Yes	No	No	Yes
8	CUUCC	No	No	Yes	No	Yes	No	No	No
9	UUCUU	No	Yes	No	No	No	No	Yes	Yes
10	UUCUC	No	Yes	No	No	No	No	No	No
11	CUCUU	No	Yes	No	No	Yes	No	Yes	Yes
12	CUCUC	No	Yes	No	No	Yes	No	No	No
13	UCUUU	No	No	Yes	Yes	No	Yes	Yes	Yes
14	UCUUC	No	No	Yes	Yes	No	No	No	No
15	CCUUU	No	No	Yes	Yes	Yes	Yes	Yes	Yes
16	CCUUC	No	No	Yes	Yes	Yes	Yes	No	No
17	UUCCU	No	Yes	No	No	No	No	No	Yes
18	UUCCC	No	Yes	No	No	No	No	No	No
19	CUCCU	No	Yes	No	No	Yes	No	No	Yes
20	CUCCC	No	Yes	No	No	Yes	No	No	No
21	UCCUU	Yes	Yes	No	No	No	No	Yes	yes
22	UCCUC	Yes	Yes	No	No	No	No	No	No
23	CCCUU	Yes	Yes	No	No	Yes	Yes	Yes	Yes
24	CCCUC	Yes	Yes	No	No	Yes	Yes	Yes	No
25	UCUCU	No	No	Yes	No	No	No	No	Yes
26	UCUCC	No	No	Yes	No	No	No	No	No
27	CCUCU	No	No	Yes	No	Yes	Yes	No	Yes
28	CCUCC	No	No	Yes	No	Yes	Yes	No	No
29	UCCCU	Yes	Yes	No	No	No	No	No	Yes
30	UCCCC	Yes	Yes	No	No	No	No	No	No
31	CCCCU	Yes	Yes	No	No	Yes	Yes	Yes	Yes
32	CCCCC	Yes	Yes	No	No	Yes	Yes	Yes	Yes

to the reconstruction of an estimate $\hat{\underline{\rho}}_l(k)$ of the actual state $\underline{\rho}_l(k)$ based on the knowledge of the output and input signals $y_l(k)$ and $u_l(k)$.

By taking into account (7.18) and (7.8), and by defining the estimation error as $e_l(k) = \hat{\underline{\rho}}_l(k) - \underline{\rho}_l(k)$, it is possible to write the error dynamics of the system as

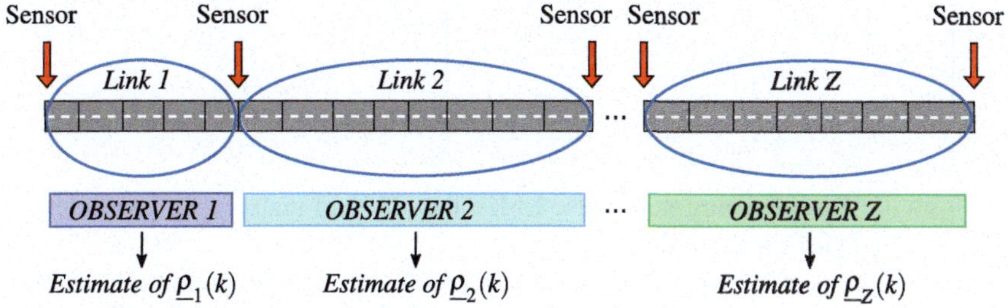

Fig. 7.5 Luenberger-like observers

$$e_l(k+1) = \left[A_l^{s_l(k)} + L_l^{s_l(k)} C_l^{s_l(k)}\right] e_l(k) \tag{7.19}$$

As discussed in Sect. 7.2.2, the switched model describing the dynamics of the freeway system may present some unobservable states depending on the operation mode $s_l(k)$. In order to identify such unobservable part of the system, it is possible to express the error dynamics in observer Kalman form by adopting a change of coordinates, i.e.

$$e_l(k+1) = \left(T_l^{s_l(k)\mathrm{T}} \begin{bmatrix} A_{l,\bar{\alpha}}^{s_l(k)} & A_{l,\bar{\alpha}\alpha}^{s_l(k)} \\ 0 & A_{l,\alpha}^{s_l(k)} \end{bmatrix} T_l^{s_l(k)} + T_l^{s_l(k)\mathrm{T}} \begin{bmatrix} 0 \\ L_{l,\alpha}^{s_l(k)} \end{bmatrix} \begin{bmatrix} 0 \ C_{l,\alpha}^{s_l(k)} \end{bmatrix} T_l^{s_l(k)} \right) e_l(k) \tag{7.20}$$

where subscripts α and $\bar{\alpha}$ denote the observable and unobservable parts, respectively, and matrix $T_l^{s_l(k)}$ allows the change of base from the standard form to the Kalman form. The remaining matrices represent the block structure separating observable and unobservable modes according to the Kalman decomposition.

Considering that only the observable part of the state can be reconstructed, the observer in (7.18) can be written as

$$\begin{aligned} \hat{\underline{\rho}}_{l,\alpha}(k+1) = {} & A_{l,\alpha}^{s_l(k)} \hat{\underline{\rho}}_l(k) + B_{l,\alpha}^{s_l(k)} u_l(k) + D_{l,\alpha}^{s_l(k)} \underline{\phi}_l(k) + E_{l,\alpha}^{s_l(k)} \underline{\rho}_l^{\max} \\ & + L_{l,\alpha}^{s_l(k)} \left[C_{l,\alpha}^{s_l(k)} \hat{\underline{\rho}}_{l,\alpha}(k) + F_{l,\alpha}^{s_l(k)} \underline{\rho}_l^{\max} - y_l(k) \right] \end{aligned} \tag{7.21}$$

where only the observable part has been extracted, with straightforward definitions of the new matrices with subscript α.

The reader is referred to [20] for the discussion and the proof of a result useful not only to analyse the stability of the proposed observer structure but also to compute the gain matrix $L_{l,\alpha}^{s_l(k)}$. According to the theorem reported in [20], the switched observer system (7.21) asymptotically converges to the observable subspace of the dynamics described in (7.8) if there exist matrices $P_{i,\alpha} > 0$, $P_{i,\bar{\alpha}} > 0$, $P_{j,\alpha} > 0$, $P_{j,\bar{\alpha}} > 0$ and Y_i satisfying the following Linear Matrix Inequality (LMI):

$$\left[\begin{array}{cc|cc} P_{i,\bar{\alpha}} & 0 & & \\ 0 & P_{i,\alpha} & & * \\ \hline P_{j,\bar{\alpha}} A^i_{l,\bar{\alpha}} & P_{i,\bar{\alpha}} A^i_{l,12} & P_{j,\bar{\alpha}} & 0 \\ 0 & P_{i,\alpha} A^i_{l,\alpha} + Y^i C^i_{l,\alpha} & 0 & P_{j,\alpha} \end{array}\right] > 0 \quad \forall (i,j) \in \mathscr{S}_l \times \mathscr{S}_l \tag{7.22}$$

From the feasible solution of these LMIs, the observer matrix gain is obtained as $L^i_{l,\alpha} = P^{-1}_{i,\alpha} Y_i$ for every mode $i \in \mathscr{S}_l$.

The theorem in [20] provides sufficient conditions for the asymptotic stability of these Luenberger-like observers. Note that (7.22) is a *set of LMIs*, since all the inequalities must be satisfied for all possible transitions between any two pairs $(i, j) \in \mathscr{S}_l \times \mathscr{S}_l$ of operation modes. The computational complexity associated with the LMIs could be drastically alleviated by considering suitable assumptions which reduce the number of possible mode transitions.

7.4 Consensus-Based Traffic State Estimation

This section presents a more general traffic state estimation scheme for a freeway stretch in which there is one observer per freeway link, as in Sect. 7.3. The main difference compared with the Luenberger-like observer scheme described before is that, in the present scheme, each observer provides an estimate of the entire freeway state. Such an estimate is done by each observer on the basis of its partial information on the freeway system related to the flow measurements at the link boundaries, and on some information that each observer can exchange with neighbouring observers (see Fig. 7.6).

Compared with a conventional centralised all-to-all communication scheme, the considered distributed estimation scheme, in which only neighbouring observers

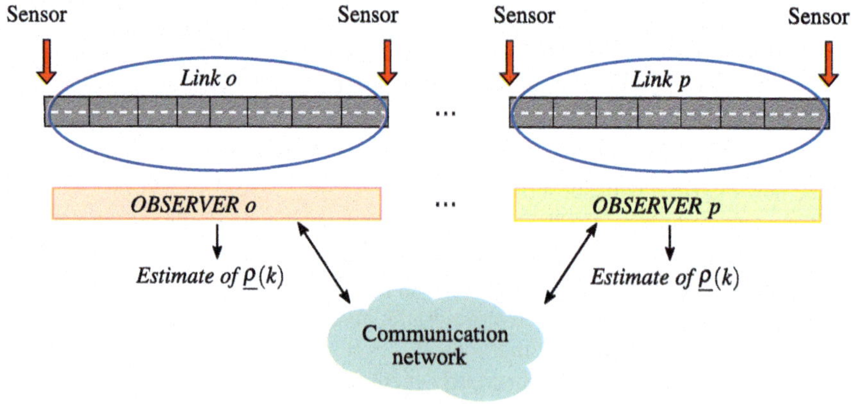

Fig. 7.6 Distributed consensus-based observers

communicate, has high advantages in terms of reduction of the communication costs, but requires that efficient *consensus reaching strategies* are implemented. This kind of strategies, which have been introduced to deal with distributed systems [29], has been applied also in traffic state estimation and control [21, 23, 24, 30]. The basic idea is to design a distributed coordination scheme which relies on observers, in case of state estimation, and on local controllers, in case of control, in which observers and controllers are regarded as *agents*. The coordination strategy is aimed at attaining a consensus among them complying with problem objectives and constraints, as well as taking into account the information exchange topology [31–34].

As shown in Fig. 7.6, the observers are connected by means of a communication network, which is represented with a directed graph $\mathcal{G} = (\mathcal{V}, \mathcal{E})$, where $\mathcal{V} = \{1, \ldots, Z\}$ is the set of observers of the network and $\mathcal{E} \subset \mathcal{V} \times \mathcal{V}$ is the set of links connecting them. The set of observers connected to observer o, i.e. the neighbourhood of o, is denoted as $\mathcal{N}_o$. Oriented communications are considered, i.e. if a link (o, p) exists, this implies that observer o receives information from observer p.

In this scheme, each observer o estimates the entire freeway system, denoted with $\underline{\hat{x}}_o(k)$, on the basis of the measured output $y_o(k)$ and on the basis of some outputs $\underline{\hat{y}}_{o,p}(k) = C^{s(k)}_{o,p}\underline{\hat{x}}_p(k)$ estimated by neighbour p and communicated to observer o. Moreover, it is assumed that observer o knows matrices $C^{s(k)}_{o,p}$, $p \in \mathcal{N}_o$.

In the following, the consensus-based traffic state estimation scheme is distinguished for the cases without and with ramps.

7.4.1 Distributed Observers for a Freeway Stretch Without Ramps

Let us consider a freeway system divided into links without ramps, modelled according to (7.4) and in which an observer is associated with each link. Each observer o runs an estimator of the system state given by

$$\begin{aligned}\underline{\hat{x}}_o(k+1) &= A^{s(k)}\underline{\hat{x}}_o(k) + D^{s(k)}\underline{d}(k) + E^{s(k)}\underline{\pi}\\ &+M^{s(k)}_o\left[C^{s(k)}\underline{\hat{x}}_o(k) + F^{s(k)}\underline{\pi} - \underline{y}_o(k)\right]\\ &+\sum_{p\in\mathcal{N}_o} N^{s(k)}_{o,p}\left[C^{s(k)}_{o,p}\underline{\hat{x}}_p(k) - C^{s(k)}_{o,p}\underline{\hat{x}}_o(k)\right]\end{aligned} \tag{7.23}$$

According to (7.23), the observer o is structured in two main parts, i.e.

- a local *Luenberger-like observer*, weighted with matrices $M^{s(k)}_o$, used to correct the estimated state based on the output $\underline{y}_o(k)$ measured by observer o;
- a *consensus-based observer*, weighted with matrices $N^{s(k)}_{o,p}$, which takes into account the information received from neighbouring observers $p \in \mathcal{N}_o$.

Let us consider now the observation error of a generic observer o defined as $\underline{e}_o(k) = \underline{\hat{x}}_o(k) - \underline{x}(k)$. Taking into account the model of the freeway stretch (7.4) and (7.23), the dynamics of the observation error can be written as

$$\underline{e}_o(k+1) = \left[A^{s(k)} + M_o^{s(k)} C^{s(k)}\right] \underline{e}_o(k) + \sum_{p \in \mathscr{N}_o} N_{o,p}^{s(k)} \left[C_{o,p}^{s(k)} \underline{e}_p(k) - C_{o,p}^{s(k)} \underline{e}_o(k)\right] \tag{7.24}$$

Considering all the Z observers, the dynamic equation of the observation errors can be written in compact form by defining a stacked error vector as $\underline{e}^{\mathrm{T}}(k) = [\underline{e}_1^{\mathrm{T}}(k)\ \underline{e}_2^{\mathrm{T}}(k)\ \ldots\ \underline{e}_Z^{\mathrm{T}}(k)]$, as follows:

$$\underline{e}(k+1) = \Gamma^{s(k)}(\mathscr{M}, \mathscr{N})\underline{e}(k) \tag{7.25}$$

where

$$\Gamma^{s(k)}(\mathscr{M}, \mathscr{N}) = \Phi^{s(k)}(\mathscr{M}) + \Lambda^{s(k)}(\mathscr{N}) \tag{7.26}$$

in which $\Phi^{s(k)}(\mathscr{M})$ and $\Lambda^{s(k)}(\mathscr{N})$ are matrices depending on the sets of gain matrices $\mathscr{M} = \{M_o^{s(k)}, s(k) \in \mathscr{S}, o \in \mathscr{V}\}$ and $\mathscr{N} = \{N_{o,p}^{s}, s(k) \in \mathscr{S}, o \in \mathscr{V}, p \in \mathscr{N}_o\}$, whose structure can be easily obtained (see [23] for further details).

In order to account for possible model uncertainties in the observers dynamics and for measurement errors of the flow sensors at the link interfaces, it is possible to add to the switched autonomous error dynamics (7.25) an additional term $\underline{\omega}(k)$, by obtaining

$$\underline{e}(k+1) = \Gamma^{s(k)}(\mathscr{M}, \mathscr{N})\underline{e}(k) + B_\omega^{s(k)}\underline{\omega}(k) \tag{7.27}$$

where the disturbance terms $\underline{\omega}(k) \in \mathscr{L}_2$ are $\mathscr{L}_2$-bounded, i.e. $\sum_{k=0}^{\infty} \underline{\omega}^{\mathrm{T}}(k)\underline{\omega}(k) < \infty$ and $B_\omega^{s(k)}$ is a matrix of suitable dimensions. Let us define also the following performance function $\underline{z}(k)$

$$\underline{z}(k) = C_e^{s(k)}\underline{e}(k) + C_\omega^{s(k)}\underline{\omega}(k) \tag{7.28}$$

where $C_\omega^{s(k)}$ is a suitable matrix.

In [23], it is proved that system (7.27)–(7.28) has γ-performance[1] if there exist matrices $P_i > 0$ and M_i, $N_{i,j}$, $i \in \mathscr{S}$, satisfying the non-linear matrix inequality

[1] An autonomous switched linear system with switching function $s(k)$ expressed as

$$\underline{x}(k+1) = A^{s(k)}\underline{x}(k) + B_\omega^{s(k)}\underline{\omega}(k)$$
$$\underline{z}(k) = C_x^{s(k)}\underline{x}(k) + C_\omega^{s(k)}\underline{\omega}(k)$$

is said to have γ-performance if the undisturbed system (i.e. with $\underline{\omega}(k) = 0$) is asymptotically stable and, under zero initial conditions, the following relation is verified

$$\sum_{k=0}^{\infty} \underline{z}^{\mathrm{T}}(k)\underline{z}(k) < \gamma^2 \sum_{k=0}^{\infty} \underline{\omega}^{\mathrm{T}}(k)\underline{\omega}(k) \quad \forall \underline{\omega}(k) \in \mathscr{L}_2$$

$$\begin{bmatrix} -P_i - C_e^{i\,\mathrm{T}} C_e^i & -C_e^{i\,\mathrm{T}} C_\omega^i & \Gamma^i(\mathscr{M},\mathscr{N})^{\mathrm{T}} P_j \\ * & -\gamma^2 I - C_\omega^{i\,\mathrm{T}} C_\omega^i & B_\omega^{i\,\mathrm{T}} P_j \\ * & * & -P_j \end{bmatrix} < 0 \quad \forall (i,j) \in \mathscr{S} \times \mathscr{S} \tag{7.29}$$

The matrix inequalities in (7.29) are non-linear due to the presence of cross terms in the block $\Gamma^i(\mathscr{M}\mathscr{N})^{\mathrm{T}} P_j$. By properly substituting the products with new unknown terms, (7.29) can be written as a set of LMIs. The interested reader can find further details on this aspect in [23], where it is also shown how to set the weighting matrices $N_{o,p}^{s(k)}$ and $M_o^{s(k)}$.

Note that inequality (7.29) is applied to all allowed transition pairs (i, j). Under the assumption of a single-congestion wave present in the links as assumed in [16, 18], only a subset of transitions between operation modes $s(k) \in \mathscr{S}$ is allowed. This surely decreases the number of inequalities to be satisfied in (7.29), allowing a reduction of the associated computational complexity.

It is also worth reminding that the adopted traffic model can present some unobservable modes, as reported in Sect. 7.2.2. When the system operates in an unobservable mode, there is no way to incorporate corrective actions on the observer dynamics, hence the correcting matrices in $M_o^{s(k)}$ and $N_{o,p}^{s(k)}$ are fixed to null values. In these cases, the observer preserves the same accuracy of the adopted traffic model.

It is finally important to underline that the considered distributed consensus-based observers, when applying (7.23), need to know the operation mode of the plant $s(k)$. To this aim, every observer o runs a local estimator of the system operation mode $\hat{s}_o(k)$, replicating the equation for $\sigma(\cdot, \cdot)$ in (7.4) as

$$\hat{s}_o(k) = \sigma\left(\underline{\hat{x}}_o(k), \underline{d}(k)\right) \tag{7.30}$$

The estimation in (7.30) uses the information available for observer o, that is the local state estimation $\underline{\hat{x}}_o(k)$, and the freeway upstream and downstream flows $\underline{d}(k)$.

7.4.2 Distributed Observers for a Freeway Stretch with Ramps

Analysing the more general case of a freeway stretch with on-ramps where traffic flows can be regulated, it is necessary to consider how traffic state estimation issues relate to traffic control, since the presence of metered input flows from the on-ramps significantly affects the state dynamics. In other terms, giving that the observer is used inside a freeway traffic control scheme, the traffic state estimation has to be performed in closed loop, explicitly taking into account the control effects. Notice that, in a closed-loop observer-controller scheme, the estimates produced by the traffic state estimation module have to be transferred to the controller, so that they can be *merged* or *fused* with sensor measurements.

For purely illustrative purposes, let us consider a *ramp metering controller* which relies on the use of suitable state estimates in place of measurements (typically, of the densities close to the on-ramps). This case is representative of all those real situations in which the traffic network is not equipped with state measurements in the proximity of the on-ramps.

In particular, a switching control scheme is considered which makes use of different control laws depending on the operating mode of the system. Referring to the freeway system described by (7.9), let us express the switched control law as follows

$$\underline{u}(k) = F^{s(k)}\underline{u}(k-1) + G^{s(k)}\underline{x}(k) + H^{s(k)} \tag{7.31}$$

where $F^{s(k)}$, $G^{s(k)}$ and $H^{s(k)}$ collect the gains associated with the control laws according to the operation mode $s(k)$. A possible example of (7.31) can be found in [24], where the adopted control scheme switches among different control laws depending on whether the density estimates upstream and/or downstream the on-ramps are observable or not. More specifically, in [24], when the density downstream the on-ramp of link l is observable, a control law based on such a density is adopted for that link; when instead link l is in a mode in which the downstream density is not observable but the upstream one is observable, a second control law relying on the downstream density is used. Finally, if link l is in a mode in which neither the upstream nor the downstream densities are observable, the metered flow in link l is given by a pre-computed control law depending on the mode $s(k)$.

Analogously to the observer described in Sect. 7.4.1, the observer o is structured in two main parts, i.e. a local Luenberger-like observer and a consensus-based observer, with gain matrices $M_o^{s(k)}$ and $N_{o,p}^{s(k)}$, respectively. More precisely, each observer o runs an estimator of the system state given by

$$\begin{aligned}\hat{\underline{x}}_o(k+1) = A^{s(k)}\hat{\underline{x}}_o(k) + B^{s(k)}\underline{u}_o(k) + D^{s(k)}\underline{d}(k) + E^{s(k)}\underline{\pi} \\ +M_o^{s(k)}\left[C^{s(k)}\hat{\underline{x}}_o(k) + F^{s(k)}\underline{\pi} - \underline{y}_o(k)\right] \\ +\sum_{p\in\mathcal{N}_o} N_{o,p}^{s(k)}\left[C_{o,p}^{s(k)}\hat{\underline{x}}_p(k) - C_{o,p}^{s(k)}\hat{\underline{x}}_o(k)\right]\end{aligned} \tag{7.32}$$

where the vector of control variables $\underline{u}_o(k)$ used by observer o depends on its traffic state estimates, i.e.

$$\underline{u}_o(k) = F^{s(k)}\underline{u}_o(k-1) + G^{s(k)}\hat{\underline{x}}_o(k) + H^{s(k)} \tag{7.33}$$

By introducing new complementary states $\underline{\xi}^{s(k)}(k)$ and $\hat{\underline{\xi}}_o^{s(k)}(k)$ and by defining expanded state vectors $\underline{\underline{x}}^{s(k)}(k)$ and $\hat{\underline{\underline{x}}}_o^{s(k)}(k)$ as

$$\underline{\underline{x}}^{s(k)}(k) = \left[\underline{x}^{s(k)}(k)^{\mathrm{T}}\ \underline{\xi}^{s(k)}(k)^{\mathrm{T}}\right]^{\mathrm{T}} \qquad \hat{\underline{\underline{x}}}_o^{s(k)}(k) = \left[\hat{\underline{x}}_o^{s(k)}(k)^{\mathrm{T}}\ \hat{\underline{\xi}}_o^{s(k)}(k)^{\mathrm{T}}\right]^{\mathrm{T}} \tag{7.34}$$

the closed-loop dynamics of the expanded state can be written as

$$\underline{\underline{x}}(k+1) = \underline{A}^{s(k)}\underline{\underline{x}}(k) + \underline{B}^{s(k)}\underline{\underline{\hat{x}}}(k) + \underline{D}^{s(k)}\underline{d}(k) + \underline{E}^{s(k)} \tag{7.35}$$

with suitable definitions of the matrices $\underline{A}^{s(k)}$, $\underline{B}^{s(k)}$, $\underline{D}^{s(k)}$ and $\underline{E}^{s(k)}$ (see [24] for further details).

Analogously, taking into account the expanded vector, the observer dynamics in (7.32) becomes

$$\begin{aligned}\underline{\underline{\hat{x}}}_o(k+1) = {} & \underline{A}^{s(k)}\underline{\underline{\hat{x}}}_o(k) + \underline{B}^{s(k)}\underline{\underline{\hat{x}}}_o(k) + \underline{D}^{s(k)}\underline{d}(k) + \underline{E}^{s(k)} \\ & + \underline{M}_o^{s(k)}\left[\underline{C}_o^{s(k)}\underline{\underline{\hat{x}}}_o(k) + \underline{F}_o^{s(k)}\underline{\pi} - \underline{y}_o(k)\right] \\ & + \sum_{p \in \mathscr{N}_o} \underline{N}_{o,p}^{s(k)}\left[\underline{C}_{o,p}^{s(k)}\underline{\underline{\hat{x}}}_p(k) - \underline{C}_{o,p}^{s(k)}\underline{\underline{\hat{x}}}_o(k)\right]\end{aligned} \tag{7.36}$$

Let us now consider the observation error of a generic observer o defined as $\underline{\underline{e}}_o(k) = \underline{\underline{\hat{x}}}_o(k) - \underline{\underline{x}}(k)$. Analysing all the Z observers, the dynamic equations of the observation errors can be written in compact form defining a stacked error vector $\underline{\underline{e}}^{\mathrm{T}}(k) = [\underline{\underline{e}}_1^{\mathrm{T}}(k)\ \underline{\underline{e}}_2^{\mathrm{T}}(k)\ \ldots\ \underline{\underline{e}}_Z^{\mathrm{T}}(k)]$, as

$$\underline{\underline{e}}(k+1) = \Xi^{s(k)}(\mathscr{M}, \mathscr{N})\underline{\underline{e}}(k) \tag{7.37}$$

where

$$\Xi^{s(k)}(\mathscr{M}, \mathscr{N}) = \Phi^{s(k)}(\mathscr{M}) + \Lambda^{s(k)}(\mathscr{N}) + \Psi^{s(k)} \tag{7.38}$$

in which matrices $\Phi^{s(k)}(\mathscr{M})$, $\Lambda^{s(k)}(\mathscr{N})$ and $\Psi^{s(k)}$ depend on the sets of observers weighting matrices to be designed $\mathscr{M} = \{\underline{M}_o^s, s \in \mathscr{S}, o \in \mathscr{V}\}$ and $\mathscr{N} = \{\underline{N}_{o,p}^s, s \in \mathscr{S}, o \in \mathscr{V}, p \in \mathscr{N}_o\}$.

In [24], it is proved that the switched error dynamics (7.37) exhibits asymptotic stability if there exist matrices $P_i > 0$, F_i, G_i, $i \in \mathscr{S}$, satisfying the following LMI optimisation problem:
Maximise ν subject to

$$\nu I < P_i < I \tag{7.39}$$

$$\begin{bmatrix} \Xi_i^{\mathrm{T}} F_i^{\mathrm{T}} + F_i \Xi_i - P_i + \nu I & \Xi_i^{\mathrm{T}} G_i - F_i \\ * & P_j - G_i - G_i^{\mathrm{T}} \end{bmatrix} < 0 \quad \forall (i, j) \in \mathscr{S} \times \mathscr{S} \tag{7.40}$$

with $\nu > 0$.

Note that the resulting matrix inequalities are non-linear, but these non-linearities can be settled by substitution with new unknown terms, so that (7.40) becomes a set of LMIs (see [24] for further details).

7.5 Traffic State Estimation with New Data Sources

As aforementioned, this chapter has not provided a thorough discussion of freeway traffic state estimation, but has illustrated some possible estimation methods that seem promising for practical implementations. It is unquestionable that, in the future, these methods will need to be adapted to the new types of data that can be acquired on freeways. Data will be collected not only through conventional sensors but also through the so-called *probe vehicles* or *floating cars*, capable to make measurements of the traffic state along their trajectories by adopting GPS technologies as well as radar and camera-based devices. Other traffic data will come from mobile phones of the drivers, from information exchanged by connected vehicles with other vehicles or with the infrastructure, from social networks and so on [35]. These data will be more and more disaggregated, inhomogeneous and asynchronous, implying the necessity of developing suitable *data fusion techniques* in order to obtain reliable information by properly respecting the users' privacy.

Another important issue associated with these mobile data is that they will contain much more information than the simple traffic state, hence advanced techniques to extract the relevant information from these data will be required. In addition, if the penetration rate of probe vehicles becomes high and the data transmission becomes very frequent, the issues of *big data* will be a reality in freeway traffic estimation and control. This will require a meditated re-adjustment of the methods used for *data mining* in order to make them scalable and efficiently applicable in the new scenario. But, more important, it will be crucial to develop data processing methods capable of manipulating data from different sources, possibly merging them also with historical and conventional data.

All these novelties will also create a challenge for the design of the models to be used to perform traffic state estimation. Certainly, in case of model-driven approaches, it will be crucial to use scalable models which allow for online computation, by respecting the constraints of real-time implementation, while being sufficiently accurate. As seen, most model-driven approaches are based on macroscopic models. It will be unavoidable that, by increasing the amount of data and their ability to capture detailed phenomena of traffic dynamics, more accurate models should be used, even within the estimation methods. This will entail an increasing trend towards distributed and decentralised state observation methods in order to ensure the practical implementation of the methods and their joint use with real-time traffic control strategies.

Some recent works in the literature deal with these aspects. For instance, in [36, 37] the traffic state estimation problem using data coming from probe vehicles is addressed. Traffic state estimation for freeways with a mixed flow of conventional and connected/automated vehicles is studied in [38–40]. The work in [41] considers the case of flow and density estimation on the basis of both standard fixed sensors and floating car data, and proposes a data fusion algorithm to merge these two sources of information. Similarly, density estimation is addressed in [42] and in [43] for cases in which measurements are given by both probe vehicles and fixed-location loop detectors.

References

1. Herrera JC, Work DB, Herring R, Ban X, Jacobson Q, Bayen AM (2010) Evaluation of traffic data obtained via GPS-enabled mobile phones: the mobile century field experiment. Transp Res Part C 18:568–583
2. Seo T, Bayen AM, Kusakabe T, Asakura Y (2017) Traffic state estimation on highway: a comprehensive survey. Ann Rev Control 43:128–151
3. Gazis DC, Knapp CH (1971) On-line estimation of traffic densities from time-series of flow and speed data. Transp Sci 5:283–301
4. Nahi NE, Trivedi AN (1973) Recursive estimation of traffic variables: section density and average speed. Transp Sci 7:269–286
5. Wang Y, Papageorgiou M (2005) Real-time freeway traffic state estimation based on extended Kalman filter: a general approach. Transp Res Part B 39:141–167
6. Wang Y, Papageorgiou M, Messmer A (2006) RENAISSANCE - a unified macroscopic model-based approach to real-time freeway network traffic surveillance. Transp Res Part C 14:190–212
7. Wang Y, Papageorgiou M, Messmer A, Coppola P, Tzimitsi A, Nuzzolo A (2009) An adaptive freeway traffic state estimator. Automatica 45:10–24
8. Tampère CMJ, Immers LH (2007) An extended Kalman filter application for traffic state estimation using CTM with implicit mode switching and dynamic parameters. In: Proceedings of the IEEE international transportation systems conference, pp 209-216
9. Kurzhanskiy AA, Varaiya P (2012) Guaranteed prediction and estimation of the state of a road network. Transp Res Part C 21:163–180
10. Hegyi A, Girimonte D, Babuska R, De Schutter B (2006) A comparison of filter configurations for freeway traffic state estimation. In: Proceedings of the IEEE intelligent transportation systems conference, pp 1029–1034
11. Work DB, Tossavainen OP, Blandin S, Bayen AM, Iwuchukwu T, Tracton K (2008) An ensemble Kalman filtering approach to highway traffic estimation using GPS enabled mobile devices. In: Proceedings of the 47th IEEE conference on decision and control, pp 5062–5068
12. Mihaylova L, Boel R, Hegyi A (2007) Freeway traffic estimation within particle filtering framework. Automatica 43:290–300
13. Pascale A, Gomes G, Nicoli M (2013) Estimation of highway traffic from sparse sensors: stochastic modeling and particle filtering. In: Proceedings of the IEEE international conference on acoustics, speech and signal processing, pp 6158–6162
14. van Lint JWC, Hoogendoorn SP (2010) A robust and efficient method for fusing heterogeneous data from traffic sensors on freeways. Comput-Aided Civ Infrastruct Eng 25:596–612
15. Treiber M, Kesting A (2011) Reconstructing the traffic state by fusion of heterogeneous data. Comput-Aided Civ Infrastruct Eng 26:408–419
16. Canudas de Wit C, Leon Ojeda L, Kibangou AY (2012) Graph constrained-CTM observer design for the Grenoble south ring. In: Proceedings of the 13th IFAC symposium on control of transportation systems, pp 197–202
17. Morbidi P, Leon Ojeda L, Canudas de Wit C, Bellicot I (2014) A new robust approach for highway traffic density estimation. In: Proceedings of the European control conference, pp 2575–2580
18. Muñoz L, Sun X, Horowitz R, Alvarez L (2003) Traffic density estimation with the cell transmission model. In: Proceedings of the American control conference, pp 3750–3755

19. Canudas De Wit C, Morbidi F, Léon Ojeda L, Kibangou AY, Bellicot I, Bellemain P (2015) Grenoble traffic lab: an experimental platform for advanced traffic monitoring and forecasting. IEEE Control Syst Mag 35:23–39
20. Brandi A, Ferrara A, Sacone S, Siri S, Vivas C, Rubio FR (2017) Model predictive control with state estimation for freeway systems. In Proceedings of the American control conference, pp 3536–3541
21. Sun Y, Work DB (2014) A distributed local Kalman consensus filter for traffic estimation. In: Proceedings of the 53rd conference on decision and control, pp 6484–6491
22. Sun Y, Work DB (2017) Scaling the Kalman filter for large-scale traffic estimation. IEEE Trans Control Netw Syst. https://doi.org/10.1109/TCNS.2017.2668898
23. Vivas C, Siri S, Ferrara A, Sacone S, Cavanna G, Rubio FR (2015) Distributed consensus-based switched observers for freeway traffic density estimation. In: Proceedings of the 54th IEEE conference on decision and control, pp 3445–3450
24. Ferrara A, Sacone S, Siri S, Vivas C, Rubio FR (2016) Switched observer-based ramp metering controllers for freeway systems. In: Proceedings of the 55th conference on decision and control, pp 6777–6782
25. Christophersen FJ (2007) Optimal control of constrained piecewise affine systems. Springer, Berlin
26. Antsaklis PJ, Michel AN (2006) Linear systems. Birkhäuser, Basel
27. Luenberger DG (1971) An intoduction to observers. IEEE Trans Autom Control 16:596–602
28. Kerner A, Isidori A (1983) Linearization by output injection and nonlinear observers. Syst Control Lett 3:47–52
29. Coulouris G, Dollimore J, Kindberg T (2001) Distributed systems: concepts and design, 3rd edn. Addison-Wesley, Boston
30. Kim BY, Ahn HS (2016) Distributed coordination and control for a freeway traffic network using consensus algorithms. IEEE Syst J 10:162–168
31. Borkar V, Varaiya P (1982) Asymptotic agreement in distributed estimation. IEEE Trans Autom Control 27:650–655
32. Jadbabaie A, Lin J, Morse AS (2003) Coordination of groups of mobile autonomous agents using nearest neighbor rules. IEEE Trans Autom Control 48:988–1001
33. Cao M, Morse AS, Anderson BDO (2008) Agreeing asynchronously. IEEE Trans Autom Control 53:1826–1838
34. Nedić A, Ozdaglar A, Parrilo PA (2010) Constrained consensus and optimization in multi-agent networks. IEEE Trans Autom Control 55:922–938
35. Patire AD, Wright M, Prodhomme B, Bayen AM (2015) How much GPS data do we need? Transp Res Part C 58:325–342
36. Seo T, Kusakabe T, Asakura Y (2015) Estimation of flow and density using probe vehicles with spacing measurement equipment. Transp Res Part C 53:134–150
37. Seo T, Kusakabe T (2015) Probe vehicle-based traffic flow estimation method without fundamental diagram. Transp Res Procedia 9:149–163
38. Bekiaris-Liberis N, Roncoli C, Papageorgiou M (2016) Highway traffic state estimation with mixed connected and conventional vehicles. IEEE Trans Intell Transp Syst 17:3484–3497
39. Fountoulakis M, Bekiaris-Liberis N, Roncoli C, Papamichail I, Papageorgiou M (2017) Highway traffic state estimation with mixed connected and conventional vehicles: microscopic simulation-based testing. Transp Res Part C 78:13–33
40. Wang R, Li Y, Work DB (2017) Comparing traffic state estimators for mixed human and automated traffic flows. Transp Res Part C 78:95–110
41. Lovisari E, Canudas de Wit C, Kibangou AY (2016) Density/flow reconstruction via heterogeneous sources and optimal sensor placement in road networks. Transp Res Part C 69:451–476
42. Wright M, Horowitz R (2016) Fusing loop and GPS probe measurements to estimate freeway density. IEEE Trans Intell Transp Syst 17:3577–3590
43. Canepa ES, Claudel CG (2017) Networked traffic state estimation involving mixed fixed-mobile sensor data using Hamilton-Jacobi equations. Transp Res Part B 104:686–709

Part III
Freeway Traffic Control

Chapter 8
An Overview of Traffic Control Schemes for Freeway Systems

8.1 Freeway Traffic Management and Control

The need for the development of surveillance and control strategies for freeway traffic networks has increased in the past decades because of the persistent growth of traffic congestion and the resulting negative effects on people and on the ecosystem. Freeway networks, although designed to meet the mobility needs of high traffic flows, have suffered in recent years the increasing demand which can be rarely solved with proper infrastructure interventions (see Chap. 1 for a more detailed discussion on these aspects). Consequently, the adoption of specific control measures represents, in many cases, the only possible answer to improve the performance of freeway traffic systems.

Moreover, in recent years, the development of information systems supporting the drivers when travelling along freeways has strongly increased thanks to the progress in detection, transmission and data processing technologies. In fact, an important aspect in efficiently managing a freeway network is the implementation of a reliable *traffic monitoring system* or, analogously, a *traffic surveillance system*, able to elaborate the information coming from sensors located throughout the network, to detect possible critical situations and to provide, both to controllers and to road users, useful information about the current state of the system and, even in some cases, a prediction of its evolution in the short–medium term.

Besides monitoring the traffic state, a further advancement in the management of a freeway traffic system consists in controlling and regulating traffic flows in order to improve the performance of the system itself. *Freeway traffic control systems* have been developed and are still under investigation by scientists, in order to act on the system in real time, depending on the present system state and, in some cases, also on its predicted evolution. One of the main objectives of a freeway traffic control tool is the reduction of congestion, i.e. the reduction of the travel times for drivers (see Fig. 8.1). Clearly, reducing congestion and delays for travellers often entails the reduction of other negative effects of traffic, more related to sustainability and quality of life of citizens. Nevertheless, in some recent freeway traffic control systems, these

A. Ferrara et al., *Freeway Traffic Modelling and Control*, Advances in Industrial Control, https://doi.org/10.1007/978-3-319-75961-6_8

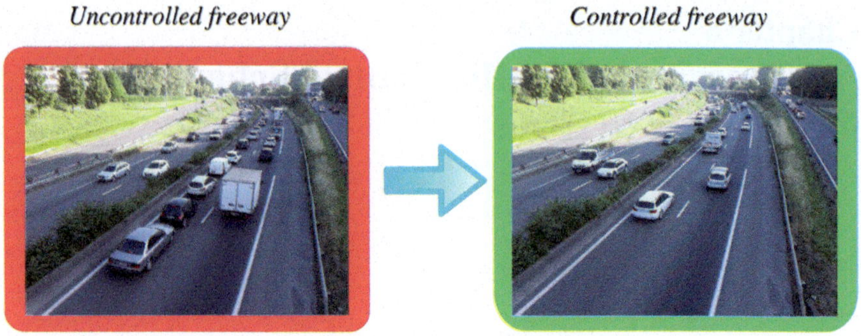

Fig. 8.1 Congestion reduction as a main objective of traffic control

related objectives are explicitly taken into account, i.e. some traffic controllers are specifically devised to reduce traffic emissions, noise, accidents and so on.

8.1.1 Traffic Control Strategies

Freeway traffic can be controlled in different ways. The most traditional way is the implementation of *road-based traffic control*, which is realised by regulating traffic at a macroscopic level. In particular, referring to road-based traffic control, it is possible to regulate the access of traffic flows to the freeway, by implementing *ramp management* policies, or to control the movement of vehicles inside the freeway, via *mainstream control*, or to route vehicles on specific paths, implementing suitable *route guidance* strategies. Of course, these control strategies can be properly combined via *integrated control*, in order to achieve better performance for the freeway system.

An example of a freeway regulated via road-based traffic control strategies is depicted in Fig. 8.2, showing two controlled on-ramps, two installations of mainstream control and one junction in which route guidance indications are provided. Note that road-based traffic control strategies act at a system level, e.g. the traffic lights at the on-ramps regulate the access to the freeway of the whole flow of incoming vehicles, as well as variable speed limits or routing indications displayed on Variable Message Signs (VMSs) are the same for all the drivers passing in front of them.

The technological development of electronic devices present on board of vehicles is allowing and will allow in the near future the wide diffusion of control policies specifically devised for each driver, according to a *vehicle-based traffic control* logic. Vehicle-based traffic control is a new concept, surely promising for the next years, but on which very few research results have been developed so far. Analysing the literature on this topic, it is worth mentioning works referring to control schemes in which the control actions are determined considering the whole traffic system but are

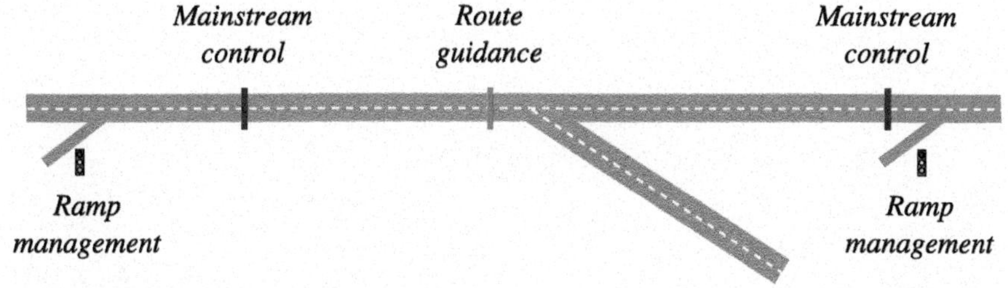

Fig. 8.2 Road-based traffic control strategies

transferred to vehicles via Vehicle-to-Infrastructure (V2I) communication systems (see e.g. [1–6]), and other works addressing the issue of coordination mechanisms for Connected and Automated Vehicles (CAVs) (see e.g. [7, 8] and the references therein).

In this chapter and in the following ones, we will mainly focus on road-based traffic control strategies, which researchers have studied for some decades and which still are the most interesting traffic control options for real implementations.

Ramp Management Ramp management strategies are applied in order to control the flow of vehicles entering the freeway mainstream. The most widespread strategy belonging to this category is surely *ramp metering*, which regulates the access of traffic flows to the mainstream through traffic signals installed at the on-ramps. The idea of controlling traffic flows by metering the on-ramps has been exploited since the 60s in the U.S., with very simple control strategies based on historical data [9, 10]. The very first implementation of ramp metering occurred in 1963 on the Eisenhower Expressway in Chicago, U.S., where a police officer was in charge of regulating traffic at the on-ramp in order to allow a safer and smoother merging into the freeway.

Most of the ramp metering strategies are devoted to reduce the *onset of congestion* phenomena, by managing the amount of traffic flows entering the freeway and by facilitating the merge of the on-ramp flows with the mainstream. Some traffic phenomena associated with merging areas are discussed in [11], based on real observations in the U.S., in which the effects of ramp metering strategies are analysed in detail.

Besides the prevention of traffic breakdowns, further benefits of ramp metering applications are widely documented in the literature (see e.g. [12, 13], which also refer to field implementations). For instance, an important phenomenon associated with congestion is the *blockage of off-ramps*, i.e. the fact that vehicles which would like to exit the freeway are delayed because they are stuck in the traffic jam, hence further increasing congestion. If a ramp metering strategy acts effectively in reducing the congestion, this can also reduce or eliminate the off-ramp blockage phenomenon, and such amelioration is more relevant in case the congestion would involve more than one off-ramp (see Fig. 8.3).

Fig. 8.3 Off-ramps in A1 freeway, close to Florence, Italy (courtesy of Autostrade per l'Italia SpA, photo from Archivio Videofotografico Autostrade per l'Italia)

In addition, the use of traffic lights at the on-ramps may increase *safety* during the merge phases. This is due to the fact that ramp metering prevents the entry of long platoons of vehicles, and, also, vehicles are induced to enter the freeway with lower speeds, reducing the risk of collisions. A better merging behaviour by vehicles can be also translated in a reduction of *pollutant emissions* in the environment.

Despite the clear positive effects achieved on the efficiency of the freeway infrastructure when implementing ramp metering policies, some critical issues may arise. One of the main drawbacks due to the application of ramp metering is the creation of long queues at the entering on-ramps. In strongly congested scenarios, the presence of *limited storage space* (which is very common, especially in urban freeways) may induce a queue spillback that can compromise the functionality of the adjacent infrastructure. In addition, the formation of long queues may generate dissatisfaction in the road users.

Ramp metering has been applied successfully for some decades and still is very widespread worldwide. Many real applications may be found in the United States, in Europe (especially in the Netherlands) and in Australia.

Mainstream Control Mainstream control is used to regulate traffic flows of vehicles travelling in the mainstream, generally showing proper indications to drivers through VMSs or with mainstream traffic lights. It is indeed proven that the operability and safety of freeway traffic may be potentially improved through control actions on the mainline (see for instance Fig. 8.4, representing a forming congestion in a freeway controlled with variable speed limits).

Fig. 8.4 Variable speed limits in A20 freeway, close to Rotterdam, the Netherlands (courtesy of Rijkswaterstaat, Photo: Essencia Communication/Rob de Voogd)

At a general level, these mainstream control actions have the aim of homogenising the traffic conditions, preventing the formation of recurrent congestions and reducing the probability of collisions among vehicles. An additional objective is to face the formation of phenomena of non-recurrent congestion, by increasing the efficiency of the system under conditions of limited capacity.

One of the most widespread mainstream control measures in freeway networks is represented by *variable speed limits*, widely applied in Northern Europe. This methodology aims to improve *mobility* and *safety* conditions in freeways by suggesting or imposing appropriate speed limits, displayed by means of VMSs. The development of V2I technologies, enabling the communication of specific messages, including also speed limits, on board of vehicles, could increase the effectiveness of this methodology in the next future.

The basic underlying idea is to reduce the speed of vehicles travelling upstream the congested area in order to *homogenise* the overall traffic conditions. Note that homogenisation means reduction of the speed differences among the vehicles composing the traffic flow, thereby limiting the onset phenomena of stop-and-go waves, that often cause accidents and traffic breakdowns (see e.g. [14, 15] for a detailed analysis on the main effects of variable speed limits). The presence of variable speed limits can have an impact also on the distribution of vehicles among the different lanes. This aspect is investigated in [16], referring to a real setup in the Netherlands, where the change in lane distribution due to variable speed limits is analysed, with reference to the merging process due to traffic flows coming from on-ramps.

Another relevant implementation of mainstream control is *mainline metering*, which involves the use of traffic lights along the mainstream. This control action is often actuated before bottlenecks in order to avoid their activations, and the associated negative consequences of system performance degradation. Mainline metering was experienced for the first time in the U.S. in the late 50s, to increase the throughput of the tunnels under the Hudson River, connecting New York City with New Jersey. The tunnel was controlled through an inflow traffic control system using real-time traffic measurements from the bottleneck location [17]. Another relevant example of mainline metering is the entrance control system with traffic lights, that had been implemented at the San Francisco–Oakland Bay Bridge for more than 35 years [18].

Another way to control the traffic flow in the mainstream is via *lane control*, in order to warn the drivers about the presence of possible queues (that may be caused by adverse weather conditions, accidents, work zones and so on) or to redirect the vehicle flows to different lanes. In Northern European countries, in particular in the Netherlands and in Germany, a widespread lane control measure is the temporary use of the *shoulder lane*. During peak hours, in order to increase the vehicle throughput, shoulder lanes are utilised as extra-lanes, and their opening or closure is communicated via VMSs. Another form of lane control consists in the exclusive use of the shoulder lane for specific classes of vehicles, such as heavy vehicles or public transport means. Further policies frequently adopted to control drivers in the mainstream are the 'keep your lane' strategy, forcing drivers to maintain their lane, or the 'early merge' strategy, encouraging drivers to merge into the open lane before the lane closure.

Finally, among the mainstream control strategies, it is possible to include also *section control*, often called also *average speed enforcement* or *point-to-point speed enforcement*. It is a speed control system, which measures the travel time of vehicles between different positions (normally with cameras) to verify the speed limit compliance. The effects of section control on freeway traffic are of several types, the most relevant ones being related to more homogenised traffic flow, increased traffic capacity, and, above all, a consequent reduction of accidents [19]. The effectiveness of section control is verified, for instance in [20], on the basis of floating car data.

Route Guidance In freeway traffic networks, drivers have often to face routing decisions, in case there are different *alternative paths* to reach their destinations (see Fig. 8.5). Among these alternatives, drivers would like to choose the most convenient path, which can correspond to the shortest, fastest or cheapest choice, depending on their preferences. Since traffic conditions vary over time, the most effective route guidance systems are the *dynamic* ones, i.e. those which are based on real-time measurements coming from the freeway network.

Route Guidance and Information Systems (RGISs) are devised in order to provide the users with information about the current state of the system (such as the presence of congestion, traffic incidents, working zones and so on) in the alternative routes or, in some cases, to give specific routing indications to the drivers. Such information can be communicated to drivers by displaying messages on VMSs or by providing them with specific (and even personalised) information by using special in-car communication devices. Even though in the future this latter option will probably become

Fig. 8.5 Alternative paths in A20 freeway, close to Rotterdam, the Netherlands (courtesy of Rijkswaterstaat, Photo: Essencia Communication/Rob de Voogd)

the most frequent, actual RGISs basically rely on the use of VMSs to communicate routing indications to the drivers. The main scientific approaches analysed in this chapter will refer to this communication option.

Different route guidance systems have been developed all over the world, simply indicating estimated travel times for alternative paths or directly suggesting paths to drivers (see e.g. [21, 22] for a survey and classification of route guidance systems).

Integrated Control Strategies Phenomena of recurrent and non-recurrent congestion in freeway systems can be relieved more efficiently if different control strategies are integrated and combined towards a common objective. It is quite evident, indeed, that the best achievements in controlling traffic in a freeway network are obtained if traffic is regulated exploiting all the possible control actions. Applying, for instance, ramp metering can provide effective results in reducing congestion phenomena but it is undeniable that, for some specific traffic scenarios, acting on the system only by regulating the access of vehicles from the on-ramps can be a limitation, while controlling also the mainstream flow or routing vehicles through alternative paths can make the overall control action more effective.

On the other hand, it is apparent that a control scheme which combines different control strategies is more challenging from the design point of view and good performance results can be obtained only if the different control strategies are properly integrated in order to achieve the same objective for the controlled system.

8.1.2 Freeway Traffic Control Schemes

The road-based traffic control strategies described in Sect. 8.1.1 are of different types and act on the freeway by intervening on different parts of the system, i.e. on the on-ramps, on the mainstream or on the diverging junctions. Regardless of their different natures, all these control strategies should act according to suitably devised traffic control algorithms to be applied online, on the basis of real-time measurements coming from the freeway network. In this sense, we are dealing with *feedback* (*closed-loop*) control schemes, since the values of the control inputs depend on the measurements of the system state. A recent work dealing with feedback control laws applied to general acyclic traffic networks can be found in [23], where robust global exponential stabilisation is proven (the robustness is referred to any uncertainty related to the Fundamental Diagram, as well as the uncertain nature of the traffic model in the congested case).

Few and very old-fashioned control schemes represent an exception to feedback control strategies, i.e. they are not based on real-time measurements but, instead, are derived off-line on the basis of historical data. Examples of this type of controllers are fixed-time ramp metering strategies (see e.g. [9, 10]), dating back to the 60s, which rely on simple static models and on past traffic data. In this book, we only consider feedback control schemes, since the control strategies that are computed off-line and are applied to the system independently from the real system state are no more of interest for real applications.

Figure 8.6 reports a very general scheme of a feedback loop for a controlled freeway traffic system. The dynamics of the freeway traffic system is affected by two different types of inputs:

- the *control inputs* are computed by a traffic controller and transferred to the real system through proper *actuators*. For instance, in case of ramp metering, the control inputs are the flows that should enter the mainstream from the on-ramp and the actuators are the traffic lights (see e.g. [24] for a discussion about how ramp metering control inputs can be translated into specific traffic light settings

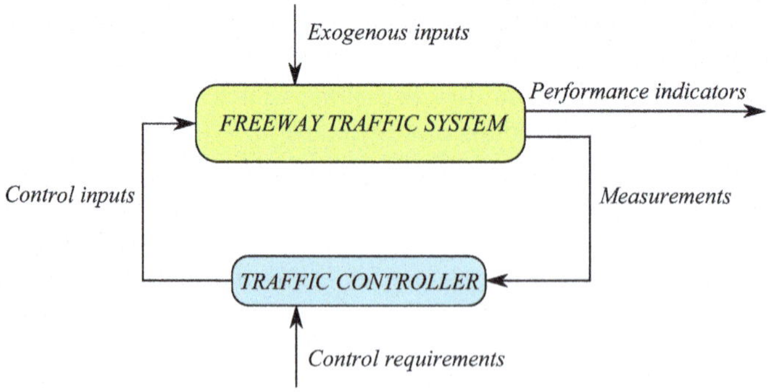

Fig. 8.6 A controlled freeway traffic system

according to the applied metering policy). In case of variable speed limits, the control inputs are the speeds of vehicles and the actuators are normally VMSs; these actuators are used also in case of route guidance control, in which the control variables are generally the splitting rates of vehicles at junctions;

- the *exogenous inputs* represent external conditions which influence the traffic system state. Typical examples of exogenous inputs affecting a freeway traffic system are the external demand (vehicles that require to enter the considered freeway), weather conditions, accidents and turning rates: some of these exogenous inputs are measurable, detectable or predictable, but they cannot be manipulated nor controlled.

The control inputs are computed by a *traffic controller*, which includes a control algorithm, that can vary from a very simple control law to highly sophisticated control frameworks. In any case, traffic controllers base the computation of the control inputs on real-time *measurements* (e.g. measurements of flows, densities, mean speeds, queue lengths), which are collected through proper *sensors*. A discussion about the possibility of measuring traffic variables in freeway systems or to estimate such variables all over the freeway network can be found in Chap. 7.

The effectiveness of the traffic controller is defined according to pre-defined *control requirements*, indicating specific functions or characteristics of the controller, as well as suitable behaviours desirable for the controlled system. Control requirements may regard, for instance, the computational time necessary to determine the control law, the use of specific types of measurements/estimates, the use (or not) of prediction models, as well as the definition of ad hoc control objectives. Strongly correlated with the control requirements, suitable *performance indicators* are defined for the freeway traffic system under investigation, e.g. the total time spent by the drivers in the system, the total delay in queues, the overall emissions, which can be referred to the entire freeway or to specific road portions. Performance indicators can be used to assess the behaviour of the system in real time, but also such indicators can be employed, via simulation, to verify the effectiveness of a given control approach, normally compared with the uncontrolled case or with other control schemes.

8.1.3 Classification of Freeway Traffic Control Schemes

The first freeway traffic control systems were developed and implemented in the U.S. in the 60s [25]. Since then, a very wide literature on freeway traffic control has been developed (see e.g. the survey papers [26, 27]). In recent years, the technological developments, especially in sensors, communication devices and processors, have allowed the actual transfer of many research results from a theoretical to a practical level. Also, the technological innovation in the context of traffic management, surveillance and control has put into evidence in some cases that it is necessary to revise conventional algorithms and control schemes in order to fully exploit the potential of new technologies.

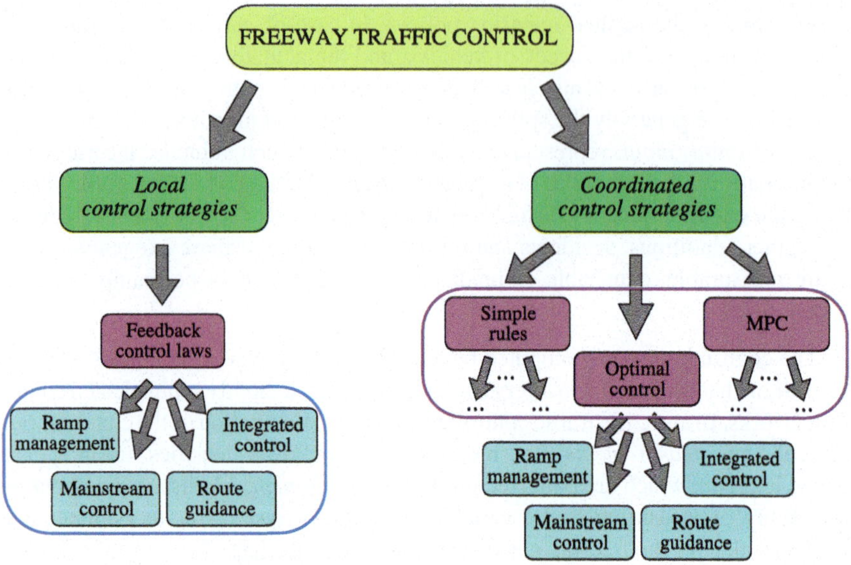

Fig. 8.7 Classification of conventional freeway traffic control approaches

This chapter is devoted to report and classify the main *conventional approaches* for freeway traffic control, according to the scheme reported in Fig. 8.7. In particular, these conventional approaches are first divided in two main categories:

- *local traffic control strategies*: they are the simplest feedback strategies, in which the control action of each controller depends on local measurements of the system state, normally coming from sensors placed in the vicinity of the corresponding actuators;
- *coordinated traffic control strategies*: the control actions actuated in different portions of the freeway are not independent and are computed taking into account measurements of the whole system state. Coordinated strategies are, in general, more effective than local ones to regulate traffic flows in a freeway network but more difficult to be designed and managed.

Both local and coordinated traffic control strategies can be further subdivided according to different criteria. The two most relevant criteria for this classification are

- the considered control methodology;
- the adopted control action.

As shown in Fig. 8.7, *local control strategies* do not differ too much in terms of control methodology, since they are mainly based on feedback control laws or on more sophisticated schemes (e.g. hierarchical), anyhow relying on feedback control concepts. Local control strategies are instead strongly differentiated on the basis of the adopted control action, i.e. ramp management, mainstream control, route guidance and integrated control. Hence, in this book, and in particular in Sect. 8.3, local control strategies are classified according to the type of control action.

Coordinated control strategies, also, can be classified depending on the type of control method and the adopted control action, as shown in Fig. 8.7. Differently from local strategies, the most meaningful classification for coordinated traffic control schemes seems related to the control methodology, and this is the criterion used in this book for their categorisation. Specifically, in Sect. 8.4, coordinated traffic control schemes are divided in schemes resulting from the coordination of simple feedback strategies, control schemes relying on optimal control approaches and Model Predictive Control (MPC) frameworks.

Chapter 9 will investigate some new concepts of traffic management, related to the *implementability* of freeway traffic control systems, i.e. the computational efficiency of the control algorithms so as to make them suitable for real-time use in possibly large freeway networks. Hence, Chap. 9 will include an overview of innovative approaches in this direction, also including event-triggered control frameworks, as well as decentralised and distributed control schemes.

Chapter 10, instead, will be focused on a new vision for freeway management and control, related to the system *sustainability*, i.e. the improvement of the quality of life of citizens as well as the efficient use of the natural resources. According to this vision, freeway traffic needs to be controlled not only for guaranteeing a significant efficiency in using the road network capacity and the improvement of global mobility, but also to limit emissions and reduce fuel consumptions. Moreover, it is particularly relevant in this context to distinguish different typologies of vehicles, leading to multi-class traffic control schemes. Hence, Chap. 10 will include an overview of the most innovative approaches including sustainability-related factors in the design of the control schemes.

8.2 Objectives of Traffic Controllers

The objectives of traffic controllers are strictly related to the improvement of the traffic conditions in the freeways, i.e. to the reduction of congestions and to the mitigation of the associated negative effects. The main goal pursued by traffic controllers is surely the reduction of the travelling times, since this is the most direct impact for travellers. As aforementioned, more recently, other control objectives have been introduced by researchers, in order to take into account environmental issues, safety aspects and, more in general, factors related to the quality of life of citizens.

In order to achieve the aforementioned benefits for freeway traffic systems, traffic controllers must be properly designed and implemented, also taking into account the requisites coming from the real application context. Among the wide variety of control approaches for freeway traffic present in the scientific literature, it is not straightforward to properly categorise all the objectives of the different freeway traffic control schemes. In the following, these objectives are classified into four main groups, respectively, corresponding to

- tracking of set-point values;
- improvement of the system performance in terms of congestion reduction;
- improvement of the system performance in terms of emission reduction;
- balancing of some system variables.

8.2.1 Tracking of Set-Point Values

A large number of traffic controllers for regulating freeway traffic have been devised in order to track some specified set-point values for the traffic variables. As it is widely applied in control theory, a *set-point* is the desired or target value for a variable of the system.

In freeway traffic, the most common choice is to fix reference values for the *traffic densities* and to design the traffic control schemes in order to track these set-points. Let us refer to a macroscopic discrete-time traffic flow model for a freeway stretch (see Sects. 3.3.1 and 4.2.1, respectively, for the CTM and for METANET), in which the stretch is composed of N road sections and the time horizon is discretised into K time intervals, where $\rho_i(k)$ is the traffic density in section i at time kT. Let us denote with $\rho_i^*(k)$ the set-point value for the traffic density in section i at time kT, $i = 1, \dots, N$, $k = 0, \dots, K$.

Set-point values for the traffic density which are different in each road section and are time-varying surely represent the most general case. Indeed, in sophisticated control schemes, the set-points can be defined according to the present traffic conditions; this is particularly suitable in hierarchical control schemes in which a supervisor computes the set-points in real time. On the opposite side, the simplest choice for these density target values is to maintain them as fixed values. In many cases, the desired density is set equal to the critical density, i.e. $\rho_i^*(k) = \rho_i^{\mathrm{cr}}$, $i = 1, \dots, N$, $k = 0, \dots K$. Designing a traffic controller in order to track the critical density is equivalent to maximise the flow, i.e. to exploit the road capacity as much as possible.

In case a set-point is fixed, it is very useful to define the *error signal*, given by the difference between the set-point and the dynamic variable and generally denoted as $e_i(k)$, referred to section i at time kT. Such error is computed as $e_i(k) = \rho_i^*(k) - \rho_i(k)$ in case the reference value is associated with traffic density. The basic idea is that, as in a standard tracking control problem, the tracking error should go to zero, hence implying stability concepts that are investigated in some research papers, as discussed later on. Note that similar considerations can be made also in case the set-points are defined for other traffic variables, such as the mean speed or the traffic flow.

A concept similar to the tracking of set-point values is related with the definition of proper *thresholds* for the traffic variables. This choice can be motivated by the fact that the real traffic control problem is not related to track a given value but to avoid specific critical situations. For instance, considering again the traffic density, the essential goal of a traffic controller is, when the density is high, to reduce it to the critical density (or another value defined according to the traffic conditions). In the opposite case in which the density is lower than the critical value and the traffic is flowing freely, there is no interest (and, often, no chance) to increase the density to the critical value. Note that the case of lower densities often corresponds to a situation in which the system does not need even to be controlled. Similar arguments can be used if a threshold is defined for the mean speed but, in that case, the traffic controller acts in the opposite way, i.e. it aims to avoid that the mean speed becomes lower than the threshold. In case thresholds are considered, it is no more relevant to define an error, but it is more useful to compute and to penalise the cases in which the threshold is overcome.

8.2.2 Improvement of the System Performance: Congestion Reduction

Instead of considering set-point or threshold values for the traffic variables, another possibility is to design the traffic controller in order to explicitly improve the performance of the freeway system, by defining suitable performance indices. The most relevant and common performance indicators are associated with *congestion reduction*. In this context, let us introduce the three most widespread indices, very often used in freeway traffic control schemes, that are the Total Time Spent, the Total Travel Distance and the Mean Speed [28].

The *Total Time Spent* (TTS) represents the time spent in the freeway by all the vehicles [veh h] in the considered time horizon. It is computed as the sum of two terms, that are the *Total Travel Time* (TTT), i.e. the total time spent by all vehicles [veh h] in the mainstream, and the *Total Waiting Time* (TWT), i.e. the total time spent by all vehicles [veh h] waiting at the on-ramps. Reducing the TTS is equivalent to reduce congestion and, equivalently, to increase the throughput exiting from the network [12]. This is due to the fact that reducing the delays suffered by vehicles implies that they will reach their destination in shorter times, i.e. improving the level of service of the infrastructure.

In addition, the *Total Travel Distance* (TTD) is the total distance [veh km] covered by all the vehicles in the considered time horizon. On the basis of the TTS and the TTD, it is possible to compute the Mean Speed (MS) [km/h] of the vehicles travelling in the considered system in the whole time horizon.

Let us refer to the METANET model for a freeway stretch with on-ramps described in Sect. 4.2.2. In this model, the freeway stretch is composed of N road sections, each one with length L_i, $i = 1, \ldots, N$, the time horizon is discretised into K time intervals with sample time T, $\rho_i(k)$ is the traffic density in section i at time kT, $q_i(k)$ is the traffic flow leaving section i during time interval $[kT, (k+1)T)$, $l_i(k)$ is the queue length of vehicles waiting in the on-ramp of section i at time kT. According to this model, the cited indices are computed as

$$TTS = TTT + TWT = T\sum_{k=0}^{K}\sum_{i=1}^{N}\rho_i(k)L_i + T\sum_{k=0}^{K}\sum_{i=1}^{N}l_i(k) \tag{8.1}$$

$$TTD = \sum_{k=0}^{K}\sum_{i=1}^{N}L_i q_i(k)T \tag{8.2}$$

$$MS = \frac{TTD}{TTS} \tag{8.3}$$

Let us report also how these indices are computed in case a multi-class traffic model is adopted, referring specifically to the multi-class METANET model for freeway stretches described in Sect. 4.3.1. In this model, again, the freeway stretch is divided into N road sections, with length L_i, $i = 1, \ldots, N$, the time horizon is discretised in K time intervals with sample time T, and, in addition, C classes of vehicles are explicitly modelled. To account for different vehicle classes, the parameter η^c, $c = 1, \ldots, C$, is used, being a conversion factor of vehicles of class c into cars. Moreover, $\rho_i^c(k)$ is the traffic density of class c in section i at time kT, $q_i^c(k)$ is the traffic flow of class c leaving section i during time interval $[kT, (k+1)T)$, $l_i^c(k)$ is the queue length of vehicles of class c waiting in the on-ramp of section i at time kT. In the multi-class case, the previous indices are computed as follows:

$$TTS = TTT + TWT = T\sum_{k=0}^{K}\sum_{i=1}^{N}\sum_{c=1}^{C}\eta^c\rho_i^c(k)L_i + T\sum_{k=0}^{K}\sum_{i=1}^{N}\sum_{c=1}^{C}\eta^c l_i^c(k) \tag{8.4}$$

$$TTD = \sum_{k=0}^{K}\sum_{i=1}^{N}\sum_{c=1}^{C}L_i\eta^c q_i^c(k)T \tag{8.5}$$

while the MS is still given by (8.3). Note that in the multi-class case, the TTS is expressed in [PCE h] and the TTD in [PCE km].

The computation of the same indices in case the CTM is used or macroscopic traffic models for freeway networks are adopted is very similar to the presented one, with only slight differences in the notation.

8.2.3 Improvement of the System Performance: Emission Reduction

The reduction of congestions is not the unique performance index to be considered in a freeway traffic system to be controlled. Many other aspects can be taken into account, such as the reduction of noise, pollution, as well the increase of safety. The performance indices associated with emission reductions are of particular interest, especially for the purposes of the present book, and will be detailed below, referring to the two emission models, COPERT and VERSIT+, described in Chap. 6.

As motivated in Chap. 6, when adopting emission models, it is useful to consider multi-class traffic flow models, allowing to explicitly consider the different emission factors of the multiple vehicle classes. In particular, let us refer to the multi-class METANET model for a freeway stretch described in Sect. 4.3.1, in which the emissions are computed, on the basis of COPERT and VERSIT+ models, as described in Sects. 6.3.2 and 6.4.2, respectively. In both cases, $E_i^M(k)$ represents the mainstream emissions in section i at time step k, and $E_i^R(k)$ indicates the on-ramp emissions in section i at time step k. Note that these emissions are given in [g/km] if COPERT model is applied, while they are expressed in [kg/s] for VERSIT+.

Analogously to the TTS previously described, a performance index associated with the *Total Emissions* (TE) in the freeway system in the whole time horizon can be defined. The TE are given by the sum of the *Mainstream Emissions* (ME) and the *Ramp Emissions* (RE), and are computed as follows:

$$TE = ME + RE = \sum_{k=0}^{K} \sum_{i=1}^{N} E_i^M(k) + \sum_{k=0}^{K} \sum_{i=1}^{N} E_i^R(k). \tag{8.6}$$

Note that this performance index can be applied also in case emission models different from COPERT and VERSIT+ are considered, provided that $E_i^M(k)$ and $E_i^R(k)$ are properly computed.

8.2.4 Balancing of the System Variables

Another possible objective of a traffic controller is to homogenise and balance traffic variables. This balancing approach can be applied following different concepts.

A first option is to design the traffic controller in order to balance the state variables along the freeway. This *space-balancing* is normally applied to the *traffic densities*, in order to obtain a homogenisation of the traffic conditions along the freeway, or to the *on-ramp queues*, in order to make the ramp metering actions more fair for drivers entering from different on-ramps. Another type of space-balancing is in some cases associated with the *control variables*: when applying variable speed limits, for instance, it is desirable that successive VMSs encountered by drivers display speed limits that are not too oscillating.

A concept of balancing can be applied also in time, to equalise the values of variables in consecutive time steps. The *time-balancing* is often applied to the *control variables*, in order to reduce oscillations in time, which could reduce the performance of the control actions. Consider for instance the control variables of ramp metering controllers, i.e. the entering flows from on-ramps: if these values have high variations from one time step to another, it is hard and often ineffective to actuate them through red and green phases of the traffic lights present at the on-ramps.

8.3 Local Control Strategies

Referring to the general scheme depicted in Fig. 8.7, let us start by describing local control strategies, that are the simplest feedback strategies, in which the control action of each controller depends on *local measurements*, i.e. measurements in the vicinity of the corresponding actuators. In particular, considering a generic freeway traffic system represented in a discrete-time framework with k indicating the time step, let us denote with $u(k)$ the generic control action computed by a given traffic controller at time step k. According to a local control strategy, the control action $u(k)$ depends on local measurements of the system state, denoted in general as $x^{\text{loc}}(k)$, which can include only one variable or more than one, depending on the specific case.

Let us refer to the example of controlled freeway depicted in Fig. 8.2, with two controlled on-ramps, two installations of mainstream control and one junction in which route guidance indications are provided. Figure 8.8 shows a scheme of local control strategies applied to that freeway stretch, in which four traffic controllers are present, denoted with A, B, C and D. In particular, traffic controller A applies a ramp management action, controller B implements mainstream control, controller C realises a route guidance strategy, and controller D implements an integrated control action, by combining ramp management and mainstream control.

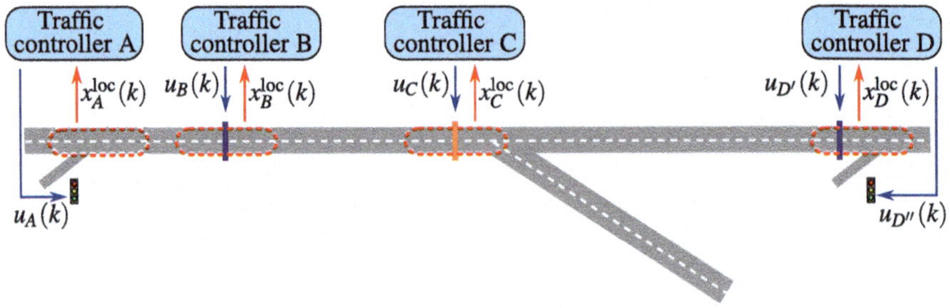

Fig. 8.8 Local control strategies

Different local strategies for freeway traffic control have been developed by researchers during the years and can be found in the literature. Among these control strategies, a very relevant class of controllers often exploited for freeway traffic is given by regulators of proportional and integral type, properly designed in order to track a specified *reference value*. Let us denote this reference value of the local system state as $x^{\mathrm{loc},*}(k)$, referred to time step k. The error at time step k, denoted as $e^{\mathrm{loc}}(k)$, can be then computed as $e^{\mathrm{loc}}(k) = x^{\mathrm{loc},*}(k) - x^{\mathrm{loc}}(k)$.

In a *proportional* controller (often called regulator of P-type), the control action is proportional to the error, i.e.

$$u(k) = K_P e^{\mathrm{loc}}(k) \tag{8.7}$$

where K_P is the proportional gain.

An *integral* controller (often called regulator of I-type) is characterised by a control action computed as

$$u(k) = u(k-1) + K_I e^{\mathrm{loc}}(k) \tag{8.8}$$

where K_I is the integral gain.

A *proportional-integral* controller (often called regulator of PI-type) is characterised by a control action which includes the proportional and the integral actions, i.e.

$$u(k) = u(k-1) + K_P \left[e^{\mathrm{loc}}(k) - e^{\mathrm{loc}}(k-1)\right] + K_I e^{\mathrm{loc}}(k) \tag{8.9}$$

Note that, in most cases, the control variables are bounded for physical reasons and must belong to a range $[u^{\min}, u^{\max}]$. Hence, the value of $u(k)$ resulting from the control laws (8.7)–(8.9) should be truncated if it is out of the requested range.

The control strategies of local type can be categorised according to different criteria. In this book, we have classified them according to the type of control action adopted, since the proposals made by researchers strongly differ according to this aspect. Hence, local control strategies are divided in:

- local ramp management strategies;
- local mainstream control strategies;
- local route guidance strategies;
- local integrated control strategies.

8.3.1 Local Ramp Management Strategies

Different feedback ramp metering control strategies of local type have been proposed in the literature and have been applied in real cases. The main objective of local ramp metering is to properly regulate the inflow from one on-ramp in order to reduce congestion in the mainstream downstream the on-ramp, which corresponds to reduce undesirable phenomena such as capacity drop or blockage of off-ramps.

The local ramp metering control strategies developed by researchers in the last decades differentiate both for the type of control law and for the type and number of local measurements needed to the traffic controller. Specifically, a common distinction is between local measurements taken *upstream* the on-ramp and measurements taken *downstream*.

Let us report in the following some of the most widespread ramp metering local strategies, by referring to the METANET model for a freeway stretch with on-ramps described in Sect. 4.2.2, in which the generic ramp metering control variable is $r_i^{\mathrm{C}}(k) \in [r_i^{\min}, r_i^{\max}]$, representing the flow that should enter section i from the on-ramp during time interval $[kT, (k+1)T)$.

One of the earliest feedback ramp metering strategies is the *demand-capacity* strategy [29], which is an open-loop disturbance-rejection policy in which the local control action depends on the flow measured upstream the on-ramp and the occupancy measurement downstream the on-ramp. Specifically, the flow that should enter section i from the on-ramp during time interval $[kT, (k+1)T)$ is given by

$$r_i^{\mathrm{C}}(k) = \begin{cases} q_i^{\max} - q_i^{\mathrm{up}}(k-1) & \text{if } o_i^{\mathrm{down}}(k) \leq o_i^{\mathrm{cr}} \\ r_i^{\min} & \text{otherwise} \end{cases} \tag{8.10}$$

where the two required measurements are $q_i^{\mathrm{up}}(k)$, i.e. the flow measured upstream the on-ramp, and $o_i^{\mathrm{down}}(k)$, i.e. the occupancy measured downstream the on-ramp. Moreover, in (8.10), $q_i^{\max}$ is the mainstream capacity downstream the on-ramp, and o_i^{cr} is the critical occupancy (at which the flow reaches its maximum value). The basic philosophy of this strategy is, in case of under-critical traffic conditions, to allow to enter in the mainstream an on-ramp flow such that the downstream freeway capacity is reached; if instead the mainstream situation is congested, only a minimum on-ramp flow is allowed to enter.

Another simple and very common local ramp metering strategy is the so-called *percent-occupancy* strategy, which is one of the most widespread ramp metering schemes in the U.S., due to its simplicity of implementation and observed effectiveness [13]. The percent-occupancy strategy provides a proportional control of the occupancy measurement and depends on the occupancy measurement taken upstream of the on-ramp. Specifically, according to the percent-occupancy strategy, the on-ramp flow of section i for time interval $[kT, (k+1)T)$ is computed as

$$r_i^{\mathrm{C}}(k) = r^{\mathrm{H}} - \frac{r^{\mathrm{H}} - r^{\mathrm{L}}}{o^{\mathrm{H}} - o^{\mathrm{L}}} \left[o_i^{\mathrm{up}}(k) - o^{\mathrm{L}} \right] \tag{8.11}$$

where the only measurement is $o_i^{\mathrm{up}}(k)$, i.e. the the occupancy measured upstream the on-ramp, while r^{H} and r^{L} are parameters (corresponding to a high and a low threshold for the on-ramp flow) and, analogously, o^{H} and o^{L} are other parameters (corresponding to a high and a low threshold for the occupancy). According to (8.11), the on-ramp flow is a decreasing linear function of the mainline occupancy, with $r_i^{\mathrm{C}}(k) = r^{\mathrm{H}}$ when $o_i^{\mathrm{up}}(k) = o^{\mathrm{L}}$, and, vice versa, $r_i^{\mathrm{C}}(k) = r^{\mathrm{L}}$ when $o_i^{\mathrm{up}}(k) = o^{\mathrm{H}}$.

One of the most well-known local ramp metering strategies is *ALINEA* [30], that is a control law of I-type, in which the flow entering from the on-ramp is computed according to an error signal expressed in terms of difference between a set-point value and the occupancy measured downstream the on-ramp. In particular, the flow that should enter section i from the on-ramp during time interval $[kT, (k+1)T)$ is computed according to the following control law

$$r_i^{\mathrm{C}}(k) = r_i^{\mathrm{C}}(k-1) + K_R\left[o_i^* - o_i^{\mathrm{down}}(k)\right] \tag{8.12}$$

where $o_i^{\mathrm{down}}(k)$ is the occupancy measured downstream the on-ramp, o_i^* is a set-point value for the downstream occupancy, and K_R is the integral gain. Note that, in case the main objective of the traffic controller is to reduce congestion and to maximise the throughput, a good choice for the set-point is $o_i^* = o_i^{\mathrm{cr}}$. In some papers, the control law of ALINEA is expressed similarly to (8.12) but in terms of density instead of occupancy. The ALINEA controller is able to react to differences of $o_i^* - o_i^{\mathrm{down}}(k)$ in a less abrupt way compared with the demand-capacity strategy, as discussed in [12]. Note that the set-point could be time-varying and, in a hierarchical control scheme, it could be communicated to the controller by a supervisor. ALINEA has been applied in real cases for some decades, especially in Europe, to maximise the freeway throughput [28, 31].

A specific case of ramp metering installation is addressed in [32], where *dual-branch on-ramps* controlled with ALINEA are considered. In on-ramps of this type, it is very important to take account of balancing concepts, both in terms of queue lengths and in terms of waiting times experienced in the two branches. Different balancing policies are analysed in [32], referring to the real ramp metering system of the Monash Freeway, in Melbourne, Australia.

Different versions of ALINEA have been proposed in the literature. In [33], three versions are analysed, which are the flow-based strategy called *FL-ALINEA*, the version relying on the upstream occupancy called *UP-ALINEA*, and the one based on the upstream flow named *UF-ALINEA*. In [34], the adaptive *AD-ALINEA* strategy is proposed, being suitable for cases in which the critical occupancy cannot be a priori estimated. Indeed, AD-ALINEA includes an estimation algorithm based on the Kalman filter, which uses real-time measurements to estimate the critical occupancy that guarantees throughput maximisation according to the present traffic conditions. Again in [34], an upstream-measurement based version of the AD-ALINEA is investigated, called *AU-ALINEA*. AU-ALINEA may be useful in real cases in which no measurement devices are present downstream the on-ramp.

Another relevant extension of ALINEA is the so-called *PI-ALINEA*, in which a proportional term is added to the integrative term, resulting in a PI regulator. The PI-ALINEA control law assumes the following form:

$$r_i^{\mathrm{C}}(k) = r_i^{\mathrm{C}}(k-1) - K_P\left[o_i^{\mathrm{down}}(k) - o_i^{\mathrm{down}}(k-1)\right] + K_R\left[o_i^* - o_i^{\mathrm{down}}(k)\right] \tag{8.13}$$

where K_P is another regulator parameter.

A comparison between ALINEA and PI-ALINEA was carried out in [35], based on a theoretical stability analysis. In particular, [35] addresses the case of *distant downstream bottlenecks*, i.e. bottlenecks with smaller capacity than the merging area which are present further downstream the on-ramp. For that case, it is argued in the paper that it is advisable to use measurements from these downstream bottlenecks rather than from the merging area. According to the stability analysis of the closed-loop ramp metering system reported in [35], it can be stated that PI-ALINEA is able to guarantee a better control performance than ALINEA. A similar case of distant downstream bottlenecks is analysed in [36], where ALINEA and PI-ALINEA are compared via a simulation analysis, considering three different types of bottlenecks, i.e. an uphill, a lane drop and an uncontrolled on-ramp, showing again a better performance of PI-ALINEA compared with ALINEA.

Another interesting and realistic case is associated with many *bottlenecks with random location*, that can form downstream the metered on-ramp. This aspect has been addressed for instance in [37], considering incidents or lane changes in merge areas as possible causes of random-location bottlenecks and referring to a real case in Melbourne, Australia. In that work, the authors propose a generalisation of PI-ALINEA: a PI-ALINEA controller is defined for each possible bottleneck and a properly defined decision policy selects the controller corresponding to the most critical situation, in order to be actuated by the traffic light at the on-ramp.

A different methodological approach has been proposed in [38, 39] for ramp metering control. This approach is based on *iterative learning control*, which is a simple and robust feed-forward control method particularly suitable for addressing modelling uncertainties and non-linear dynamics (very common in the traffic case) and which exploits the repetitiveness of traffic phenomena to learn and improve the performance of the traffic controller. Indeed, traffic patterns are in general repeated similarly every day, and it is possible to find also similarities on monthly and yearly bases. According to the authors of [38, 39], this learning mechanism can allow the controller to improve its performance over time, differently from standard feedback regulators. On the other hand, compared with more sophisticated approaches as neural networks or fuzzy logic, iterative learning control has some advantages, as discussed in [39]. In particular, the control scheme proposed in [38] has the objective of driving the traffic density to converge to a desired density value, by combining the iterative learning control law with a generic feedback control law. The specific case in which the iterative learning control law is combined with ALINEA is addressed in [39].

Note that most of the ramp metering controllers should act in connection with *queue control* strategies, since the on-ramps are normally characterised by a limited space, that in some real cases can be quite large but, in others, can be very restrictive. In addition, it is important to point out that a long queue formed at an on-ramp may cause traffic problems to the adjacent streets, also affecting the possibly close urban traffic network. This means that, in case of maximum queue limits, the ramp metering controller should take into account the queue upper bound and regulate the on-ramp flow accordingly. This aspect can strongly limit the performance of ramp metering actions that has been proven, instead, to be very high in ideal conditions, i.e. without on-ramp storage space limitations. Queue control strategies can be implemented in different ways; for instance, the authors of [33] propose a proportional regulator which could work in conjunction with feedback ramp metering policies, such as ALINEA or PI-ALINEA.

8.3.2 Local Mainstream Control Strategies

Local mainstream control strategies regulate the mainstream flow of vehicles according to local measurements of the system state. Mainstream control can be actuated in different ways, and the most general concept of this type of control action has been proposed in [40, 41], where *mainstream traffic flow control* is defined as a general tool to regulate traffic in the mainstream, adopting different types of actuators, such as mainstream traffic lights, variable speed limits, or more advanced systems according to which suitable indications are provided directly to drivers on board of vehicles.

The objective of mainstream control strategies can be of different types. Safety increase has been one of the first goals of mainstream control, but, over the years, the most common objective has become to homogenise traffic conditions along the mainstream and to avoid the activation of bottlenecks, in order to mitigate all the negative effects of such phenomena, such as capacity drop, off-ramp blockage, and stop-and-go waves. Mainstream control can be realised in different ways, by imposing or suggesting variable speed limits to the drivers (normally by means of VMSs), by applying mainline metering actions or with lane control policies.

One of the first feedback local strategies for variable *speed limits* can be found in [42], where an ALINEA-like mainstream regulator is proposed, along with a switched activation-deactivation mechanism for the controller.

An interesting dynamic speed limit control algorithm, named *SPECIALIST*, was proposed in [43]. It is a feed-forward control scheme based on the shock wave theory aiming at maximising the discharge rate of vehicles. The main goal of this algorithm is to eliminate moving shock waves, i.e. short moving jams that propagate upstream causing an increase of travel times and unsafe situations for drivers, as well as a rise in noise and pollution for the environment.

Starting from the SPECIALIST algorithm, a more sophisticated variable speed limit control scheme was presented in [44], with the twofold objective of maximising the discharge rate at the bottleneck and to reduce speed variations upstream. To reach this result, a variable speed limit control is applied upstream the bottleneck in order to dissipate the possible forming queue. In addition, another variable speed limit control is applied further upstream to solve the queue generated by the first variable speed limit and to better regulate the inflow to the bottleneck.

A local mainstream traffic flow feedback controller, enabled via variable speed limits, is discussed in [45]. It is a controller of I-type, designed in order to prevent the congestion formation at an active bottleneck, hence eliminating the negative effects of capacity drop and blocking of off-ramps. Let us refer to the METANET model for a freeway network described in Sect. 4.2.3, where the control variable is given by

$b_m(k) \in [b^{\min}, 1]$, representing the variable speed limit rate to be displayed in each section of link m during time interval $[kT, (k+1)T)$. In [45], the variable speed limit rate $b_m(k)$ is computed according to the following control law

$$b_m(k) = b_m(k-1) + K_I \left[o_m^* - o_m^{\text{down}}(k) \right] \tag{8.14}$$

where K_I is the integral gain, $o_m^{\text{down}}(k)$ is the occupancy measured at the bottleneck downstream, and o_m^* is the occupancy set-point value, typically set equal to the critical occupancy, to maximise the flow. In [45], this variable speed limit controller is tested via micro-simulation for the case of an on-ramp merge bottleneck, showing very high performance results.

Controllers of I-type and PI-type are investigated also in [46] for application of variable speed limits. In particular, in [46], continuous-time traffic flow models are adopted and the effectiveness of the controllers is shown both through analytical results and with numerical evaluations.

A more sophisticated control scheme is discussed in [47], where a *cascade control* framework is devised, with two nested control loops. According to the cascade control scheme proposed in [47], the variable speed limit rate $b_m(k)$ is computed with the *secondary loop* controller of I-type, as

$$b_m(k) = b_m(k-1) + K_I \left[q_m^{\text{C},*}(k) - q_m^{\text{C}}(k) \right] \tag{8.15}$$

where $q_m^{\text{C}}(k)$ is the controlled mainstream flow measured downstream the VMS location, and $q_m^{\text{C},*}(k)$ is the reference value. Such value is computed according to the control law of PI-type of the *primary loop* given by

$$q_m^{\text{C},*}(k) = q_m^{\text{C},*}(k-1) + K_P' \left[\rho_m^{\text{down}}(k-1) - \rho_m^{\text{down}}(k) \right] + K_I' \left[\rho_m^* - \rho_m^{\text{down}}(k) \right] \tag{8.16}$$

where K_P' and K_I' are gains, while $\rho_m^{\text{down}}(k)$ is the measured downstream density and ρ_m^* is the set-point value for this density.

In [47], an interesting analysis of practical application aspects regarding the implementation of variable speed limits is reported. Indeed, in practice, there are several constraints and practical limitations for speed limits to be displayed on VMSs, that controllers computing variable speed limits must take into account. First of all, variable speed limits can assume discrete values belonging to a pre-defined set. Also, the speed limits to be shown on a given VMS cannot change too fast in time and, analogously, there is a limited space variation of speed limits displayed on consecutive VMSs. The interested reader can find more details on these practical issues in [47].

The cascade controller described in [47] was extended in [48] to account for the case of *multiple bottleneck locations*. In particular, in [48], the variable speed limit control is used to deal with multiple bottlenecks in different downstream locations. At each of these locations, a suitable sensor is placed to measure the density and a controller of PI-type similar to (8.16) is applied for each location. A decision logic is used to identify the most critical downstream bottleneck, in order to compute the reference value $q_m^{C,*}(k)$ for the secondary loop controller given by (8.15).

An integral regulator is used in [49] for *mainline metering*, with reference to the case of a generic merge area, i.e. a freeway infrastructure characterised by a high number of lanes merging into a lower number of lanes. This is the case for instance of the merging of two highways, toll plazas, or working zones which reduce the number of available lanes. In case the incoming flow exceeds the capacity, the capacity-drop phenomenon occurs, with consequent delays for the drivers. As outlined in [49], there are different real configurations of merging areas, for instance lane changing may be allowed or not, the merging can be asymmetric or symmetric, i.e. some lanes may have priorities over others or not, and so on. Each of these cases should be addressed in a specific way, with proper policies. The control law adopted in [49] is analogous to ALINEA, but the control variable is the flow entering the merge area. Once this flow is computed, it is necessary to define how to distribute such a flow among the controllable lanes. Depending on the type of merging, different distribution policies are suggested in [49].

A similar case of mainline metering for merging zones is analysed in [50], specifically addressing the case of work zones. A regulator of PI-type is applied and particular attention is paid to the location of the traffic lights. In [50], it is shown that choosing the right location for the traffic lights, at a given distance from the merge area, has a relevant impact on the performance of the traffic controller.

8.3.3 Local Route Guidance Strategies

Route guidance systems aim at routing vehicles along alternative paths either by providing the drivers with specific information about these paths (e.g. expected travel time, presence of work zones, accidents, and so on) or by directly suggesting the path to follow. Since in this chapter we are considering freeway traffic control schemes, the latter option is taken into consideration, i.e. the case in which a traffic controller computes the *splitting rates* at a given junction, corresponding to the portions of vehicles which should choose each alternative path. Then, these control inputs are transferred to the system with proper actuators, that are normally VMSs before the junction displaying appropriate messages to the drivers. Alternatively, it is possible to adopt proper interfaces on board of vehicles in which specific indications are provided to the drivers. In the following, we will refer to the case of VMSs, being at present the most conventional solution, but most of the reported results could also be generalised to different actuator devices.

For route guidance, local control strategies are those in which the splitting rates computed by the traffic controller associated with a given road junction and displayed on a VMS placed before that junction are based on local information and are not correlated with other splitting rates referred to other junctions. It is worth noting that, in the context of route guidance, the concept of *local state* should be distinguished from the cases just analysed and related to ramp management and mainstream control. If for these latter control actions the local state is generally a measurement of occupancy, flow or density close to the actuator, for route guidance the corresponding 'local measurement' is mostly referred to *travel times* along the alternative paths to reach a common destination. This is, in fact, the most relevant information on the basis of which the control action is derived.

It is important to make a relevant distinction between

- instantaneous travel times;
- predictive travel times.

The easiest way to account for travel times along alternative paths is to measure *instantaneous travel times* (also called reactive travel times). The instantaneous travel time can be defined as the travel time of a virtual vehicle travelling along a given path facing the current traffic conditions. This variable can be measured in real time, assuming that the freeway stretches are equipped with sensors providing the mean speed. Of course, instantaneous travel times can be misleading or inappropriate for a traffic controller in case of traffic conditions changing fast in the considered freeway stretch, because the driver that will follow a given path will experience traffic conditions that are different from those that are present on those links when he leaves the junction.

For these reasons, in many cases, it can be more appropriate to consider *predictive travel times* (often called also experienced travel times). Experienced travel times can be known only after completion of the corresponding trip, hence it is necessary to predict them in real time in order to properly feed a route guidance traffic controller. This is normally done by running online a traffic model in order to predict future traffic conditions in the alternative paths.

Another very relevant distinction when discussing route guidance strategies is related to the main principles adopted by the traffic controller in terms of route choice. Two main alternative approaches are available, coming from the theory of traffic assignment for transport networks, that are the Wardrop's principles of user equilibrium and system optimum [51]. In particular,

- the *user equilibrium* corresponds to a purely selfish behaviour of drivers who want to minimise their travel times; in the resulting equilibrium condition, the alternative routes that are actually used are characterised by the same travel times, which are lower than those of the unused routes;
- the *system optimum* concept, instead, is related to a social behaviour of users that allows to minimise the total travel times, according to a system perspective.

These two principles correspond to two different points of view, respectively, the one of road users, aiming at minimising their travel times, and the one of authorities, trying to improve the global performance of the system [52]. These two different views, the individual and the system perspectives, should be accurately analysed and studied for real implementations, taking into account also the rationality and selfishness levels of the drivers [53].

Analogously to ramp management and mainstream control, researchers have exploited the possibility of applying *feedback control strategies* for route guidance as

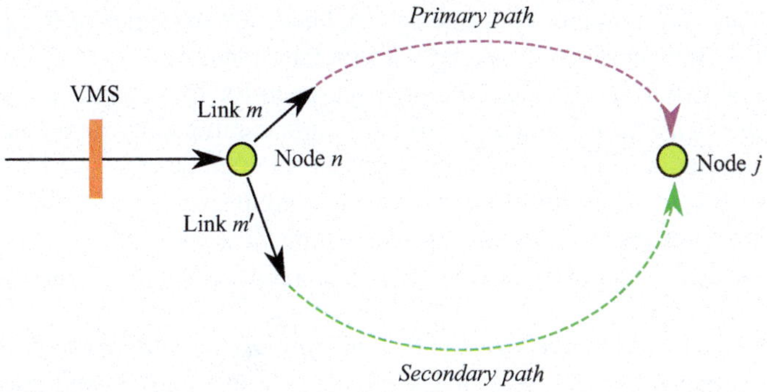

Fig. 8.9 Routing choices at a junction

well. In this case, the control law is computed taking into account the instantaneous travel times along the alternative paths, originated from the considered junction.

Let us refer to the METANET model for a freeway network described in Sect. 4.2.3, in which the control variable is the splitting rate at a given node. Considering the simple case of only two alternative paths originating from node n, let us denote with m and m' the two links exiting node n, corresponding respectively to the primary and secondary path (see Fig. 8.9). The primary path is the one characterised by the shortest travel time, in case of regular traffic conditions. In particular, the control variable is the splitting rate $\beta^{\mathrm{C}}_{m,n,j}(k) \in [0, 1]$, representing the portion of flow present in node n at time instant kT which should choose link m to reach destination j. The other control variable is $\beta^{\mathrm{C}}_{m',n,j}(k)$, referred to link m', but it is easily computed from $\beta^{\mathrm{C}}_{m,n,j}(k)$, since $\beta^{\mathrm{C}}_{m',n,j}(k) = 1 - \beta^{\mathrm{C}}_{m,n,j}(k)$.

Feedback regulators of P-type or PI-type have been proposed for route guidance systems [54, 55]. According to a *proportional control law*, the portion of flow present in node n at time instant kT which should choose link m to reach destination j is computed as

$$\beta^{\mathrm{C}}_{m,n,j}(k) = \beta^{\mathrm{N}}_{m,n,j}(k) + K_P \Delta\tau_{n,j}(k) \tag{8.17}$$

where $\beta^{\mathrm{N}}_{m,n,j}(k)$ is the nominal splitting rate, K_P is a gain, $\Delta\tau_{n,j}(k)$ is the instantaneous travel time difference between the secondary and primary direction from n to j. Note that $\beta^{\mathrm{C}}_{m,n,j}(k)$ is bounded and should be truncated in the interval [0, 1]. According to (8.17), the splitting rate for the primary path is decreased in case the instantaneous travel time difference becomes negative, i.e. in case the secondary path is characterised by a lower travel time. In this way, the traffic controller aims at equalising the travel times along the two alternative paths, in accordance with the user equilibrium principle.

In proportional-integral regulators, the splitting rate is instead computed as

$$\beta^{\mathrm{C}}_{m,n,j}(k) = \beta^{\mathrm{C}}_{m,n,j}(k-1) + K_P \left[\Delta\tau_{n,j}(k) - \Delta\tau_{n,j}(k-1)\right] + K_I \Delta\tau_{n,j}(k) \tag{8.18}$$

where K_P and K_I are other controller gains. Feedback strategies of P-type and PI-type for route guidance are compared via simulation in [56].

An alternative to the feedback control approach for route guidance is given by *iterative strategies*, in which the control action is computed by iteratively running different simulations in real time with different route guidance, in order to achieve conditions of either user equilibrium or system optimum (see e.g. [57, 58]). Although iterative strategies are very efficient in establishing these ideal conditions (more than feedback strategies that only approximate such conditions), they require a very high computational effort.

For this reason, the authors of [59] proposed a *predictive feedback* approach which incorporates the advantages of feedback and iterative strategies. In particular, in [59], the METANET model is run in real time, at each predicting time step, in order to forecast the travel times of vehicles which leave node n until they reach destination j. Note that the simulation model is initialised with the traffic state measured when the simulation is run and, for the whole prediction horizon, the splitting rates are assumed to be constant and equal to those implemented in the real system when the simulation starts. The predicted travel times computed by the simulation model for the alternative paths are used to calculate the time difference $\Delta\tau_{n,j}(k)$, which is applied as input for feedback regulators of I-type or PI-type, as those given by (8.17) and (8.18). A similar feedback approach, relying on predictive capabilities, was described in [60], referring to a real application in the Scottish freeway network.

8.3.4 Local Integrated Control Strategies

Local integrated control strategies are based on the principle that different control actions can be combined to obtain higher performance for the freeway system.

The integration of *ramp metering* with *variable speed limits* is discussed in [61], where the algorithm SPECIALIST proposed in [43] is extended to include also on-ramps controlled via ramp metering strategies, resulting in the so-called *SPECIALIST-RM*. According to the same logic of SPECIALIST, the proposed integrated control algorithm relies on shock wave theory and primarily aims at mitigating moving jams in freeways.

Another work dealing with ramp metering integrated with variable speed limits is proposed in [62]. As argued in [62], this integration is applied in order to overcome the limit of ramp metering strategies, which is related to the possible restrictions on queue lengths at the on-ramps. The basic idea developed in that paper is to exploit ramp metering as much as possible until the on-ramp is completely full of vehicles or the minimum metering lower bound is reached, and, when these limits are met, to activate variable speed limit control integrated with ramp metering. This integration is achieved by extending the *cascade control* law proposed in [47], through the application of a split-range-like scheme. A relevant practical issue is also addressed in [62], regarding the possible different control periods for the implementation of ramp metering and variable speed limits. If these periods are equal, the implementation is straightforward, whereas it is more elaborate in case of different periods. For this latter case, often occurring in real implementations, the authors propose an appropriate methodology and provide some insights.

The integration of ramp metering control and variable speed limits is addressed also in [63], where iterative learning control techniques are exploited. As already explained above for local ramp metering strategies, iterative learning control can be effective for freeway traffic, since it requires a low modelling knowledge and can be applied also in case of model uncertainties and disturbances to the system difficult to estimate, thanks to the capability of the traffic controller to learn from previous executions.

The integration of *variable speed limits* with *lane change control* is investigated in [64], with specific reference to a bottleneck or an accident causing a reduction of the number of lanes. The authors claim that, in a situation of that type, the only application of variable speed limits can provide very limited results, while the integration with lane change recommendations can highly improve the performance of the system, avoiding the capacity drop. The variable speed limit controller proposed in [64] is designed according to an analytical method based on the CTM, by exploiting feedback linearisation techniques. Stability properties are also proven in that paper under the assumption of speed limits varying continuously. Since instead in real cases the speed limits displayed on VMSs must comply with practical constraints of discretisation, space and time variation limitations, the authors of [64] also discuss these aspects and propose an integrated control strategy respecting these constraints.

8.4 Coordinated Control Strategies

In contrast with local control strategies described in Sect. 8.3, coordinated control strategies compute the control law taking into account *measurements* of the system state that are *not local*, but are related to a wider area (see Fig. 8.7 for the classification scheme adopted in this chapter). Also, the control actions applied on different actuators placed in different locations are not independent, as in local control strategies, but are in some way related and synchronised. Thanks to these coordination mechanisms and to information on the traffic state of an entire region of the network, a coordinated strategy is in general more effective than the combination of multiple independent local strategies. It goes without saying that designing coordinated control strategies requires more complex control schemes, in which a large number of information should be dealt with and more complicated control algorithms are needed.

Referring again to the freeway depicted in Fig. 8.2 with two controlled on-ramps, two installations of mainstream control and one junction with route guidance indications, a scheme of coordinated control strategies is reported in Fig. 8.10, to be compared with local control strategies shown in Fig. 8.8. According to the example reported in Fig. 8.10, the different control laws are computed by a centralised traffic controller on the basis of the knowledge of the whole system state $\underline{x}(k)$.

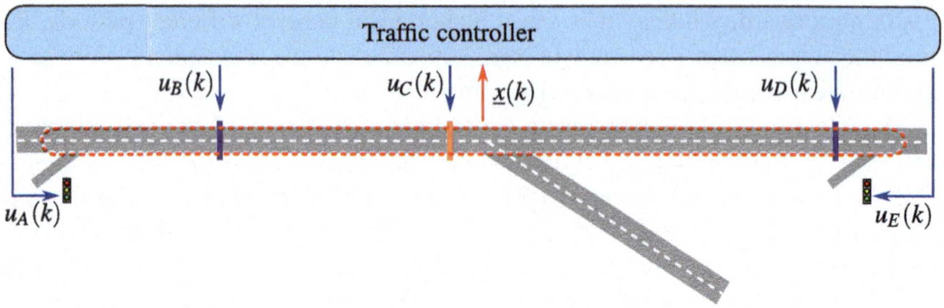

Fig. 8.10 Coordinated control strategies

The scheme shown in Fig. 8.10 represents the most general case of coordinated control strategies in which different control actions are combined, i.e. it is an example of *coordinated and integrated control strategies*. There are also many cases in which control strategies are not integrated but are coordinated. For instance, in coordinated ramp metering, the control actions implemented in different on-ramps are properly synchronised and based on traffic measurements coming from the whole region of the traffic network these on-ramps belong to. Analogously, it is possible to define coordinated mainstream control and coordinated route guidance strategies.

Traffic engineers have studied different approaches for designing coordinated control strategies, which are still under development in order to find the correct compromise between effectiveness and ease of implementation. Differently from local control strategies, which have been classified according to the type of control action (see Sect. 8.3), coordinated control strategies are categorised according to the control method adopted, since this represents the main differentiation feature. In particular, coordinated control strategies are classified, according to the coordination method, in:

- coordination of simple feedback control strategies;
- coordination via optimal control;
- coordination via Model Predictive Control.

8.4.1 Coordination of Simple Feedback Control Strategies

Different approaches to coordinate traffic control actions can be found in the literature. Most of them are based on optimisation or optimal control techniques (see Sects. 8.4.2 and 8.4.3), which can lead to efficient but very complicated control schemes. A less sophisticated but more practice-oriented way to coordinate different traffic control actions is to adopt simple regulators or heuristic rules, the main of which are summarised in this section.

Coordinated strategies have been developed especially for *ramp metering* controllers. One of the first attempts to design coordinated ramp metering strategies was the generalisation and extension of ALINEA in a multivariable regulator strategy called *METALINE* [65]. According to this feedback strategy, the controlled flow of one on-ramp depends not only on occupancy measurements immediately downstream the on-ramp but on more measurements referred to a larger area.

After these early approaches, most of the literature developed later for simple coordinated ramp metering control strategies included heuristic and rule-based algorithms. For instance, heuristic coordinated ramp metering algorithms, such as the *Zone* and the *Stratified Zone* algorithms, the *Bottleneck* algorithm, and the *Helper* algorithm, were developed and applied in some freeways in the U.S. [66, 67].

More recently, the rule-based coordinated ramp metering strategy *HERO* was developed and described in [68]. It is based on a control scheme adopting ALINEA at the local level and some coordination mechanisms for the control actions of different on-ramps, particularly useful in case of limited on-ramp space. In [68], it is shown that the effectiveness of local ALINEA controllers is low in case of limited queues, and it is exactly in this case that a coordination mechanism is useful to improve the performance of the freeway traffic system. In order to preserve easy applicability to real freeway networks, HERO is a simple reactive and rule-based strategy which decides in real time whether to activate or not the coordination mechanism among the on-ramps depending on real-time measurements.

Another research work dealing with coordinated ramp metering is [69], in which the main question addressed by the authors is when to start the control of on-ramps, having in mind that the final goals of coordinated ramp metering are to increase the throughputs and, at the same time, to mitigate traffic instabilities which characterise high flow traffic states, often improving the probability of crashes. To this aim, a production stability indicator is defined on the basis of a Macroscopic Fundamental Diagram (MFD) characterising the system at a network level and relying on a risk assessment technique, since traffic breakdown is seen as a risk. This approach is validated with empirical traffic data of the city of Shanghai, China.

A comparison between local and coordinated ramp metering strategies is reported in [70], with reference to field results based on traffic data of the A6W freeway in Paris, France. In particular, the considered local strategy is ALINEA, whereas the coordinated strategy is *CORDIN*, a new strategy proposed in that paper, which is based on a heuristic approach adopting ALINEA control laws that are properly corrected in order to achieve a better coordination. The comparative tests are devoted to evaluate the system performance, not only in terms of total travel times, but also in terms of travel time reliability. In that paper, it is shown that CORDIN performs better than ALINEA in terms of travel time reductions, but the two strategies are rather similar from a reliability point of view.

A quite general framework for coordinating different actuators is proposed in [71], referring in particular to *integrated traffic control*, combining ramp metering and variable speed limits. By exploiting some ideas developed by the same authors in previous papers (see e.g. [37, 48]), the feedback integrated control scheme proposed in [71] aims at controlling a freeway stretch in which multiple bottlenecks are present, in order to maximise the throughput. A set of controllers of PI-type is used, each associated with a measurement from a possible bottleneck (all the bottlenecks are downstream the actuators); a decision algorithm computes the smallest smoothed flow to be implemented, after distributing the flows among the available actuators in order to balance the delays upstream them.

8.4.2 *Coordination via Optimal Control*

The most advanced coordinated control strategies are based on the solution of an optimisation problem. In some cases, this problem is assumed to be solved off line, leading to optimal control approaches as the ones treated in this section, while in other cases it needs to be solved online, by applying MPC schemes, which are dealt with in Sect. 8.4.3.

Compared with the local control strategies or the simple heuristic coordination schemes described above, optimisation-based control strategies allow to obtain results that are in general more efficient. Indeed, the control actions are computed considering the overall freeway system and its dynamic evolution over a given time horizon, leading to optimal (or suboptimal) solutions to be implemented in the network.

Generally speaking, *optimal control* theory is concerned with the control of a dynamic system in the optimal way (a very detailed review of optimal control techniques can be found in the books [72–74]). The dynamic process is described by differential or difference equations, and the objective is to optimise an objective function (normally a cost function to be minimised) related to state and control variables. The problem constraints represent the dynamics of the system and bounds on the control variables.

Most of the research works on traffic control use discrete-time macroscopic models to describe the dynamic evolution of freeway systems. Two of the most widespread traffic flow models are the CTM of first-order type (see Sect. 3.3) and the more complex second-order model METANET (see Sect. 4.2), both represented in a discrete-time framework with k indicating the time step. These models are non-linear and can be in general written in the following form:

$$\underline{x}(k+1) = \underline{f}\left[\underline{x}(k), \underline{u}(k), \underline{d}(k)\right] \tag{8.19}$$

where $\underline{x}(k)$ is the vector gathering all the system state variables, $\underline{u}(k)$ is the vector of control inputs, and $\underline{d}(k)$ is the vector including exogenous inputs.

The *system state* is represented only by traffic densities when the CTM is used, and by both traffic densities and mean speeds in case METANET is instead applied. In both cases, conservation equations for the on-ramp queue lengths can be added to the model, so that the state vector may also include queue lengths.

The *control variables* used in the optimal control formulation depend on the choice of the control strategy to be applied to the system, as already discussed in the previous sections. In case a ramp metering control scheme is defined, the control variables are associated with the vehicles entering the freeway from the on-ramps, in terms of traffic flows or control rates. If instead variable speed limits are taken into account, the control variables are the values of speed suggested to the drivers via VMSs or equivalent control rates. In case route guidance is chosen, the control variables are normally the splitting rates, i.e. the ratios of vehicles arriving in a node and choosing a given path or direction. A combination of these variables is considered in case of integrated control strategies.

The *exogenous inputs* of the traffic model include all the external uncontrollable variables affecting the system. These exogenous inputs can be referred to external input signals (e.g. traffic demands or turning rates), as well as to modelling uncertainties and measurement noises.

In order to properly state a freeway traffic optimal control problem, a suitable *objective function* must be defined. This function is related to the final goal of the designed traffic controller, as described in Sect. 8.2. In particular,

- if the goal of the traffic controller is to track specific set-points (see Sect. 8.2.1), the cost function in the optimal control problem normally takes into account the *quadratic deviations* of the considered variables from their *reference values*, i.e. the quadratic errors (the choice of the quadratic form is not compulsory but it is the most common in optimal control problems); in case the goal is to avoid that given variables exceed threshold values, suitable *penalty terms* are added in the cost function;
- if the goal is associated with the improvement of the system performance, e.g. congestion reduction or emission reduction (see Sects. 8.2.2 and 8.2.3), the relative *performance indices*, i.e. TTS and TE, are minimised in the optimal control problem;
- if the goal is to balance system or control variables over space, *quadratic deviations* of these variables between consecutive road sections are minimised in the cost function and, analogously, quadratic deviations between consecutive time steps are minimised in case of time-balancing (see Sect. 8.2.4).

Note that the cost function can be also a combination of more terms, in order to take into account different goals. This is often dealt with by minimising a cost function given by a *weighted sum* of the different terms. Such terms can be conflicting or not; in the former case, the optimal control problem has a multi-objective nature and different Pareto-optimal solutions can be analysed in order to better understand which traffic control strategies are more suitable to be implemented in practice.

The general formulation of the optimal control problem over a finite horizon of K time steps is the following.

Problem 8.1 Given the system initial conditions $\underline{x}(0) = \underline{x}_0$ and the estimated sequence of exogenous inputs $\underline{d}(k)$, $k = 0, \dots, K-1$, find the optimal control sequence $\underline{u}(k)$, $k = 0, \dots, K-1$, that minimises

$$J = \vartheta[\underline{x}(K)] + \sum_{k=0}^{K-1} \varphi\left[\underline{x}(k), \underline{u}(k), \underline{d}(k)\right] \tag{8.20}$$

subject to the system dynamics expressed by (8.19), with $k = 0, \ldots, K-1$, and

$$\underline{u}^{\min} \leq \underline{u}(k) \leq \underline{u}^{\max} \qquad k = 0, \ldots, K-1 \tag{8.21}$$

□

In (8.20) $\vartheta[\cdot]$ represents the final cost, $\varphi[\cdot]$ is the stage cost, while vectors $\underline{u}^{\min}$, $\underline{u}^{\max}$ include, respectively, lower and upper bounds for the control variables. Note that the formulation of Problem 8.1 is suitable for discrete-time models. An analogous problem statement can be provided for continuous traffic models, but it is not reported here, since they are less common in freeway traffic control approaches (see e.g. [75, 76]).

It is worth noting that the structure of the problem to be solved depends both on the type of traffic model chosen for (8.19) and on the type of objective function. In most cases, the optimal control problem has a *non-linear* nature, since the most common traffic models are non-linear, e.g. CTM and METANET. In these cases, the numerical solution is often hard to find, because of the problem dimensions and complexity, and, also, there are generally no guarantees about the global optimality of the obtained solution. For this reason, some of the efforts made by researchers are still devoted to find efficient solution algorithms for non-linear optimal control problems for traffic networks. In other cases, the adopted traffic models are simplified or rewritten in suitable forms (see e.g. the Switching Mode Model (SMM) described in Sect. 3.3.6) or properly linearised, in order to obtain more tractable formulations.

The scientific literature on traffic control in freeway networks via the solution of an optimal control problem in the form of Problem 8.1 is very wide. This literature may be classified on the basis of the type of problem to be solved, on the adopted solution method or on how specific aspects of the problem (e.g. exogenous inputs) are treated.

In some works, first-order models are adopted to represent the dynamic evolution of traffic. For instance, in [77, 78], a discretised version of the LWR is considered for coordinated ramp metering and the resulting non-linear optimal control problem is solved with *gradient-based* techniques. The same first-order model is also applied in [79], in which the proposed non-linear feedback control law is obtained via neural networks.

Another type of first-order models is considered in [80], where a modified version of the CTM, i.e. the Asymmetric Cell Transmission Model (ACTM) (see Sect. 3.3.6), is used for a coordinated ramp metering case. The resulting non-linear optimal control problem is proven to be equivalent to a linear formulation, under specific conditions. Similarly, in [81], the adoption of another version of the CTM, i.e. the LN-CTM (see Sect. 3.3.6), leads to a non-linear problem to be solved. The authors of [81] show that, under some conditions, the solution of a linear problem is equivalent to the one of the original problems for a freeway in which variable speed limits and ramp metering are applied.

A very efficient numerical solution algorithm has been adopted in the optimal freeway traffic control tool *AMOC* [82, 83], based on the second-order METANET to represent the traffic system dynamics. AMOC is a general framework conceived to control traffic networks with different topology and with different types of control measures, such as ramp metering, mainstream control and route guidance [40, 84]. AMOC includes different solution methods, all gradient-based, which are compared in [83]. From this comparison, it results that the most promising choice is given by the *feasible direction algorithm* applying the derivative backpropagation method *RPROP* [85] (the original form of the same algorithm was proposed in [86]). This algorithm has been used also in recent works for more complicated optimal control problems, for instance for multi-class traffic regulation [87] or applications to urban networks [88].

Other solution methods to Problem 8.1 have been proposed in the literature. For instance, in [89], the non-linear optimal control problem is solved in a receding-horizon scheme in which a parameterised control law is found by using multi-layer feed-forward neural networks. In other works, the discrete adjoint method is used for solving the arising non-linear optimal control problem, e.g. in a coordinated ramp metering traffic case [90] and in a dynamic traffic assignment problem [91]. The adjoint method is also employed in [92], where a traffic control problem is solved for cybersecurity applications, i.e. in order to evaluate the potential for an adversary having access to control the freeway infrastructure.

An alternative to gradient-based algorithms for solving the non-linear optimal control problems arising in traffic systems is represented by *derivative-free* algorithms, that can be useful for complex cases, in which the gradient computation is very time consuming or even impossible (e.g. if the objective function is not differentiable). A comparative analysis between gradient-based and derivative-free algorithms is reported in [93] for a specific traffic control problem dealing with a coordinated ramp metering strategy to reduce congestion and emissions in a two-class flow environment.

While in these cited works the exogenous inputs are assumed to be known over the considered time horizon, there are other works in which such inputs are seen as *disturbances* or *uncertainties*, explicitly modelled and taken into account in the determination of the control law. For instance, stochastic disturbances, acting on the system and on the measurement channel, are considered and modelled as noises in [94]. Moreover, in [95], the traffic control problem is formulated as an H_∞ control problem which accounts for uncertainties of different natures associated with the macroscopic traffic model, while a robust control approach is proposed in [96], taking into account both model uncertainties and disturbances in the freeway network. In [97], the control law is determined by exploiting the Pontryagin maximum principle and the robustness of the controlled system with respect to uncertainties in the input parameters is analysed.

8.4.3 Coordination via Model Predictive Control

A further possibility to coordinate different control actions for traffic regulation is via *Model Predictive Control* (the interested reader can find theoretical and practical issues on MPC in the books [98–100]). Analogously to optimal control methods described in Sect. 8.4.2, MPC aims at controlling a dynamic system in an optimal way, by taking into account its dynamic evolution and by considering a suitable objective function to be optimised. The difference is that MPC is an *online* control scheme, i.e. it is applied to the system in real time by iteratively solving a *Finite-Horizon Optimal Control Problem* (FHOCP) that is updated on the basis of real system measurements.

Considering a generic MPC scheme in a discrete-time framework, a FHOCP is solved at each time step k, $k = 0, \ldots, K$. This optimisation problem is characterised by an objective function and some constraints. Among them, the system model equations are included, in the form of (8.19), thus allowing the prediction of the system evolution over a given *prediction horizon* of K_p time steps. It is worth noting that the prediction model is initialised with the system state $\underline{x}(k)$ measured at time step k.

In particular, the FHOCP to be solved at time step k can be stated as follows.

Problem 8.2 Given the initial conditions on the system state $\underline{x}(k)$ and the estimated sequence of exogenous inputs $\underline{d}(h)$, $h = k, \ldots, k + K_p - 1$, find the optimal control sequence $\underline{u}(h)$, $h = k, \ldots, k + K_p - 1$, that minimises

$$J(k) = \vartheta[\underline{x}(k + K_p)] + \sum_{h=k}^{k+K_p-1} \varphi\left[\underline{x}(h), \underline{u}(h), \underline{d}(h)\right] \tag{8.22}$$

subject to

$$\underline{x}(h + 1) = \underline{f}\left[\underline{x}(h), \underline{u}(h), \underline{d}(h)\right] \qquad h = k, \ldots, k + K_p - 1 \tag{8.23}$$

$$\underline{u}^{\min} \le \underline{u}(h) \le \underline{u}^{\max} \qquad h = k, \ldots, k + K_p - 1 \tag{8.24}$$

and further possible additional constraints on the state variables $\underline{x}(h)$, $h = k + 1, \ldots, k + K_p$ and control variables $\underline{u}(h)$, $h = k, \ldots, k + K_p - 1$. □

By solving the FHOCP at time step k, the optimal state trajectory $\underline{x}^{\circ}(h|k)$, $h = k + 1, \ldots, k + K_p$, and the optimal control sequence $\underline{u}^{\circ}(h|k)$, $h = k, \ldots, k + K_p - 1$, are found over the whole prediction horizon. Note that the notation $f^{\circ}(h|k)$ means that this is the optimal value of the generic variable f, referred to time step h and obtained by solving the FHOCP at time step k.

MPC relies on a *receding-horizon* framework. This means that, after solving the FHOCP at time step k, only the first element of the whole optimal control sequence is implemented in the system, i.e. $\underline{u}^{\circ}(k|k)$, and at the subsequent time step $k + 1$

the optimisation procedure is repeated again, updated with the system state values $\underline{x}(k+1)$, over a prediction horizon that is shifted one time step ahead.

The objective function (8.22) can be used to pursue different goals of the traffic controller, e.g. tracking set-points, improving the system performance, or balancing the system variables, as discussed in Sect. 8.4.2. Moreover, according to the adopted traffic model, the type of constraints and objective function, Problem 8.2 can have a different structure, e.g. linear, quadratic or non-linear, convex or nonconvex (more detailed considerations on these aspects can be found in [101]).

Problem 8.2 provides the basic statement of a generic FHOCP, but some modifications can often be found. In some cases, in order to reduce the number of variables, the prediction horizon is distinguished from the *control horizon* of K_c time steps (with $K_c < K_p$). The idea is that, after the control horizon has passed, the control actions are forced to be constant. The parameters K_c and K_p should be carefully chosen, to account for the trade-off between accuracy of the controller and computational complexity.

Another possible variation in the formulation of the FHOCP is given by the distinction between the simulation time step index k referred to the model sample time T and the controller time step index k_c referred to the *controller sample time* T_c (with T_c integer multiple of T), this latter representing the rate at which the control actions are updated. This difference is normally related to implementation issues, since the model sample time is chosen small enough to correctly capture the traffic dynamic evolution, while the controller sample time is dictated by the type of actuator.

As widely recognised, MPC has many *advantages*. First of all, it has prediction capabilities, so that the control action computed at time step k is optimal not only for the present situation but also for the future evolution of the system. Second, MPC allows to explicitly handle constraints on the system, and, finally, it has a closed-loop nature, since the control action is determined on the basis of real-time measurements. The main *drawbacks* of MPC are related with computational issues: since MPC is applied in real time, the FHOCP must be solved very fast. This represents a big challenge for real-case traffic networks, because the FHOCP often has a non-linear form and a large number of variables is involved in the optimisation. Some implementation-oriented strategies to deal with MPC schemes for real applications are discussed in Chap. 9.

The first works regarding MPC frameworks for freeway traffic control are [102–104], respectively for variable speed limits, ramp metering, and their coordinated action. In those works, the adopted prediction model is METANET and the resulting optimisation problem is, then, non-linear. A similar framework is considered in [105], where a traffic simulator is used to represent the real freeway, causing a model mismatch between the simulation and the prediction model.

Different research works have dealt with MPC for freeway traffic in the last decade, also considering specific and peculiar cases. For instance, a *mixed traffic network* with two urban regions and a freeway is addressed in [106], where the MFD and the ACTM are used, respectively, to model urban and freeway traffic. The control measures are represented by perimeter controllers and ramp metering. The case of *discrete variable speed limits* is addressed in [107], resulting in discrete control variables and, consequently, in a non-linear mixed-integer problem to solve. Since it is very computationally demanding to solve this type of problem in real time, some methods are proposed to find a reasonable solution in acceptable computational times. Freeway control via *reversible lanes* is studied in [108], where METANET is modified to take into account this type of control action. A search-tree method is used to solve the resulting mixed-integer non-linear problem.

A more advanced control scheme including MPC is reported in [109] for coordinated ramp metering. This is a *hierarchical control* scheme composed of three layers: the upper estimation/prediction level provides estimates and predictions of the future system disturbances; the intermediate layer includes the optimal control tool AMOC [83] used in a receding-horizon framework; the lower level adopts local feedback ALINEA-based controllers, in which the optimal state trajectory found at the upper level is used to fix the set-point values.

A two-level hybrid control scheme, represented with the formalism of discrete-time discrete-event automata, is described in [110] for regulating traffic conditions in freeway systems. In that paper, both ramp metering and variable speed limits can be applied but the MPC regulator to be used at the lower level is chosen by the higher level of the control scheme, on the basis of the present operating conditions. A similar concept applied to a multi-class traffic case is developed in [111].

In order to deal with more tractable problem formulations, some model simplifications or relaxations have been studied in some papers. For instance, heuristic restrictions and relaxations are applied to solve the nonconvex optimisation problem in [112], while a simple first-order model but extended to account for the capacity drop is adopted in [113], enabling a good prediction accuracy and, at the same time, a fast computation of the optimal solution.

In other research works on MPC for freeway traffic, the exogenous inputs are explicitly modelled as *disturbances* or *uncertainties*. In [114], the exogenous inputs affecting the traffic demands are explicitly considered as additive and bounded quantities and the input-to-state practical stability of the system controlled via MPC is proven. In a framework similar to the one considered in [114], a new concept of 'natural robustness' is introduced for a traffic system subject to these types of disturbances and controlled with MPC [115, 116]. A robust control approach is proposed in [117] to handle uncertainties in freeway traffic networks via a min–max scheme. In particular, to reduce the computational complexity, a scenario-based receding horizon parametrised approach is adopted.

References

1. Chien C-C, Zhang Y, Ioannou PA (1997) Traffic density control for automated highway systems. Automatica 33:1273–1285
2. Alvarez L, Horowitz R, Li P (1999) Traffic flow control in automated highway systems. Control Eng Pract 7:1071–1078
3. Baskar LD, De Schutter B, Hellendoorn H (2013) Optimal routing for automated highway systems. Transp Res Part C 30:1–22

4. Roncoli C, Papageorgiou M, Papamichail I (2015) Traffic flow optimisation in presence of vehicle automation and communication systems - part II: optimal control for multi-lane motorways. Transp Res Part C 57:260–275
5. Roncoli C, Papamichail I, Papageorgiou M (2016) Hierarchical model predictive control for multi-lane motorways in presence of vehicle automation and communication systems. Transp Res Part C 62:117–132
6. Spiliopoulou A, Perraki G, Papageorgiou M, Roncoli C (2017) Exploitation of ACC systems towards improved traffic flow efficiency on motorways. In: Proceedings of the 5th IEEE international conference on models and technologies for intelligent transportation systems, pp 37–43
7. Rios-Torres J, Malikopoulos AA (2017) Automated and cooperative vehicle merging at highway on-ramps. IEEE Trans Intell Transp Syst 18:780–789
8. Rios-Torres J, Malikopoulos AA (2017) A survey on the coordination of connected and automated vehicles at intersections and merging at highway on-ramps. IEEE Trans Intell Transp Syst 18:1066–1077
9. Wattleworth JA (1965) Peak-period analysis and control of a freeway system. Highway Res Rec 157 (1965)
10. Yuan LS, Kreer JB (1971) Adjustment of freeway ramp metering rates to balance entrance ramp queues. Transp Res 5:127–133
11. Kim K, Cassidy MJ (2012) A capacity-increasing mechanism in freeway traffic. Transp Res Part B 46:1260–1272
12. Papageorgiou M, Kotsialos A (2002) Freeway ramp metering: an overview. IEEE Trans Intell Transp Syst 3:271–281
13. Horowitz R, May A, Skabardonis A, Varaiya P, Zhang M, Gomes G, Muñoz L, Sun X, Sun D (2005) Design, field implementation and evaluation of adaptive ramp metering algorithms. California PATH Research Report UCB-ITS-PRR-2005-2. University of California, Berkeley
14. Papageorgiou M, Kosmatopoulos E, Papamichail I (2008) Effects of variable speed limits on motorway traffic flow. Transp Res Rec 2047
15. Soriguera F, Martínez I, Sala M, Menéndez M (2017) Effects of low speed limits on freeway traffic flow. Transp Res Part C 77:257–274
16. Knoop VL, Duret A, Buisson C, van Arem B (2010) Lane distribution of traffic near merging zones influence of variable speed limits. In: Proceedings of the 13th international IEEE annual conference on intelligent transportation systems, pp 485–490
17. Gazis DC, Foote RS (1969) Surveillance and control of tunnel traffic by an on-line digital computer. Transp Sci 3:255–275
18. Mc Calden MS (1984) A traffic management system for the San Francisco-Oakland Bay Bridge. ITE J 54:46–51
19. Soole DW, Watson BC, Fleiter JJ (2013) Effects of average speed enforcement on speed compliance and crashes: a review of the literature. Accid Anal Prev 54:46–56
20. Vanlommel M, Houbraken M, Audenaert P, Logghe S, Pickavet M, De Maeyer P (2015) An evaluation of section control based on floating car data. Transp Res Part C 58:617–627
21. Ben-Akiva M, Bottom J, Ramming MS (2001) Route guidance and information systems. Proc Inst Mech Eng Part I J Syst Control Eng 215:317–324
22. Schmitt EJ, Jula H (2006) Vehicle route guidance systems: classification and comparison. In: Proceedings of the IEEE intelligent transportation systems conference, pp 242–247
23. Kontorinaki M, Karafyllis I, Papageorgiou M (2017) Global exponential stabilisation of acyclic traffic networks. Int J Control (2017). Published on line, https://doi.org/10.1080/00207179.2017.1362114
24. Papageorgiou M, Papamichail I (2008) Overview of traffic signal operation policies for ramp metering. Transp Res Rec 2047:28–36
25. Isaksen L, Payne HJ (1973) Freeway traffic surveillance and control. Proc IEEE 61:526–536
26. Papageorgiou M, Diakaki C, Dinopoulou V, Kotsialos A, Wang Y (2003) Review of road traffic control strategies. Proc IEEE 91:2043–2067

27. Papamichail I, Papageorgiou M, Wang Y (2007) Motorway traffic surveillance and control. Eur J Control 13:297–319
28. Haj-Salem H, Papageorgiou M (1995) Ramp metering impact on urban corridor traffic: field results. Transp Res Part A 29:303–319
29. Masher DP, Ross DW, Wong PJ, Tuan PL, Zeidler HM, Petracek S (1975) Guidelines for design and operation of ramp control systems. Standford Research Institute Report NCHRP 3-22, SRI Project 3340, California
30. Papageorgiou M, Hadj-Salem H, Blosseville J-M (1991) ALINEA: a local feedback control law for on-ramp metering. Transp Res Rec 1320:58–64
31. Papageorgiou M, Kosmatopoulos E, Papamichail I, Wang Y (2007) ALINEA maximises motorway throughput - an answer to flawed criticism. TEC Mag 271–276
32. Papamichail I, Papageorgiou M (2011) Balancing of queues or waiting times on metered dual-branch on-ramps. IEEE Trans Intell Transp Syst 12:438–452
33. Smaragdis E, Papageorgiou M (1856) Series of new local ramp metering strategies. Transp Res Rec 2003:74–86
34. Smaragdis E, Papageorgiou M, Kosmatopoulos E (2004) A flow-maximizing adaptive local ramp metering strategy. Transp Res Part B 38:251–270
35. Wang Y, Kosmatopoulos EB, Papageorgiou M, Papamichail I (2014) Local ramp metering in the presence of a distant downstream bottleneck: theoretical analysis and simulation study. IEEE Trans Intell Transp Syst 15:2024–2039
36. Kan Y, Wang Y, Papageorgiou M, Papamichail I (2016) Local ramp metering with distant downstream bottlenecks: a comparative study. Transp Res Part C 62:149–170
37. Wang Y, Papageorgiou M, Gaffney J, Papamichail I, Guo J (2010) Local ramp metering in the presence of random-location bottlenecks downstream of a metered on-ramp. In: Proceedings of the 13th international IEEE conference on intelligent transportation systems, pp 1462–1467
38. Hou Z, Xu J-X, Yan J (2008) An iterative learning approach for density control of freeway traffic flow via ramp metering. Transp Res Part C 16:71–97
39. Hou Z, Xu X, Yan J, Xu J-X, Xioung G (2011) A complementary modularized ramp metering approach based on iterative learning control and ALINEA. IEEE Trans Intell Transp Syst 12:1305–1318
40. Carlson RC, Papamichail I, Papageorgiou M, Messmer A (2010) Optimal mainstream traffic flow control of large-scale motorway networks. Transp Res Part C 18:193–212
41. Carlson RC, Papamichail I, Papageorgiou M, Messmer A (2010) Optimal motorway traffic flow control involving variable speed limits and ramp metering. Transp Sci 44:238–253
42. Zhang J, Chang H, Ioannou PA (2006) A simple roadway control system for freeway traffic. In: Proceedings of the American control conference, pp 4900–4905
43. Hegyi A, Hoogendoorn SP, Schreuder M, Stoelhorst H, Viti F (2008) SPECIALIST: a dynamic speed limit control algorithm based on shock wave theory. In: Proceedings of the 11th international IEEE conference on intelligent transportation systems, pp 827–832
44. Chen D, Ahn S, Hegyi A (2014) Variable speed limit control for steady and oscillatory queues at fixed freeway bottlenecks. Transp Res Part B 70:340–358
45. Müller ER, Carlson RC, Kraus W, Papageorgiou M (2015) Microsimulation analysis of practical aspects of traffic control with variable speed limits. IEEE Trans Intell Transp Syst 16:512–523
46. Jin H-Y, Jin W-L (2015) Control of a lane-drop bottleneck through variable speed limits. Transp Res Part C 58:568–584
47. Carlson RC, Papamichail I, Papageorgiou M (2011) Local feedback-based mainstream traffic flow control on motorways using variable speed limits. IEEE Trans Intell Transp Syst 12:1261–1276
48. Iordanidou G-R, Roncoli C, Papamichail I, Papageorgiou M (2015) Feedback-based mainstream traffic flow control for multiple bottlenecks on motorways. IEEE Trans Intell Transp Syst 16:610–621
49. Papageorgiou M, Papamichail I, Spiliopoulou AD, Lentzakis AF (2008) Real-time merging traffic control with applications to toll plaza and work zone management. Transp Res Part C 16:535–553

50. Tympakianaki A, Spiliopoulou A, Kouvelas A, Papamichail I, Papageorgiou M, Wang Y (2014) Real-time merging traffic control for throughput maximization at motorway work zones. Transp Res Part C 44:242–252
51. Wardrop JG (1952) Some theoretical aspects of road traffic research. Proc Inst Civ Eng 1:325–378
52. Vreeswijk JD, Landman RL, van Berkum EC, Hegyi A, Hoogendoorn SP, van Arem B (2015) Improving the road network performance with dynamic route guidance by considering the indifference band of road users. IET Intell Transp Syst 9:897–906
53. van Essen M, Thomas T, van Berkum E, Chorus C (2016) From user equilibrium to system optimum: a literature review on the role of travel information, bounded rationality and non-selfish behaviour at the network and individual levels. Transp Rev 36:527–548
54. Messmer A, Papageorgiou M (1994) Automatic control methods applied to freeway network traffic. Automatica 30:691–702
55. Pavlis Y, Papageorgiou M (1999) Simple decentralized feedback strategies for route guidance in traffic networks. Transp Sci 33:264–278
56. Wang Y, Papageorgiou M, Messmer A (2001) Feedback and iterative routing strategies for freeway networks. In: Proceedings of the IEEE international conference on control applications, pp 1162–1167
57. Mahmassani HS, Peeta S (1993) Network performance under system optimal and user equilibrium dynamic assignments: implications for advanced traveller information systems. Transp Res Rec 1408:83–93
58. Wang Y, Messmer A, Papageorgiou M (2001) Freeway network simulation and dynamic traffic assignment using METANET tools. Transp Res Rec 1776:178–188
59. Wang Y, Papageorgiou M, Messmer A (2002) A predictive feedback routing control strategy for freeway network traffic. In: Proceedings of the American control conference, pp 3606–3611
60. Messmer A, Papageorgiou M, Mackenzie N (1998) Automatic control of variable message signs in the interurban scottish highway network. Transp Res Part C 6:173–187
61. Schelling I, Hegyi A, Hoogendoorn SP (2011) SPECIALIST-RM: integrated variable speed limit control and ramp metering based on shock wave theory. In: Proceedings of the 14th international IEEE conference on intelligent transportation systems, pp 2154–2159
62. Carlson RC, Papamichaill I, Papageorgiou M (2014) Integrated feedback ramp metering and mainstream traffic flow control on motorways using variable speed limits. Transp Res Part C 46:209–221
63. Hou Z, Xu J-X, Zhong H (2007) Freeway traffic control using iterative learning control-based ramp metering and speed signaling. IEEE Trans Veh Technol 56:466–477
64. Zhang Y, Ioannou PA (2017) Combined variable speed limit and lane change control for highway traffic. IEEE Trans Intell Transp Syst 18:1812–1823
65. Papageorgiou M, Blosseville J-M, Hadj-Salem H (1990) Modeling and real-time control of traffic flow on the southern part of Boulevard Périphérique in Paris - Part II: coordinated on-ramp metering. Transp Res Part A 24:361–370
66. Bogenberger K, May AD (1999) Advanced coordinated traffic responsive ramp metering strategies. California PATH Working Paper, University of California, Berkeley
67. Hadi MA (2005) Coordinated traffic responsive ramp metering strategies - an assessment based on previous studies. In: Proceedings of the world congress on intelligent transport systems
68. Papamichail I, Papageorgiou M (2008) Traffic-responsive linked ramp-metering control. IEEE Trans Intell Transp Syst 9:111–121
69. Tu H, Li H, Wang Y, Sun L (2014) When to control the ramps on freeway corridors? A novel stability-and-MFD-based approach. IEEE Trans Intell Transp Syst 15:2572–2582
70. Bhouri N, Haj-Salem H, Kauppila J (2013) Isolated versus coordinated ramp metering: field evaluation results of travel time reliability and traffic impact. Transp Res Part C 28:155–167
71. Iordanidou G-R, Papamichail I, Roncoli C, Papageorgiou M (2017) Feedback-based integrated motorway traffic flow control with delay balancing. IEEE Trans Intell Transp Syst 18:2319–2329

72. Bryson AE, Ho Y-C (1975) Applied optimal control: optimization, estimation and control. Taylor & Francis Group, New York
73. Athans M, Falb PL (2007) Optimal control: an introduction to the theory and its applications. Dover Publications, Mineola, New York
74. Lewis FL, Vrabie DL, Syrmos VL (2012) Optimal control, 3rd edn. Wiley, Hoboken, New Jersey
75. Bayen AM, Raffard RL, Tomlin CJ (2004) Network congestion alleviation using adjoint hybrid control: application to highways. In: Alur R, Pappas GJ (eds) Hybrid systems: computation and control. Springer, Berlin, pp 95–110
76. Li Y, Canepa E, Claudel C (2014) Optimal control of scalar conservation laws using linear/quadratic programming: application to transportation networks. IEEE Trans Control Netw Syst 1:28–39
77. Zhang H, Ritchie S, Recker W (1996) Some general results on the optimal ramp metering control problem. Transp Res Part C 4:51–69
78. Zhang HM, Recker WW (1999) On optimal freeway ramp control policies for congested traffic corridors. Transp Res Part B 33:417–436
79. Zhang HM, Ritchie SG, Jayakrishnan R (2001) Coordinated traffic-responsive ramp control via nonlinear state feedback. Transp Res Part C 9:337–352
80. Gomes G, Horowitz R (2006) Optimal freeway ramp metering using the asymmetric cell transmission model. Transp Res Part C 14:244–262
81. Muralidharan A, Horowitz R (2012) Optimal control of freeway networks based on the link node cell transmission model. In: Proceedings of the American control conference, pp 5769–5774
82. Kotsialos A, Papageorgiou M, Middelham F (2001) Optimal coordinated ramp metering with AMOC. Transp Res Rec 1748:55–65
83. Papageorgiou M, Kotsialos A (2004) Nonlinear optimal control applied to coordinated ramp metering. IEEE Trans Intell Transp Syst 12:920–933
84. Kotsialos A, Papageorgiou M, Mangeas M, Haj-Salem H (2002) Coordinated and integrated control of motorway networks via non-linear optimal control. Transp Res Part C 10:65–84
85. Papageorgiou M, Marinaki M, Typaldos P, Makantasis K (2016) A feasible direction algorithm for the numerical solution of optimal control problems - extended version, internal report no 2016–26. Chania, Greece
86. Riedmiller M, Braun H (1993) A directive adaptive method for faster backpropagation learning: the RPROP algorithm. In: Proceedings of the IEEE international conference, neural networks pp 586–591
87. Pasquale C, Papamichail I, Roncoli C, Sacone S, Siri S, Papageorgiou M (2015) Two-class freeway traffic regulation to reduce congestion and emissions via nonlinear optimal control. Transp Res Part C 55:85–99
88. Jamshidnejad A, Papamichail I, Papageorgiou M, De Schutter B (2017) Sustainable model-predictive control in urban traffic networks: efficient solution based on general smoothening methods. IEEE Trans Control Syst Technol, published on line, https://doi.org/10.1109/TCST.2017.2699160
89. Di Febbraro A, Parisini T, Sacone S, Zoppoli R (2001) Neural approximations for feedback optimal control of freeway systems. IEEE Trans Veh Technol 50:302–313
90. Reilly J, Samaranayake S, Delle Monache ML, Krichene W, Goatin P, Bayen AM (2015) Adjoint-based optimization on a network of discretized scalar conservation laws with applications to coordinated ramp metering. J Optim Theory Appl 167:733–760
91. Samaranayake S, Reilly J, Krichene W, Lespiau JB, Delle Monache ML, Goatin P, Bayen A (2015) Discrete-time system optimal dynamic traffic assignment (SO-DTA) with partial control for horizontal queuing networks. In: Proceedings of the American control conference, pp 663–670
92. Reilly J, Martin S, Payer M, Bayen AM (2016) Creating complex congestion patterns via multi-objective optimal freeway traffic control with application to cyber-security. Transp Res Part B 91:366–382

93. Pasquale C, Anghinolfi D, Sacone S, Siri S, Papageorgiou M (2016) A comparative analysis of solution algorithms for nonlinear freeway traffic control problems. In: Proceedings of the 19th IEEE intelligent transportation systems conference, pp 1773–1778
94. Alessandri A, Di Febbraro A, Ferrara A, Punta E (1998) Optimal control of freeways via speed signalling and ramp metering. Control Eng Pract 6:771–780
95. Chiang Y-H, Juang J-C (2008) Control of freeway traffic flow in unstable phase by H_∞ theory. IEEE Trans Intell Transp Syst 9:193–208
96. Zhong RX, Sumalee A, Pan TL, Lam WHK (2014) Optimal and robust strategies for freeway traffic management under demand and supply uncertainties: an overview and general theory. Transportmetrica A: Transp Sci 10:849–877
97. Como G, Lovisari E, Savla K (2016) Convexity and robustness of dynamic traffic assignment and freeway network control. Transp Res Part B 91:446–465
98. Maciejowski J (2002) Predictive control with constraints. Prentice Hall, Harlow, UK
99. Camacho EF, Bordons C (2007) Model predictive control. Springer, London
100. Rawlings JB, Mayne DQ (2009) Model predictive control: theory and design. Nob Hill Publishing, Madison, Wisconsin
101. Burger M, van den Berg M, Hegyi A, De Schutter B, Hellendoorn J (2013) Considerations for model-based traffic control. Transp Res Part C 35:1–19
102. Hegyi A, De Schutter B, Hellendoorn J (2005) Optimal coordination of variable speed limits to suppress shock waves. IEEE Trans Intell Transp Syst 6:102–112
103. Bellemans T, De Schutter B, De Moor B (2006) Model predictive control for ramp metering of motorway traffic: a case study. Control Eng Pract 14:757–767
104. Hegyi A, De Schutter B, Hellendoorn H (2005) Model predictive control for optimal coordination of ramp metering and variable speed limits. Transp Res Part C 13:185–209
105. Hegyi A, Burger M, De Schutter B, Hellendoorn J, van den Boom TJJ (2007) Towards a practical application of model predictive control to suppress shock waves on freeways. In: Proceedings of the European control conference, pp 1764–1771
106. Haddad J, Ramezani M, Geroliminis N (2013) Cooperative traffic control of a mixed network with two urban regions and a freeway. Transp Res Part B 54:17–36
107. Frejo JRD, Núñez A, De Schutter B, Camacho EF (2014) Hybrid model predictive control for freeway traffic using discrete speed limit signals. Transp Res Part C 46:309–325
108. Frejo JRD, Papamichail I, Papageorgiou M, Camacho EF (2016) Macroscopic modeling and control of reversible lanes on freeways. IEEE Trans Intell Transp Syst 17:948–959
109. Papamichail I, Kotsialos A, Margonis I, Papageorgiou M (2010) Coordinated ramp metering for freeway networks - a model-predictive hierarchical control approach. Transp Res Part C 18:311–331
110. Sacone S, Siri S (2012) A control scheme for freeway traffic systems based on hybrid automata. Discret Event Dyn Syst 22:3–25
111. Sacone S, Siri S, Torriani F (2012) A hybrid automaton for multi-class ramp metering in freeway systems. In: Proceedings of the 4th IFAC conference on analysis and design of hybrid systems, pp 344–349
112. Muralidharan A, Horowitz R (2015) Computationally efficient model predictive control of freeway networks. Transp Res Part C 58:532–553
113. Han Y, Hegyi A, Yuan Y, Hoogendoorn S, Papageorgiou M, Roncoli C (2017) Resolving freeway jam waves by discrete first-order model-based predictive control of variable speed limits. Transp Res Part C 77:405–420
114. Ferrara A, Nai Oleari A, Sacone S, Siri S (2015) Freeways as systems of systems: a distributed model predictive control scheme. IEEE Syst J 9:312–323
115. Ferrara A, Sacone S, Siri S (2014) Simulation-based assessment of natural robustness of freeway traffic systems controlled via MPC. In Proceedings of the 22nd Mediterranean conference on control and automation, pp 1255–1260
116. Ferrara A, Sacone S, Siri S (2014) Time-varying triggering conditions for the robust control of freeway systems. In: Proceedings of the 53rd IEEE conference on decision and control, pp 1741–1746

117. Liu S, Sadowska A, Frejo JRD, Núñez A, Camacho EF, Hellendoorn H, De Schutter B (2016) Robust receding horizon parameterized control for multi-class freeway networks: a tractable scenario-based approach. Int J Robust Nonlinear Control 26:1211–1245

Chapter 9
Implementation-Oriented Freeway Traffic Control Strategies

9.1 Practical Implementability Issues in Freeway Traffic Control

A freeway traffic control scheme can be deemed as *practically implementable* only if it is compatible with the constraints typical of a real-time control system. In other terms, the control scheme for traffic regulation has to be real-time implementable and scalable.

As discussed in Sect. 8.4.3, coordinated traffic control strategies based on Model Predictive Control (MPC) are very effective, thanks to the possibility of taking a decision about the control action which relies both on the system state measurements and on a prediction of the system evolution, but they are in general highly demanding from the computational point of view. In fact, at any controller time step, it is necessary to solve a Finite-Horizon Optimal Control Problem (FHOCP) which is often complex and involves a number of variables that can be very large.

The three main issues affecting the practical applicability of MPC in freeway traffic control systems can then be summarised in the following points:

1. the FHOCP is normally a non-linear optimal control problem with a high number of variables;
2. the optimisation problem must be solved at any controller time step;
3. the solution procedure requires that the entire system state is transmitted from the sensors to the controller at any controller time step.

Depending on the available technological infrastructure and on the size of the freeway traffic network, the previous three issues can have a different impact on the applicability of the control scheme, making the online application of centralised MPC schemes hard for complex traffic networks (see Fig. 9.1).

Different possible directions can be investigated to design implementation-oriented freeway traffic control strategies, some of which are addressed in this chapter. In particular, the following possibilities may be considered towards practically implementable control frameworks:

A. Ferrara et al., *Freeway Traffic Modelling and Control*, Advances in Industrial Control, https://doi.org/10.1007/978-3-319-75961-6_9

Fig. 9.1 Junction Ridderster of the A15 and A16 freeways, the Netherlands (courtesy of Rijkswaterstaat, Photo: Essencia Communication/Rob de Voogd)

- *decomposing* the whole control problem into smaller sub-problems;
- *simplifying* the problem structure, by reformulating or approximating it;
- *controlling* only *when* and *where* necessary;
- *transmitting* only *when* and *where* necessary.

The first point consists in dividing the control problem into small sub-problems to be solved locally, leading to the so-called *distributed control*. This is a practical alternative to centralised control solutions which has been applied to different dynamic processes and real applications [1]. Generally speaking, distributed control schemes are based on the synergy of local controllers, designed to solve small-scale control problems, which contribute to regulate the overall system, by relying on specific cooperation patterns. Local controllers can either exchange information or be totally independent, giving rise to *decentralised control* schemes. The concepts of distributed and decentralised control can be applied to freeway traffic systems, in order to design practically implementable control frameworks, especially for the case of large traffic networks. Section 9.2 provides a survey of some research papers on distributed and decentralised control schemes for freeway traffic and describes two distributed control algorithms to be applied in large-scale traffic networks.

Another very relevant issue towards implementation-oriented freeway traffic control is the reduction of the *computational effort*, both in terms of computational complexity of the FHOCP and in terms of number of times in which the control law is computed. This latter aspect is related with the concept of *event-triggered control*, in which the control sampling is event-triggered rather than time-triggered, in order to reduce the energy, computation and communication effort. This means that the control scheme includes a triggering mechanism that determines *when* the control input has to be *recomputed* [2]. Section 9.3 deals with event-triggered control for freeway traffic systems in order to reduce the computational effort. This is achieved both by reformulating the traffic model in a mixed-integer linear form and by defining suitable triggering conditions in order to decide when a new FHOCP needs to be solved.

A major feature of freeway traffic control systems which has received attention only recently is the necessity, when evaluating their efficiency and practical implementability, of considering not only the controller but also all the measurement equipments, i.e. the sensors, and the way in which all the system components are connected and communicate. These aspects are related to the concept of *Networked Control Systems* (NCSs). A control system is classified as an NCS when it is spatially distributed and the connections among feedback controllers, sensors and actuators are implemented via a communication network [3, 4]. In Sect. 9.4, we discuss control frameworks for controlling freeway traffic systems considered as NCSs, referring in particular to the cases in which the communication among the sensors, the controller and the actuators located in the traffic systems occurs through a shared digital communication network. In the design of these control schemes, specific attention is devoted not only to reduce the effort for keeping the control law updated, by *recomputing* it only *when* necessary, but also to limit as much as possible the communication effort. This latter goal is achieved by *transmitting* only *when* necessary. To this end, the number of transmissions is decreased by defining specific triggering conditions.

To provide a solution which contemporarily complies with the requirement of updating the control law and transmitting the measurements only *where* and *when* necessary, the concept of event-triggered control can be declined with the logic of distributed and decentralised control. These aspects are illustrated in Sect. 9.5, giving rise to solutions which size the complexity of the optimisation problem to solve and the frequency of the solution of the problem to the actual traffic conditions in the freeway traffic system.

9.2 Distributed and Decentralised Control for Freeway Traffic

The concept of decomposing the control problem into small sub-problems to be solved locally is quite natural in the control of large-scale systems, not only in road traffic networks. A large number of distributed and decentralised MPC algorithms have been proposed and discussed in the literature for different types of systems to be controlled. The interested reader can examine these issues more in detail in [5] (and in the references therein cited), which provides a comprehensive survey and classification of distributed, decentralised and hierarchical MPC architectures for large-scale systems.

Distributed and decentralised control schemes are normally applied when the centralised approach is too hard to be implemented in real time. This is normally true when computationally intensive control problems are faced, typically in case of MPC frameworks and large-scale applications. In distributed control schemes, local controllers solve small control problems and, in some way, cooperate and exchange information in order to control the overall system in a synergistic way. In decentralised control schemes, instead, local controllers do not communicate and are completely independent.

Compared with distributed frameworks, decentralised solutions are easier to implement and usually preferable in terms of scalability. They do not require communication between different areas of the system to control. Yet, their limited implementation cost is often accompanied by a performance lower than the one obtainable via distributed control [6].

9.2.1 Overview of Freeway Traffic Distributed Control Schemes

In order to design distributed or decentralised control schemes for freeway traffic, the basic idea is to subdivide the whole freeway system to control into *subsystems*. The most common choice is to define these subsystems as subsets of contiguous freeway portions which contain one or more actuators (e.g. traffic lights for ramp metering).

The choice of considering subsystems with only one actuator is one of the possible alternatives, which can be convenient to reformulate the centralised (multi-input) control problem as a set of single-input control problems, which is a common practice in process control. Depending on the traffic network morphology and dynamic aspects, one could of course adopt different solutions, in which the subsystems have more than one input.

Note that the most extreme case of single-input subsystems is normally different from the case of *local control strategies* described in Sect. 8.3, for two main reasons. First, when talking about distributed or decentralised control schemes, even in case of only one actuator per subsystem, the controller normally relies on an MPC scheme, hence including predictive capabilities, that are absent in simple local strategies. Second, in distributed or decentralised control schemes, the control action often depends on measurements of the system state referred to the related subsystem, while in local control strategies the control law is generally based on one measurement, taken close to the associated actuator.

Coming to the control system design, the overall freeway system is regarded as a *large-scale system* to which the control requirements refer, while local controllers, for instance, of MPC type, are associated with the subsystems which compose the whole system. In case of *distributed* control schemes (see Fig. 9.2), there is an exchange of information among local controllers of neighbouring subsystems, which communicate according to a specific network communication topology. In

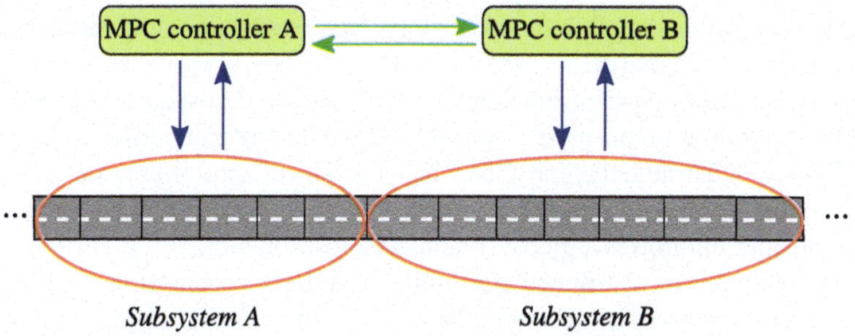

Fig. 9.2 A generic distributed MPC scheme for a freeway traffic system

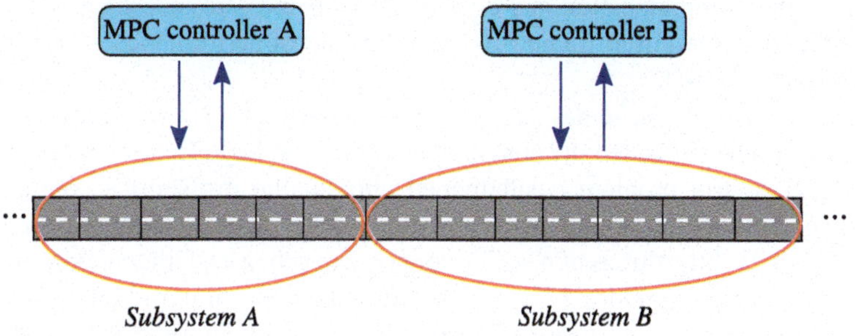

Fig. 9.3 A generic decentralised MPC scheme for a freeway traffic system

case of *decentralised* control (see Fig. 9.3), instead, local controllers operate in a completely independent way [7].

As for distributed control, it is possible to distinguish among different settings. According to the classification provided in [5], a first classification can be based on the topology of the communication network. In *fully connected algorithms*, the information is transmitted and received from any local controller to all the others, while *partially connected algorithms* are applied when information is transmitted and received from any local controller to a given subset of the others. Note that the latter type of algorithms is appropriate only when some of the subsystems of the considered large-scale system are loosely connected. As for freeway traffic systems, this could be, for instance, the case of road portions which are far apart from each other, so that the reciprocal dynamic influence is limited.

In addition to the topology of the communication network, the classification of distributed control algorithms can also be made considering the type of protocol adopted to exchange information among local controllers. In particular, in *non-iterative algorithms* the information is transmitted and received by the local controllers only once within each sample time interval, while in *iterative algorithms* the information can be transmitted and received by the local controllers more than once. Note that, in case of iterative algorithms, the iterative procedure is a kind of negotiation in order to reach, within the sample interval, a global consensus among controllers on the control actions to apply [8].

A classification of the algorithms can be made also on the basis of the cost function adopted. According to this aspect, it is possible to distinguish between *independent algorithms*, i.e. distributed algorithms where each local controller minimises a local cost function, and *cooperating algorithms*, i.e. distributed algorithms where each local controller minimises a global or nonlocal cost function.

A wide number of distributed algorithms have been proposed by researchers to control large-scale freeway networks, mainly in case of predictive traffic controllers. For instance, in [9, 10], distributed and decentralised MPC schemes are proposed in order to deal with the computational requirements of real-time applications. In those papers, some communication and cooperation schemes are investigated and compared with the centralised framework, and the control performance of the different proposals is assessed.

Distributed MPC schemes are also investigated in [11] to reduce both the communication efforts and the computational times in large freeway networks. The centralised control problem is subdivided into smaller problems according to an agent-based modelling approach. In particular, a serial partially cooperative approach is studied, with three different possible cooperative schemes. Other distributed MPC algorithms can be found in [12, 13], together with a decentralised one, all within a cluster-based architecture. These algorithms are analysed in detail in Sect. 9.2.2.

Moreover, in [14], distributed optimisation algorithms are developed for the case of shared state systems, like in coordinated ramp metering and variable speed limit control, while consensus-based coordination mechanisms are examined in [15]. A distributed ramp metering strategy is addressed in [16], in which the control actions are coordinated by sharing local variables with upstream and downstream neighbours.

In some works, distributed traffic control schemes are designed according to game theory concepts. For instance, in [17], the optimal coordination of ramp metering and variable speed limits is seen as a game in which different players (traffic controllers) decide simultaneously. In [18], the problem of optimally balancing the vehicle density in a freeway traffic system is formulated as a non-cooperative Nash game and the controllability properties of the adopted traffic model are exploited to identify the subsystems associated with local controllers.

9.2.2 *Cluster-Based Distributed MPC Algorithms*

In this section, two distributed MPC algorithms for freeway traffic control are presented, considering a cluster-based controller distribution. More specifically, the whole freeway system is subdivided into subsystems, called *clusters of cells*. A cluster of cells is a subset of contiguous freeway cells, which contains a single actuator. In case of ramp metering control, this means that each cluster contains a single traffic light placed at the on-ramp. Each cluster is regulated via an MPC controller.

This cluster-based distributed framework has been proposed in [13] for ramp metering control in a large freeway system. In [13], the MPC controllers adopt the CTM rewritten in a mixed-integer linear form (see Sect. 3.3.3) as prediction model, while the objective function is the minimisation of the quadratic deviations of the state variables from specific set-points. Nevertheless, the algorithms described hereinafter are rather general and can be applied to any distributed MPC scheme for freeway traffic control.

In both the algorithms, there is a communication among each cluster and a subset of other clusters, called *neighbours*. The simplest choice assumes that each cluster only communicates with the adjacent clusters, i.e. the neighbours of each cluster are the previous and the following cluster in the freeway. On the basis of the foregoing classification, the two cluster-based distributed MPC algorithms for freeway ramp metering described in this section can be classified as follows:

1. a *partially connected non-iterative independent algorithm* (Algorithm 1), in which each local controller computes the control action by minimising a local cost function and by exchanging information with the neighbours only once within each sample time interval;
2. a *partially connected non-iterative cooperative algorithm* (Algorithm 2), in which each local controller computes the control action by minimising a partial cost function which includes the local cost function of the cluster itself and the local cost functions of the neighbours. As in Algorithm 1, also in this case, the optimisation problem is solved once within each sample time interval.

Let us consider a discrete-time framework, with k index of the time step. Let C be the number of clusters into which the freeway stretch is subdivided and $\mathcal{N}_i$ the set of indexes of clusters to which cluster i is connected (i.e. the neighbours of cluster i).

The control variable associated with cluster i, $i = 1, \ldots, C$, at time step k is denoted as $u_i(k)$. In case of ramp metering control, this control variable is the traffic flow entering from the on-ramp of the cluster. Let $\underline{x}_i(k)$ be the vector containing all the state variables of cluster i, at time step k. In case the CTM is used, as in [13], the traffic state vector includes the traffic densities of the cells of the cluster and the queue length of the on-ramp. Moreover, let $\underline{d}_i(k)$ include the exogenous inputs affecting cluster i at time step k.

As defined in Sect. 8.4.3, the global cost function $J(k)$ associated with the centralised FHOCP of an MPC scheme, solved at time step k over a finite horizon of K_p time steps, can be written as

$$J(k) = \vartheta[\underline{x}(k + K_p)] + \sum_{h=k}^{k+K_p-1} \varphi\left[\underline{x}(h), \underline{u}(h), \underline{d}(h)\right] \tag{9.1}$$

The global cost function to solve is partitioned in the local cost functions $J_i(k)$ of each cluster i, $i = 1, \ldots, C$, i.e.

$$J(k) = \sum_{i=1}^{C} J_i(k) \tag{9.2}$$

where the local cost function can be defined as

$$J_i(k) = \vartheta[\underline{x}_i(k + K_p)] + \sum_{h=k}^{k+K_p-1} \varphi_i \left[\underline{x}_i(h), u_i(h), \underline{d}_i(h)\right] \tag{9.3}$$

In *Algorithm 1*, the local controller of cluster i solves, at time step k, its own FHOCP. Specifically, the local controller finds the optimal control sequence $u_i^\circ(h|k)$, $h = k, \dots, k + K_p - 1$, which minimises $J_i(k)$. Then, it uses the sequence $u_j^\circ(h|k)$, $h = k, \dots, k+K_p-1$, $j \in \mathcal{N}_i$, computed by the local controllers of the neighbours, to determine the *actual control sequence* for cluster i, according to the rule

$$u_i(h) = \nu_i u_i^\circ(h|k) + \sum_{j \in \mathcal{N}_i} \nu_j u_j^\circ(h|k) \tag{9.4}$$

with $h = k, \dots, k + K_p - 1$, $\nu_i + \sum_{j\in\mathcal{N}_i} \nu_j = 1$ and $\nu_i > 0$, $\nu_j > 0$, $j \in \mathcal{N}_i$. According to the MPC logic, only the first element of the control sequence, i.e. $u_i(k)$, is applied to the system. A sketch of the control scheme for Algorithm 1 is given in Fig. 9.4, in which $\mathcal{N}_i = \{i - 1, i + 1\}$.

In *Algorithm 2*, the local controller of cluster i solves, at time step k, the FHOCP by minimising the *nonlocal* cost function

$$J_i^{\text{nonloc}}(k) = \eta_i J_i(k) + \sum_{j \in \mathcal{N}_i} \eta_j J_j(k) \tag{9.5}$$

where $\eta_i + \sum_{j\in\mathcal{N}_i} \eta_j = 1$ and $\eta_i > 0$, $\eta_j > 0$, $j \in \mathcal{N}_i$. Hence, the local controller of cluster i finds the optimal control sequence related to cluster i itself $u_i^\circ(h|k)$,

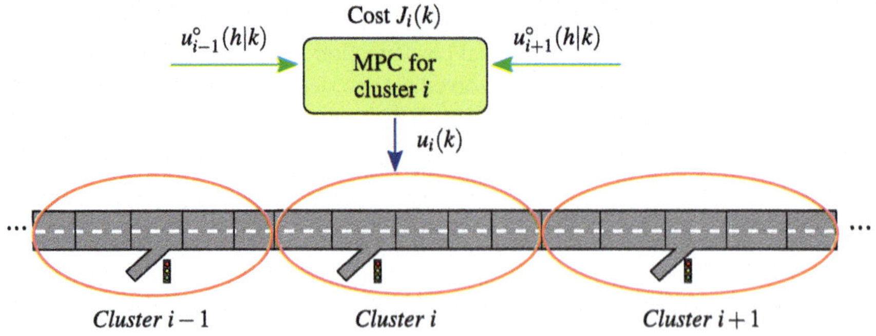

Fig. 9.4 Cluster-Based distributed MPC: Algorithm 1

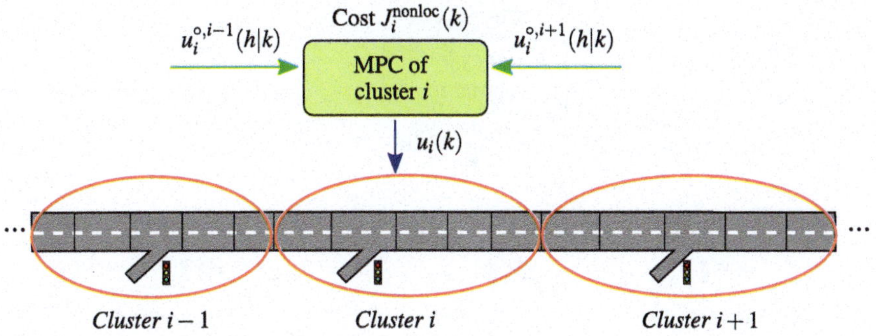

Fig. 9.5 Cluster-Based distributed MPC: Algorithm 2

$h = k, \ldots, k + K_p - 1$, and the control variables of the neighbours $u_j^{\circ,i}(h|k)$, $h = k, \ldots, k + K_p - 1$, $j \in \mathcal{N}_i$.

Then, the local controller of cluster i uses the optimal sequences $u_i^{\circ,j}(h|k)$, $h = k, \ldots, k + K_p - 1$, $j \in \mathcal{N}_i$, computed by the local controllers of the connected clusters, as well as the optimal control sequence related to itself $u_i^{\circ}(h|k)$, $h = k, \ldots, k + K_p - 1$, to determine the *actual control sequence* for cluster i, according to the rule

$$u_i(h) = \nu_i u_i^{\circ}(h|k) + \sum_{j \in \mathcal{N}_i} \nu_j u_i^{\circ,j}(h|k) \tag{9.6}$$

with $h = k, \ldots, k + K_p - 1$, where ν_i, ν_j, $j \in \mathcal{N}_i$, have the same meaning as in (9.4). Again, according to the MPC logic, the first element $u_i(k)$ of the control sequence is applied to the system. Figure 9.5 reports a sketch of the control scheme with Algorithm 2, again with $\mathcal{N}_i = \{i-1, i+1\}$.

Analogously, it is possible to consider a *cluster-based decentralised* MPC scheme for freeway systems (Fig. 9.6), in which the local controller of cluster i, at time step k, finds the optimal control sequence $u_i^{\circ}(h|k)$, $h = k, \ldots, k + K_p - 1$, which minimises $J_i(k)$. This optimal control sequence is the *actual control sequence*, the first element of which is applied to the system.

It is worth noting that, in distributed control frameworks, it is not guaranteed that the obtained solution is optimal for the global problem [5]. Using a suboptimal solution is the cost to pay to obtain a considerable simplification in the implementation, with a corresponding reduction of the computation and transmission burden. Obviously, decentralised schemes are those which allow the highest reduction in computation and transmission effort, at the expense of generally lower performance of the controlled system. A comparison among the two distributed algorithms described above, the decentralised scheme and the centralised approach can be found in [13], considering, as a show case, a stretch of the A12 Freeway in the Netherlands.

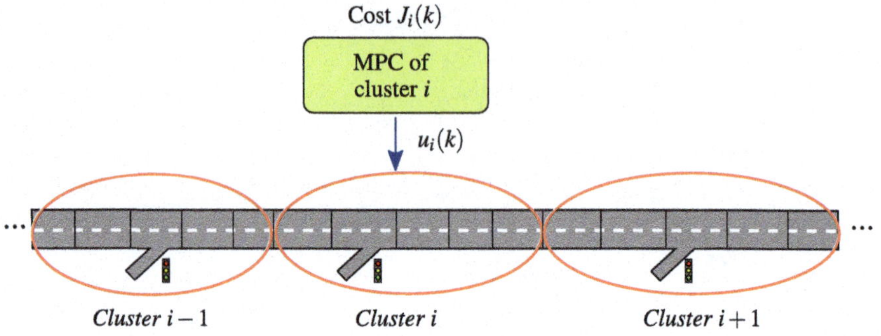

Fig. 9.6 Cluster-based decentralised MPC

9.3 Event-Triggered MPC for Freeway Traffic

The concept of *event-triggered* strategies, in which the control sampling is event-triggered rather than time-triggered, is a rather recent idea conceived in order to reduce the energy, as well as the computation and communication efforts associated with a control scheme. Event-triggered control goes in the direction of making complex control schemes implementable even in the case of large-scale systems (see, for instance, [2, 19, 20] and the references therein).

In case event-triggered concepts are integrated with MPC strategies, the main logic is to reduce the *frequency* in solving the optimal control problems. Yet, simply reducing the number of times in which the FHOCP is solved does not modify the complexity of the optimisation problem itself, which depends on several aspects such as the number of state variables involved, the number of constraints, the type and accuracy of the prediction model, and the length of the control horizon. Hence, another relevant direction in reducing the *computational load* of an MPC scheme is related to simplify the problem structure in order to be able to solve it more efficiently.

In freeway traffic, the event-triggered control paradigm can be used to comply with the requirement of updating the control only *when* necessary. This means that a *triggering condition* based on measurements of the system state is checked at any time step and, only when it is satisfied, the control law is recomputed. Some recent papers have been devoted to the reduction of the computational effort by designing event-triggered MPC schemes for traffic control. The first approach adopting event-triggered concepts for freeway traffic systems can be found in [21], where the FHOCP is based on the CTM as prediction model, but rewritten in a mixed-integer linear form in order to solve it more efficiently. Further extensions have been proposed in [22, 23], in which different triggering conditions are defined. A more sophisticated control scheme for freeway traffic can be found in [24], where a supervisory framework is adopted. In [24], at each time step, the supervisor not only decides whether the control action needs to be recomputed or not, but it also decides if the controller needs to be changed.

Other works in the literature address the problem of reducing the computational effort by simplifying the structure of the FHOCP, which is often non-linear (see Sect. 8.4.3 for further details). Most of these works are devoted to transform the problem in a more tractable formulation [25–29]. In some cases, this is a simple reformulation of the original traffic model, while in other cases this represents a relaxation or a simplification of the model. The resulting optimisation problem is generally rewritten as a linear program (often including integer or binary variables) that can be solved much more efficiently than the non-linear one.

9.3.1 Event-Triggered MPC Schemes

In this section, we refer to the ramp metering control scheme proposed in [22], based on an event-triggered MPC approach. In order to be able to cope with large freeway systems and to solve the problem in times acceptable for a real-time application, the structure of the FHOCP must be computationally efficient. This is achieved by considering a mixed-integer linear prediction model, allowing to use efficient solvers. In the following, the event-triggered scheme is described, starting from the standard MPC framework, then detailing the event-triggered scheme, and, finally, by specifying the triggering conditions. More details on this event-triggered MPC approach, as well as some simulation results, can be found in [22].

The MPC Framework The MPC scheme has been devised so that the FHOCP is computationally efficient. This is achieved by considering, as prediction model, the CTM seen as a Mixed Logical Dynamical (MLD) system, which is a mixed-integer linear model. This allows to define the FHOCP as a mixed-integer linear mathematical programming problem, for which very efficient solvers guaranteeing to find the optimal solution are available. The basic idea of this reformulation of the CTM is that the non-linear relations present in the CTM are rewritten as linear relations by adding some auxiliary variables (both binary and continuous) and some inequalities, as in the classical MLD framework (see Sect. 3.3.3 for the entire model formulation). Note that this is only a reformulation of the original CTM and not a relaxation or simplification of the model.

The considered control objective corresponds to penalise the situations in which the traffic densities exceed given threshold values and the queues at the on-ramps become positive. Note that the threshold values for the traffic density can be determined depending on the specific purpose of the considered application, since they could be devised in order to maximise the outflow (and correspondingly to minimise the total time spent by vehicles in the freeway) or in order to pursue other objectives related, for instance, to environmental or safety aspects.

It is then possible to formalise the FHOCP (see [22] for further details), in which the constraints are given by the prediction model (i.e. the CTM in MLD form), together with the definition of lower and upper bounds for the considered variables.

According to the standard MPC, the FHOCP is solved at each time step k, over a prediction horizon of K_p time steps, and the optimal control sequence is obtained. Let us denote this control sequence, given by ramp metering flows, as $r_i^{\circ}(h|k)$, $i = 1, \ldots, N$, $h = k, \ldots, k + K_p - 1$, where N is the number of cells in which the freeway stretch is subdivided. The optimal solution of the FHOCP also provides the state variables over the prediction horizon, i.e. the predicted traffic densities $\rho_i^{\circ}(h|k)$ and the predicted queue lengths $l_i^{\circ}(h|k)$, $i = 1, \ldots, N$, $h = k + 1, \ldots, k + K_p$.

In the standard MPC scheme, only the control variables corresponding to time step $h = k$ are applied, i.e. $r_i^{\circ}(k|k)$, $i = 1, \ldots, N$, and the overall procedure is repeated at the following time step $k + 1$.

The Event-Triggered MPC Framework The standard MPC scheme is computationally demanding since the FHOCP problem is solved at each time step $k = 0, \ldots, K$. This is typically necessary for cases in which either the prediction model is not completely effective or the system is affected by significant disturbances. Whenever instead the prediction of the system state is adequate and the effect of disturbances is not so relevant, the computational effort necessary to solve the FHOCP at each time step could be spared using the control sequence already computed. It is indeed worth noting that the control sequence found by solving the FHOCP at time step k is referred to a given prediction horizon of K_p time steps.

Relying on this idea, an event-triggered control scheme can be designed for freeway traffic in which the control law is not updated at each time step but only when a given set of conditions is satisfied. Let us call such a set of conditions as *triggering rule* and the time steps in which the triggering rule is verified as *triggering time steps*. Figure 9.7 schematically shows the main differences between the MPC and the event-triggered MPC logics.

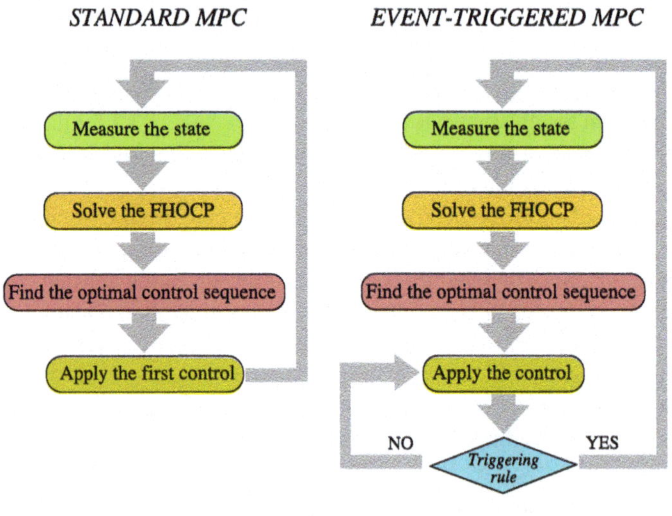

Fig. 9.7 Comparison between the standard MPC scheme and the event-triggered MPC scheme

The *event-triggered MPC mechanism* can be defined as follows:

- at time step $k = 0$, the FHOCP is solved determining the optimal control sequence $r_i^\circ(h|0)$, $i = 1, \dots, N$, $h = 0, \dots, K_p - 1$, and $r_i^\circ(0|0)$ is applied;
- at a generic time step $k > 0$, the triggering rule is verified:
  - if it is not met, the already available control sequence is applied, i.e. $r_i^\circ(k|\bar{k})$ where $\bar{k}$ is the time step in which the optimisation problem has been solved for the last time (of course it must be $k - \bar{k} < K_p$, otherwise the control sequence found in $\bar{k}$ does not cover time step k);
  - if it is met, time step k becomes a triggering time step, the FHOCP is solved, the optimal control sequence $r_i^\circ(h|k)$, $i = 1, \dots, N$, $h = k, \dots, k + K_p - 1$ is derived and $r_i^\circ(k|k)$ is applied.

According to this logic, the values of the control variables composing the last determined control sequence are applied to the system until the next triggering time step.

The Triggering Rule In order to derive the triggering rule, different logics can be used. In [22], the idea is to define a triggering rule related to the difference between the real and the predicted behaviour of the system: if such a difference becomes too large, it means that the prediction in the FHOCP is no more reliable and, consequently, the already computed control variables are no more efficient for the system.

The triggering conditions are first of all referred to the system *state variables*, i.e. the traffic densities $\rho_i(k)$ and the queue lengths $l_i(k)$, for cell i and for time step k. These values, measured in the traffic system, will be compared with the corresponding predicted values $\rho_i^\circ(k|\bar{k})$ and $l_i^\circ(k|\bar{k})$ referred to time step k found by solving the FHOCP at time step $\bar{k}$.

Another important aspect related to the traffic behaviour regards the state of traffic, i.e. free-flow or congested. The triggering condition can be based also on the evaluation of the difference between the current state of traffic in each cell and the predicted one. To express a triggering condition of this type, it is convenient to adopt the binary variables $\delta_i^{\mathrm{M}}(k)$ of the CTM in MLD form, which by definition are equal to 1 in case of free-flow conditions and equal to 0 otherwise (see Sect. 3.3.3). Such variables can be computed at any time step on the basis of the measured state variables, while the predicted values are computed by the prediction model in the FHOCP. Specifically, let $\delta_i^{\mathrm{M},\circ}(k|\bar{k})$ denote the predicted values of these binary variables referred to time step k, found by solving the FHOCP at time step $\bar{k}$.

A possible way to design the triggering rule is to compute the set of cells for which, at time step k, there is a relevant deviation of the real system behaviour from the predicted one. Let us denote this set as $\mathrm{I}(k)$. Then, the triggering rule may be based on the cardinality of this set, i.e. on the number of cells for which there is a relevant deviation between the real and the predicted state. At each time step k, considering that $\bar{k}$ is the previous triggering time step, this set is created according to the following logic:

$$\begin{array}{ll} \text{If} & \delta_i^{\mathrm{M}}(k) \neq \delta_i^{\mathrm{M},\circ}(k|\bar{k}) \ \vee \ |\rho_i(k) - \rho_i^{\circ}(k|\bar{k})| > \varepsilon^{\rho} \ \vee \ |l_i(k) - l_i^{\circ}(k|\bar{k})| > \varepsilon^{l} \\ \text{then} & i \in \mathrm{I}(k) \end{array} \tag{9.7}$$

where ε^{ρ} and ε^{l} are threshold values.

Then, the *triggering rule* to be verified at a generic time step k can be written as

$$|\mathrm{I}(k)| \geq \varepsilon^{\mathrm{I}} \tag{9.8}$$

$$\vee$$

$$k - \bar{k} \geq K_p \tag{9.9}$$

where ε^{I} is a threshold value. Note that this triggering rule includes also condition (9.9), which verifies if the number of time steps after the last triggering time step is greater than K_p. Indeed, when the control sequence has been completely used, the following time step is a triggering time step by definition.

The triggering rule given by (9.8) and (9.9), together with (9.7), is only one of the possible choices to define an event-triggered MPC control scheme (see, e.g. [21, 23] for other types of triggering conditions).

9.4 Model-Based Event-Triggered MPC for Freeway Traffic

An alternative to a purely event-triggered control scheme, as the one discussed in Sect. 9.3, is a *model-based* event-triggered control scheme (see, e.g. the general references [30, 31] illustrating the main concepts underlying model-based event-triggered control). The idea is that both the controller and the sensors can make their decisions supported by an adequately accurate model of the freeway traffic system. In particular, the controller decides if it is required to recompute the control law, while the sensors decide if to send new measurements to the controller.

The level of accuracy of the adopted traffic model depends on the computation and storage capability of the device hosting the controller, as well as on the possibility of delocalising computation power and storage volume at the level of the sensor network. In principle, in case of very high computation and storage capabilities of the hardware, the model to be selected should be very precise, e.g. of microscopic type, to counterbalance the reduced complexity and precision of the prediction model used within the MPC-based controller.

Even in case of model-based schemes, one can conceive different solutions. The scheme proposed in [32] presents the delocalisation of intelligence at the level of *access points* which should be *smart*, having processing and storage power. This control scheme is depicted in Fig. 9.8 and discussed in detail in Sect. 9.4.1.

A further possibility of implementing model-based event-triggered MPC has been discussed in [33], where it is assumed that the intelligence is delocalised at the level

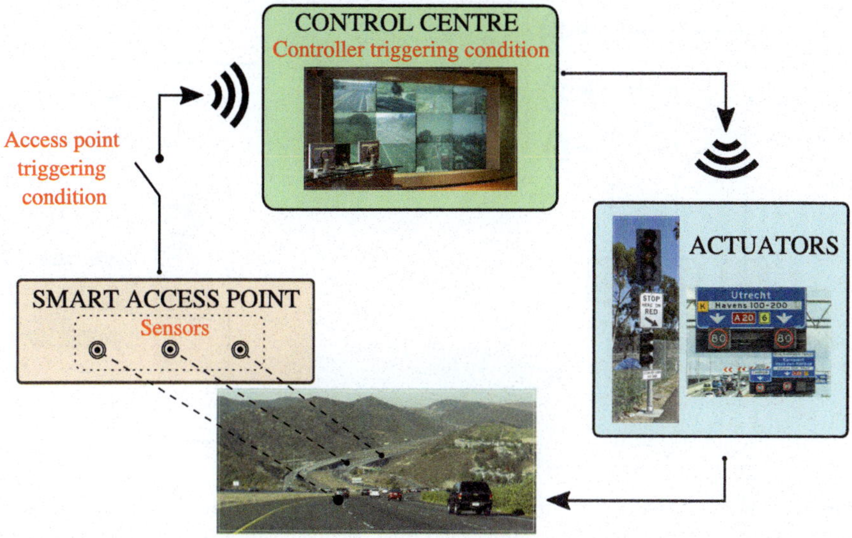

Fig. 9.8 The model-based event-triggered MPC scheme for freeway traffic with smart access points (courtesy of Autostrade per l'Italia SpA, photo from Archivio Videofotografico Autostrade per l'Italia; courtesy of Michael Ballard; courtesy of Rijkswaterstaat, Photo: Essencia Communication/Rob de Voogd)

of each single sensor. Hence, such a scheme requires the use of *smart sensors* that should have sufficient processing and storage capability. The model-based event-triggered control scheme for freeway traffic with smart sensors is depicted in Fig. 9.9 and detailed in Sect. 9.4.2.

According to the model-based event-triggered logic, the freeway system to be controlled is seen as an NCS, which can be affected by network delays and packet dropouts, generally increasing with the network overload. Hence, the explicit objective of these control schemes is not only associated with the reduction of the computational load but also with the minimisation of the communication effort. Since, in real applications of freeway traffic systems, the number of actuators is generally lower than the number of sensors used to measure the system state, the *sensors–controller link* is more critical than the controller-actuator link. Then, the main research focus is on reducing the transmissions of the state measurements from the sensors to the controller. This aspect is dealt with by both the model-based event-triggered MPC schemes described below.

In case instead the research focus is on the *controller–actuator link*, delay and packet dropouts could be faced by adopting *packetised MPC* techniques (see [34] as a general reference on such method and [35] for an application to freeway traffic). According to the packetised logic, the entire optimal control sequence found by the MPC controller is sent to the actuator, which is smart, i.e. it is able to store the predictions received and to apply them to the system.

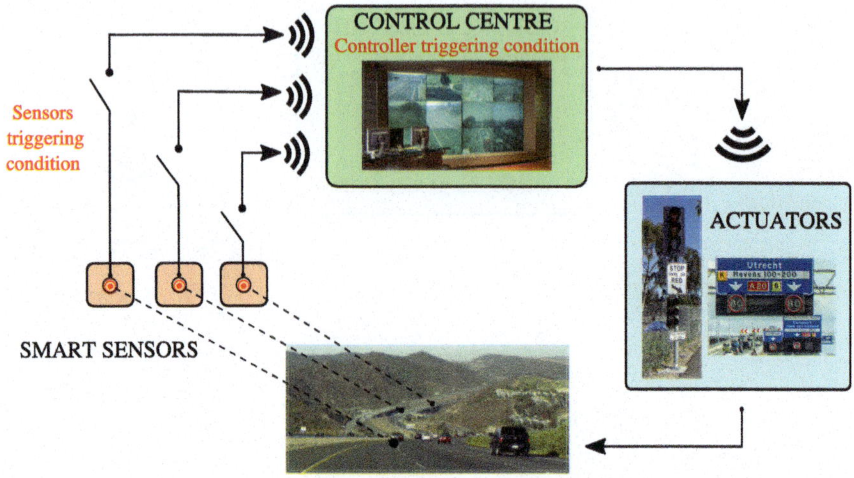

Fig. 9.9 The model-based event-triggered MPC scheme for freeway traffic with smart sensors (courtesy of Autostrade per l'Italia SpA, photo from Archivio Videofotografico Autostrade per l'Italia; courtesy of Michael Ballard; courtesy of Rijkswaterstaat, Photo: Essencia Communication/Rob de Voogd)

9.4.1 Access Point Centralised Decision-Making

In case a model-based event-triggered control architecture as the one depicted in Fig. 9.8 is adopted to control freeway traffic, one can regard the entire set of sensors as a sensor network partitioned into portions, which communicate with the controller, located in the control room, and possibly with other portions of the sensor network through the *access points*. The sensor network portions only acquire the traffic measurements and transmit them to the associated access points. As such, in this kind of implementation, the sensors can be of conventional type. It is at the level of the access points that some processing and storage power is needed to efficiently perform data collection, run the prediction model and check the triggering rule.

In this section, the concept underlying the scheme in Fig. 9.8 is described making reference to a freeway stretch in which all the sensors are connected to *a single access point*, where the decision-making is centralised. Clearly, if the freeway traffic system to control is large, one can assume that several access points are located along the road, each of them acting as a hub for a portion of the sensor network, and that the strategy illustrated below is replicated for any access point.

The control scheme is shown in Fig. 9.10 for a generic plant modelled as a discrete-time system, characterised by a state vector $\underline{x}(k)$ and a control vector $\underline{u}(k)$, referred to the generic time step k. The two main components are the MPC controller and the sensors, both operating according to event-triggered logics.

The objective of reducing both the computational effort of the controller and the communication effort of the transmission network (from the sensors to the controller)

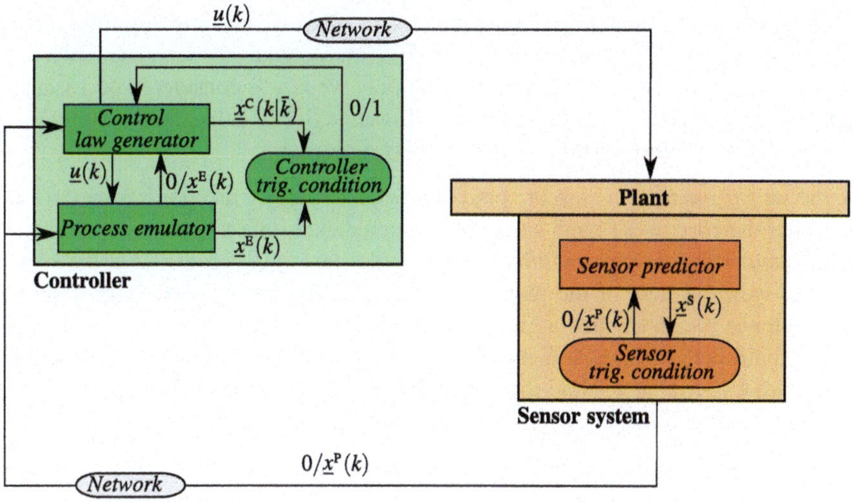

Fig. 9.10 Model-based event-triggered MPC scheme, with access point decentralised decision-making, for a generic plant

is pursued by limiting the number of times in which the control law is updated and the measurements are transmitted to cases in which it is considered useful or even necessary for the controlled system.

The *event-triggered controller* depicted in Fig. 9.10 is made of three blocks:

- the *control law generator* is an MPC controller able to solve an FHOCP in which a model for the system prediction is considered as a set of constraints; at each time step k, supposing that the control law has been computed at time step $\bar{k} < k$, the control law generator provides the control $\underline{u}(k)$ and the predicted state $\underline{x}^{\mathrm{C}}(k|\bar{k})$; note that the control law is recomputed only if activated either by the controller triggering condition (in this case, the state $\underline{x}^{\mathrm{E}}(k)$ is used as initial condition in the FHOCP) or by the sensor triggering condition (in this other case, the state $\underline{x}^{\mathrm{P}}(k)$ is used as initial condition);
- the *process emulator* provides a prediction of the system state using a suitable model, in general, more detailed than the prediction model adopted in the MPC controller; by receiving at each time step k the control $\underline{u}(k)$ computed by the control law generator, the process emulator predicts the system state evolution, by generating the state $\underline{x}^{\mathrm{E}}(k)$;
- the *controller triggering condition* is used to determine if the control law must be computed, i.e. a new FHOCP must be solved, or if the current control is still suitable to be applied. At each time step k, relying on the state $\underline{x}^{\mathrm{C}}(k|\bar{k})$ predicted by the controller predictor and the state $\underline{x}^{\mathrm{E}}(k)$ produced by the process emulator, the triggering condition is checked and a binary triggering signal is communicated to the control law generator.

The plant, and specifically all the actuators, receives at each time step k the control $\underline{u}(k)$ to be applied. The state of the plant $\underline{x}^{\mathrm{P}}(k)$ is measured by the sensors at each time step k, but it is transmitted to the controller only when it is considered necessary to guarantee a good performance of the overall system. Specifically, the *event-triggered sensor system* is represented in Fig. 9.10 by two blocks:

- the *sensor predictor* is used to predict the plant state, by receiving from the controller the control $\underline{u}(k)$ and computing the predicted state $\underline{x}^{\mathrm{S}}(k)$;
- the *sensor triggering condition* compares the state $\underline{x}^{\mathrm{S}}(k)$ predicted at the sensor level with the state of the plant $\underline{x}^{\mathrm{P}}(k)$ measured by the sensors. Only if such a difference is greater than a given threshold, the measured state is transmitted to the controller and is used as initial state both by the control law generator and by the process emulator. Note that when $\underline{x}^{\mathrm{P}}(k)$ is transmitted to the controller, it is also transmitted to the sensor predictor that uses it as a new initial condition.

The event-triggered controller and the event-triggered sensor system are customised with reference to the freeway traffic control case. They are here described, considering the ramp metering control scheme illustrated in [32], where the interested reader can find more details and some simulation results testing the performance of the control scheme.

The Event-Triggered Controller for Freeway Traffic Systems As in Sect. 9.3.1, the *control law generator* is an MPC controller adopted to perform ramp metering, which considers the CTM in MLD form as a prediction model. The objective function of the FHOCP penalises the cases in which the traffic densities exceed given thresholds and the queue lengths at the on-ramps become positive.

Let us denote the optimal control sequence, given by ramp metering flows, computed by the control law generator at time step k as $r_i^{\mathrm{C}}(h|k)$, $i = 1, \ldots, N$, $h = k, \ldots, k + K_p - 1$. Moreover, let us denote with $\rho_i^{\mathrm{C}}(h|k)$ and $l_i^{\mathrm{C}}(h|k)$ the predicted values of the traffic state, respectively, the traffic densities and queue lengths, computed by the MPC controller at time step k and referred to cell i and time step $h = k + 1, \ldots, k + K_p$.

The novelty here, compared with the event-triggered control scheme described in Sect. 9.3.1, is the element named *process emulator*. The idea is that it is realised by means of an accurate dynamic model of the traffic evolution. It can be the CTM with a sufficiently large number of cells or a more detailed traffic model, such as the second-order METANET model or, even, a microscopic or mesoscopic model.

According to the event-triggered control logics, the FHOCP is not solved at each time step k but only when a given set of conditions is satisfied. The *event-triggered MPC mechanism* is the same illustrated in Fig. 9.7, i.e.

- at time step $k = 0$, the FHOCP is solved determining the optimal control variables $r_i^{\mathrm{C}}(h|0)$, $i = 1, \ldots, N$, $h = 0, \ldots, K_p - 1$ and $r_i^{\mathrm{C}}(0|0)$ is applied;
- at a generic time step $k > 0$, the triggering rule is checked:
  - if it is not met, the already available control sequence $r_i^{\mathrm{C}}(k|\bar{k})$ is applied, where $\bar{k}$ is the last triggering time step;

– if it is met, the FHOCP is solved, the optimal control variables $r_i^{\mathrm{C}}(h|k)$, $i = 1, \ldots, N$, $h = k, \ldots, k + K_p - 1$ are derived and $r_i^{\mathrm{C}}(k|k)$ is applied.

The main difference with the event-triggered scheme described in Sect. 9.3.1 stands in the controller triggering condition. In Sect. 9.3.1, the triggering condition verifies the differences between the state predicted by the controller and the real measured system state. In the model-based event-triggered control scheme described here, the triggering rule verifies if there is a significant difference between the predicted system state and the system state obtained by the process emulator. In fact, the errors between the state predicted by the controller and the state given by the emulator (that is supposed to be more accurate than the one obtained by the controller predictor) give an indication about the effectiveness of the prediction in the FHOCP and consequently of the control variables found.

To provide an example of triggering rule, we assume, as in Sect. 9.3.1, to identify the set $\mathrm{I}(k)$ of cells for which, at time step k, there is a relevant deviation of the predicted system evolution from the one determined by the emulator. In particular, let us denote with $\delta_i^{\mathrm{M,E}}(k)$, $\rho_i^{\mathrm{E}}(k)$, $l_i^{\mathrm{E}}(k)$, respectively, the binary variable associated with the cell state, the density, the queue length computed by the emulator and referred to cell i at time step k. These values should be compared with the corresponding predicted values $\delta_i^{\mathrm{M,C}}(k|\bar{k})$, $\rho_i^{\mathrm{C}}(k|\bar{k})$, $l_i^{\mathrm{C}}(k|\bar{k})$, computed at the last triggering time step $\bar{k}$.

Then, the set $\mathrm{I}(k)$, at each time step k, is created in analogy with (9.7), i.e.

$$\begin{aligned} &\text{If} \quad \delta_i^{\mathrm{M,E}}(k) \neq \delta_i^{\mathrm{M,C}}(k|\bar{k}) \ \vee \ |\rho_i^{\mathrm{E}}(k) - \rho_i^{\mathrm{C}}(k|\bar{k})| > \varepsilon^{\rho} \ \vee \ |l_i^{\mathrm{E}}(k) - l_i^{\mathrm{C}}(k|\bar{k})| > \varepsilon^{l} \\ &\text{then} \quad i \in \mathrm{I}(k) \end{aligned} \tag{9.10}$$

where ε^{ρ} and ε^{l} are threshold values. Taking into account (9.10), the *triggering rule* is given again by (9.8) and (9.9).

The Event-Triggered Access Point for Freeway Traffic Systems In the present control scheme, the access point collects measurements from the connected sensors, getting a picture of the present system state, but also decides when a measurement packet must be transmitted to the controller.

For the sake of simplicity, we suppose that sensors are capable of providing measurements of densities and on-ramp queue lengths in each cell of the freeway to the access point. Apart from measurement storage devices, the access point relies on a predictor of the system state, here named *sensor predictor* to distinguish it from the one used by the controller.

A possible choice for the model to be adopted to realise the sensor predictor is again the CTM, since it can be easily implemented and even embedded in dedicated low-cost hardware. Obviously, if the processing and storage capability of the access point is significant (this depends on the technology used for practical implementations), the process emulator included in the controller and the sensor predictor could use the same freeway traffic model. This would be the preferable situation: even when the access point is not transmitting the actual traffic measurements over the network and the controller operates in an open-loop fashion, one could have the guarantee that the controller and the access points share a very similar information basis, i.e. the same model of the freeway traffic.

The *sensor triggering condition* is used to transmit the measured system state and to activate the computation of the control law only when it is appropriate. Then, whenever the system state determined by the sensor predictor and the measured system state differ significantly, the system state is transmitted to the controller and a new computation of the control law is forced. Specifically, let us denote with $\rho_i^{\mathrm{S}}(k)$ and $l_i^{\mathrm{S}}(k)$, respectively, the traffic density and the queue length computed by the sensor predictor, referred to cell i and time step k.

Different types of sensor triggering conditions can be devised depending on the number of sensors which present a high deviation between the predicted and measured state. In [32], six different sensor triggering conditions have been proposed, based on the deviations between the predicted state $\rho_i^{\mathrm{S}}(k)$, $l_i^{\mathrm{S}}(k)$ and the measured state $\rho_i^{\mathrm{P}}(k)$, $l_i^{\mathrm{P}}(k)$. We report them here as possible examples:

1. at least in *one sensor* there is a deviation between the predicted and the measured state overcoming a given threshold; two possible choices can be considered:
   (a) for each cell i in which the corresponding sensor has relieved such a deviation, the measured traffic density $\rho_i^{\mathrm{P}}(k)$ and queue length $l_i^{\mathrm{P}}(k)$ are transmitted to the controller;
   (b) all the measured state is transmitted to the controller, i.e. $\rho_i^{\mathrm{P}}(k)$ and $l_i^{\mathrm{P}}(k)$, $\forall i = 1, \dots, N$.
2. at least *a given number $\bar{s}$ of sensors* (even though they are not contiguous in the freeway stretch) compute a high deviation between the predicted and the measured state; two possible choices can be considered:
   (a) for the $\bar{s}$ cells in which the corresponding sensor has relieved this deviation, the measured traffic density $\rho_i^{\mathrm{P}}(k)$ and queue length $l_i^{\mathrm{P}}(k)$ are transmitted to the controller;
   (b) all the measured state is transmitted to the controller, i.e. $\rho_i^{\mathrm{P}}(k)$ and $l_i^{\mathrm{P}}(k)$, $\forall i = 1, \dots, N$;
3. at least *a given number $\bar{s}_c$ of close sensors* (i.e. sensors that are contiguous in the freeway stretch) compute a high deviation between the predicted and the measured state; two possible choices can be considered:
   (a) for the $\bar{s}_c$ close cells in which the corresponding sensor has relieved such a deviation, the measured traffic density $\rho_i^{\mathrm{P}}(k)$ and queue length $l_i^{\mathrm{P}}(k)$ are transmitted to the controller;
   (b) all the measured state is transmitted to the controller, i.e. $\rho_i^{\mathrm{P}}(k)$ and $l_i^{\mathrm{P}}(k)$, $\forall i = 1, \dots, N$.

9.4.2 Sensor Decentralised Decision-Making

Considering the available technology, it is also possible to further delocalise the intelligence of the freeway traffic control system, assuming that any sensor is equipped with sufficient computation power and storage capability to be able to host and run a freeway model (see Fig. 9.9). On the basis of the acquired measurements and the state of the simulated system, each sensor can be enabled to make an individual decision whether to transmit or not the newly measured quantities. This is the philosophy which inspired the control scheme presented in [33] and reported in Fig. 9.11 for a generic plant.

The plant is composed of N subsystems, each of which is equipped with a sensor that measures the local state of the corresponding subsystem. The plant is regulated by a feedback controller, connected to measurement and actuation devices through a communication network. As before, the goal is still to reduce both the computational and the communication efforts. Hence, both the controller and the sensors are characterised by the presence of event-triggered logics.

Since the measurements are not transmitted at each time step, both the controller and the sensors are equipped with a process emulator (which is here assumed to be the same) that is able to reproduce the plant state dynamic evolution in a very accurate way. The reduction of the number of computations is achieved by computing the control law either when there is a significant deviation between the state predicted by the emulator and the one predicted by the control law generator or when new measurements are received from the plant. Moreover, state transmissions are limited to the cases in which there is a significant deviation between the real system state and the one predicted by the emulator. Since each subsystem is provided with a sensor able to measure the local plant state, at each time step each sensor decides whether to send the measurement to the other sensors and to the controller verifying if the difference between the measured state and the one provided by the emulator exceeds a given threshold.

Referring to a generic time step k, a similar notation as in Sect. 9.4.1 is considered. In particular, $\underline{u}(k)$ is the control action, whereas the system state is represented in three different ways: $\underline{x}^{\mathrm{C}}(k|\bar{k})$ is the state predicted by the control law generator according to the last solution computed at time step $\bar{k}$, $\underline{x}^{\mathrm{E}}(k)$ is the state predicted by the process emulator and $\underline{x}^{\mathrm{P}}(k)$ is the real state of the plant.

The main difference compared to the scheme presented in Sect. 9.4.1 is that, here, the overall system is partitioned into N subsystems, and the state vector is partitioned as well. Hence, $\underline{x}_i^{\mathrm{P}}(k)$ denotes the state of the plant referred to the generic subsystem i, measured by sensor i, $i = 1, \dots, N$, while $\underline{x}_i^{\mathrm{E}}(k)$ and $\underline{x}_i^{\mathrm{C}}(k|\bar{k})$ denote the state of the plant subsystem i provided by the emulator and by the controller. Note that the total plant state $\underline{x}^{\mathrm{P}}(k)$ is obtained by putting together the states $\underline{x}_i^{\mathrm{P}}(k)$ measured in each subsystem i, and obviously the same holds for the system predicted by the emulator and by the controller. Moreover, let $\mathscr{I}(k) \subseteq \{1 \dots, N\}$ indicate the set of sensors which transmit their measurements at time step k.

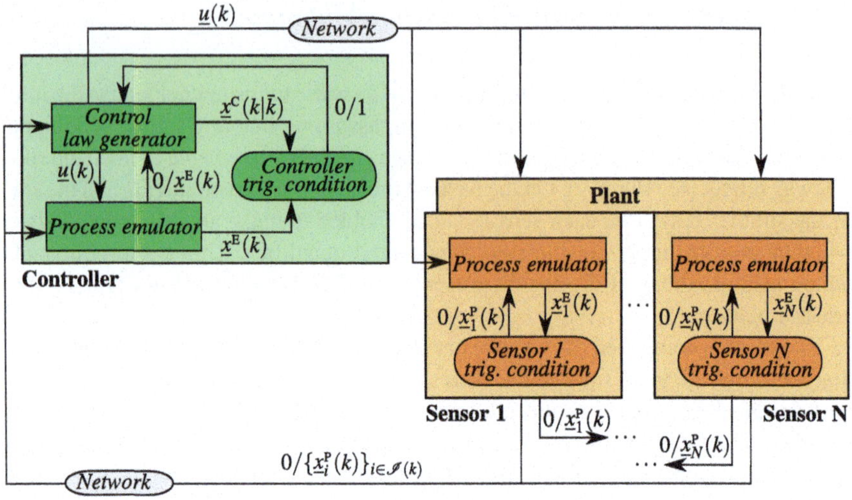

Fig. 9.11 Model-based event-triggered MPC scheme, with sensor decentralised decision-making, for a generic plant

The *event-triggered controller* depicted in Fig. 9.11 is made of three blocks:

- the *control law generator* is an MPC controller which computes the control law by solving an FHOCP; this computation is activated either by the controller triggering condition ($\underline{x}^{\mathrm{E}}(k)$ is used as initial condition) or by a new transmission of the system state ($\underline{x}_i^{\mathrm{P}}(k), i \in \mathscr{I}(k)$, are used as initial condition); at each time step k, supposing that the control law has been computed at time step $\bar{k} < k$, the control law generator provides the predicted state $\underline{x}^{\mathrm{C}}(k|\bar{k})$ and the control $\underline{u}(k)$;
- the *process emulator* provides a prediction of the system state using a model, which is in general more detailed than the prediction model adopted in the MPC controller; by receiving at each time step k the control action $\underline{u}(k)$, the process emulator generates the state $\underline{x}^{\mathrm{E}}(k)$; it updates this state when it receives new measurements $\underline{x}_i^{\mathrm{P}}(k), i \in \mathscr{I}(k)$, from the plant;
- the *controller triggering condition* is used to determine if a new FHOCP must be solved; at each time step k, on the basis of the state $\underline{x}^{\mathrm{C}}(k|\bar{k})$ predicted by the controller and the state $\underline{x}^{\mathrm{E}}(k)$ generated by the process emulator, the triggering condition is checked and a binary triggering signal is communicated to the control law generator.

Referring again to Fig. 9.11, each *event-triggered sensor* i, $i = 1, \ldots, N$, is represented with two blocks:

- the *process emulator*, at time step k, receives the control $\underline{u}(k)$ and generates the state $\underline{x}^{\mathrm{E}}(k)$;
- the *sensor triggering condition* compares the state $\underline{x}_i^{\mathrm{E}}(k)$ provided by the emulator referred to subsystem i with the measured state $\underline{x}_i^{\mathrm{P}}(k)$; if such a difference is greater

than a given threshold, the measured state is transmitted to the controller and to the other sensors, so that it is used as initial state for the control law generator and for all the process emulators.

Note that, according to the considered scheme, all the process emulators (in the sensors and in the controller) are completely synchronised since they use the same model, update the state simultaneously and consider the same control $\underline{u}(k)$.

In the following, the model-based event-triggered MPC scheme with sensor decentralised decision-making is described referring to a freeway system controlled via ramp metering, as reported in [33]. The interested reader can find, in [33], an analysis of the robustness properties of the controlled system, under the assumption of additive and bounded exogenous inputs, by proving the *input-to-state practical stability* of the system. Moreover, in [33], the effectiveness of this control scheme is assessed via simulation by analysing different scenarios.

The Event-Triggered Controller for Freeway Traffic Systems The *control law generator* is an event-triggered MPC controller, as described in Sect. 9.4.1, which uses the CTM in MLD form for prediction. In the FHOCP, the cost function has a very general form and is a weighted sum of traffic densities, queue lengths, and terms penalising the cases in which the traffic density exceeds a given threshold value. When solving the FHOCP at time step k, the optimal control sequence $r_i^{\mathrm{C}}(h|k)$, $i = 1, \ldots, N, h = k, \ldots, k+K_p-1$ is found, as well as the predicted state variables $\rho_i^{\mathrm{C}}(h|k)$ and $l_i^{\mathrm{C}}(h|k)$, for the traffic densities and queue lengths, respectively, in cell i and referred to time step $h = k+1, \ldots, k+K_p$.

The *process emulator* can be realised via a dynamic model of the freeway traffic. As observed in Sect. 9.4.1, this model can be the CTM with a sufficiently large number of cells or any other more detailed traffic model. In any case, at time step k, the process emulator determines the state variables $\rho_i^{\mathrm{E}}(k)$ and $l_i^{\mathrm{E}}(k)$, referred to cell i. Note that, if at time step k some measurements of the plant state $\rho_i^{\mathrm{P}}(k)$, $l_i^{\mathrm{P}}(k)$, $i \in \mathscr{I}(k)$, are sent to the controller, the emulator updates its state, and in particular the part of the state relative to the cells which have transmitted the measurements, i.e. $\rho_i^{\mathrm{E}}(k) = \rho_i^{\mathrm{P}}(k)$ and $l_i^{\mathrm{E}}(k) = l_i^{\mathrm{P}}(k), i \in \mathscr{I}(k)$.

According to the event-triggered control philosophy, the FHOCP is not solved at each time step but only when the *triggering rule* is met. This happens when the controller triggering conditions are satisfied or when a new measurement is received from sensors. Let us specifically refer to the triggering conditions used in [33]. Considering that $\bar{k}$ is the previous triggering time step, the controller triggering conditions to be verified at a generic time step $k \geq \bar{k}$ can be written as

$$\exists i \in \{1, \ldots, N\} : \left|\rho_i^{\mathrm{E}}(k) - \rho_i^{\mathrm{C}}(k|\bar{k})\right| > \varepsilon_i^{\rho} \tag{9.11}$$

$$\vee$$

$$\exists i \in \{1, \ldots, N\} : \left|l_i^{\mathrm{E}}(k) - l_i^{\mathrm{C}}(k|\bar{k})\right| > \varepsilon_i^{l} \tag{9.12}$$

$$\vee$$

$$k \geq \bar{k} + K_p \tag{9.13}$$

where ε_i^{ρ} and ε_i^{l} are threshold values on the density and queue length errors, respectively, for cell i. Then, the overall triggering rule which activates the solution of a new FHOCP is given by the controller triggering conditions (9.11)–(9.13) and the check if a new measurement is received from sensors, i.e.

$$\exists i \in \{1, \dots, N\} : \rho_i^{\mathrm{P}}(k) \text{ is received from sensors} \tag{9.14}$$
$$\vee$$
$$\exists i \in \{1, \dots, N\} : l_i^{\mathrm{P}}(k) \text{ is received from sensors} \tag{9.15}$$

The *event-triggered MPC mechanism* can then be defined as follows:

- at time step $k = 0$, the FHOCP is solved determining the optimal control variables $r_i^{\mathrm{C}}(h|0)$, $i = 1, \dots, N$, $h = 0, \dots, K_p - 1$, and $r_i^{\mathrm{C}}(0|0)$ is applied;
- at a generic time step $k > 0$, the triggering rule is checked:
  - if it is not met, the already available control sequence $r_i^{\mathrm{C}}(k|\bar{k})$ is applied, where $\bar{k}$ is the last triggering time step;
  - if it is met, the FHOCP is solved, the optimal control variables $r_i^{\mathrm{C}}(h|k)$, $i = 1, \dots, N$, $h = k, \dots, k + K_p - 1$ are derived and $r_i^{\mathrm{C}}(k|k)$ is applied. Note that if the triggering rule is met because triggering conditions (9.11)–(9.13) are verified, the optimisation problem is solved using as initial state the one provided by the emulator, i.e. $\rho_i^{\mathrm{C}}(k|k) = \rho_i^{\mathrm{E}}(k)$ and $l_i^{\mathrm{C}}(k|k) = l_i^{\mathrm{E}}(k)$, $i = 1, \dots, N$. If conditions (9.14) and (9.15) are verified, the initial state for the optimisation problem is given by $\rho_i^{\mathrm{C}}(k|k) = \rho_i^{\mathrm{P}}(k)$ and $l_i^{\mathrm{C}}(k|k) = l_i^{\mathrm{P}}(k)$ for cells $i \in \mathscr{I}(k)$ and $\rho_i^{\mathrm{C}}(k|k) = \rho_i^{\mathrm{E}}(k)$ and $l_i^{\mathrm{C}}(k|k) = l_i^{\mathrm{E}}(k)$ for cells $i \notin \mathscr{I}(k)$.

The Event-Triggered Sensors for Freeway Traffic Systems In order to delocalise the decision-making capability of the freeway traffic control scheme, the sensors need to be *smart*, in the sense that they not only measure the local system state but also decide when this state must be transmitted to the controller. For the sake of simplicity, we suppose that the sensors are capable of providing measurements of densities and on-ramp queue lengths along the freeway and that each cell is provided with a sensor.

As already mentioned, the *process emulator* in each sensor provides the same system state variables $\rho_i^{\mathrm{E}}(k)$ and $l_i^{\mathrm{E}}(k)$, referred to cell i, generated by the process emulator present in the controller.

The objective of the *sensor triggering condition* is to transmit the measured system state and to activate the computation of the control law only when it is actually necessary. Then, whenever the system state determined by the process emulator in each sensor and the measured system state differ significantly, the system state is transmitted (to the controller and to the other sensors), the process emulators (in the controller and in the other sensors) update the system state and a new computation of the control law is forced. More precisely, at a generic time step k, the sensor present in cell i verifies the following triggering condition:

$$\left|\rho_i^{\mathrm{P}}(k) - \rho_i^{\mathrm{E}}(k)\right| > \kappa_i^{\rho} \tag{9.16}$$

$$\vee$$

$$\left|l_i^{\mathrm{P}}(k) - l_i^{\mathrm{E}}(k)\right| > \kappa_i^{l} \tag{9.17}$$

where κ_i^{ρ}, κ_i^{l} are given threshold values. In case this condition is fulfilled, the sensor of cell i transmits the current measurement $\rho_i^{\mathrm{P}}(k)$ and $l_i^{\mathrm{P}}(k)$ to the controller and to all the other sensors.

9.5 Decentralised Event-Triggered MPC Solutions for Freeway Traffic

In case of large and very large freeway networks, the aim of reducing the communication effort as well as the computational effort is particularly important. *Resource-aware* control schemes need then to be conceived: they have to avoid an unnecessary utilisation of the available computation and communication resources which, especially when battery-powered devices are used, can decrease the practical applicability of the designed control. To cope with this kind of scenarios, *decentralised event-triggered MPC* solutions can be designed, getting the benefits that both decentralised schemes and event-triggered control solutions feature individually.

An original approach which belongs to the foregoing class of control schemes is presented in this section. In this scheme, the entire freeway is divided into *clusters of cells*, in analogy with the scheme described in Sect. 9.2.2. In order to reduce the number of state transmissions between the sensors and the controller, a set of sensor triggering conditions is defined. Moreover, the computational effort is reduced in two ways. First of all, there is a reduction in the number of times in which the control law is computed and this is achieved by adopting a model-based event-triggered MPC approach, as the one discussed in Sect. 9.4.2. Second, in each cluster, specific conditions are verified at the level of the sensors, in order to *activate the controller* of that cluster only when it is considered useful for the system performance (and to deactivate it in the opposite case), further reducing the computational effort.

9.5.1 Cluster-Based Decentralised Event-Triggered MPC

This section describes a decentralised control scheme which relies on an event-triggered logic for a generic traffic network, subdivided into C clusters. It is worth noticing that now the objective of dividing a large freeway network into clusters of cells is twofold: on the one hand, this allows to deactivate some clusters when it is not necessary to control them, and, on the other hand, it permits to solve smaller optimal control problems, with a consequent important advantage from the computational

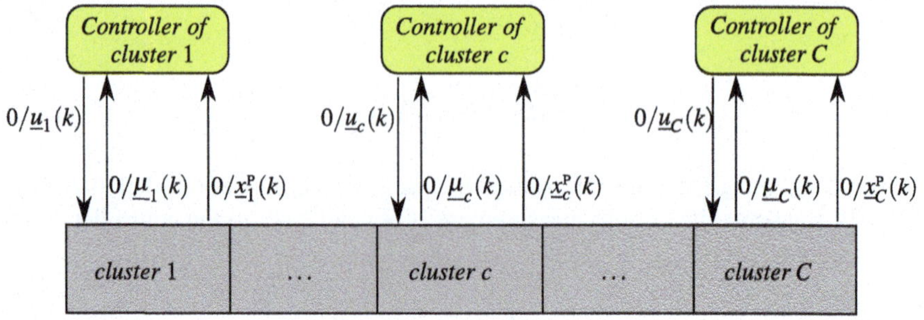

Fig. 9.12 Cluster-based decentralised event-triggered control scheme, for a generic plant

point of view. By virtue of its cluster-based nature, the decentralised cluster-based control scheme described hereafter is scalable and can be applied to large freeway networks.

The overall control scheme is composed of a set of local controllers, one for each cluster, which act in a decentralised way. Note that the concept of clusters of cells is here more general than in Sect. 9.2.2, since here a cluster can include *more than one actuator*.

The controller present in each cluster is of event-triggered MPC type, triggering conditions being present both at the sensor level and at the controller level, as described in Sect. 9.4.2. Yet, differently from the control scheme described in Sect. 9.4.2, each cluster, depending on the measurements of the system state and verifying a proper *activation/deactivation condition*, can activate or deactivate the controller.

Figure 9.12 shows the cluster-based decentralised event-triggered control scheme for a generic plant, where $\underline{x}_c^{\mathrm{P}}(k)$ indicates the state of the plant referred to cluster $c = 1, \ldots, C$ at a generic time step k. Note that the total plant state $\underline{x}^{\mathrm{P}}(k)$ is obtained by putting together the states $\underline{x}_c^{\mathrm{P}}(k)$, $c = 1, \ldots, C$.

Considering a generic cluster c and a generic time step k, the sensors, according to the event-triggered logic, can transmit the state $\underline{x}_c^{\mathrm{P}}(k)$ to the controller or not; analogously, if the controller is activated, the control $\underline{u}_c(k)$ is sent to the plant, otherwise no control is sent and the cluster works uncontrolled. If the sensors verify that the state of the controller must be changed (for instance, it is active and must be deactivated or vice versa), they transmit an *activation/deactivation signal* $\underline{\mu}_c(k)$ to the controller, otherwise they do not transmit any signal.

The control scheme for a generic cluster c is shown in Fig. 9.13. Each cluster c is composed of N_c subsystems, equipped with sensors to measure the local state. Referring to time step k and cluster c, $\underline{x}_c^{\mathrm{C}}(k|\bar{k}_c)$ is the state of cluster c predicted by the control law generator and computed at time step $\bar{k}_c$, while $\underline{x}_c^{\mathrm{E}}(k)$ and $\underline{x}_c^{\mathrm{P}}(k)$ are the states of cluster c predicted by the process emulator and measured from the plant, respectively.

Since the plant is composed of C clusters, each one divided into N_c subsystems, let $\underline{x}_{c,i}^{\mathrm{C}}(k|\bar{k}_c)$, $\underline{x}_{c,i}^{\mathrm{E}}(k)$, $\underline{x}_{c,i}^{\mathrm{P}}(k)$ and $\underline{u}_{c,i}(k)$ denote, respectively, the state predicted by the

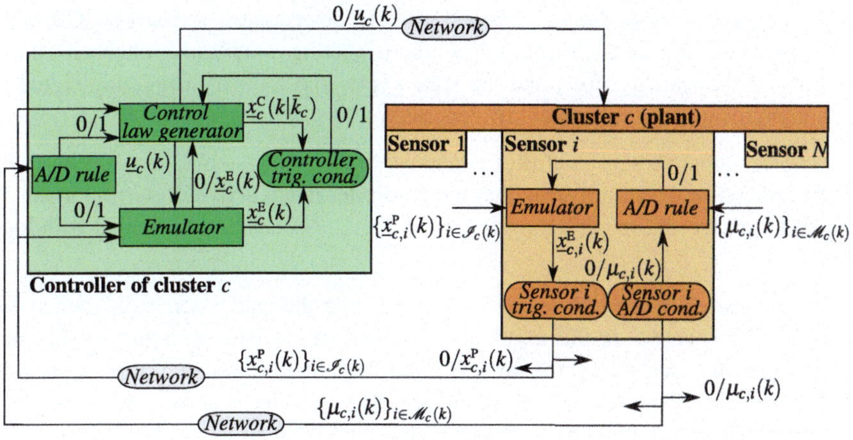

Fig. 9.13 The decentralised event-triggered MPC scheme in cluster c, for a generic plant

control law generator, the state provided by the emulator, the real state and the control of subsystem $i = 1, \ldots, N_c$ of cluster $c = 1, \ldots, C$. Moreover, let $\mu_{c,i}(k)$ represent the activation/deactivation signal sent by subsystem i of cluster c if, according to its measured state, the controller should change its state (from active to not active or vice versa). Further, let $\mathscr{I}_c(k) \subseteq \{1 \ldots, N_c\}$ indicate the set of sensors of cluster c which transmit their measurements $\underline{x}^{\mathrm{P}}_{c,i}(k)$ at time step k and $\mathscr{M}_c(k) \subseteq \{1 \ldots, N_c\}$ the set of sensors of cluster c transmitting a signal $\mu_{c,i}(k)$ at time step k, i.e. requiring a change in the controller state.

Referring to Fig. 9.13, the *event-triggered controller* of cluster c is made of four blocks:

- the *activation/deactivation rule* receives the signals $\mu_{c,i}(k)$ from the sensors $i \in \mathscr{M}_c(k)$ and decides if activating/deactivating the controller;
- the *control law generator*, of MPC type, works only if activated by the activation rule. When it is activated, it receives the measurements of the state from all the sensors $\underline{x}^{\mathrm{P}}_{c,i}(k)$, $i = 1, \ldots, N_c$. The FHOCP is solved either when the controller triggering condition is verified (the controller uses as initial conditions the state $\underline{x}^{\mathrm{E}}_c(k)$) or when a new transmission of the real state of the cluster is received (the controller initial conditions include the measurements $\underline{x}^{\mathrm{P}}_{c,i}(k)$, $i \in \mathscr{I}_c(k)$). At each time step k, the control law generator computes the predicted state $\underline{x}^{\mathrm{C}}_c(k|\bar{k}_c)$ and the control $\underline{u}_c(k)$, denoting with $\bar{k}_c < k$ the last triggering time step;
- the *process emulator* works only if activated by the activation rule; when activated, it receives the measurements from all the sensors $\underline{x}^{\mathrm{P}}_{c,i}(k)$, $i = 1, \ldots, N_c$ and provides a prediction of the plant state using a dynamic model, in general, more detailed than the prediction model in the FHOCP. By receiving at each time step k the control action $\underline{u}_c(k)$, the process emulator generates the state $\underline{x}^{\mathrm{E}}_c(k)$; such a state is updated when new measurements from the plant $\underline{x}^{\mathrm{P}}_{c,i}(k)$, $i \in \mathscr{I}_c(k)$ are received;

- the *controller triggering condition* is used to determine if a new FHOCP must be solved or not; at each time step k, the triggering condition is checked on the basis of the state $\underline{x}_c^{\mathrm{C}}(k|\bar{k}_c)$ and the state $\underline{x}_c^{\mathrm{E}}(k)$, and a binary triggering signal is communicated to the control law generator.

The actuators present in cluster c receive, at each time step k, the control $\underline{u}_c(k)$ to be applied or the information that the controller is not active. Referring again to Fig. 9.13, each *event-triggered sensor* i, $i = 1, \ldots, N_c$, can be represented with four blocks:

- the *process emulator* is the same used in the controller and is active only when the controller is active; in this case, at time step k, it receives the control $\underline{u}_c(k)$ and generates the state $\underline{x}_c^{\mathrm{E}}(k)$; the process emulator updates the state if it receives new measurements $\underline{x}_{c,i}^{\mathrm{P}}(k)$, $i \in \mathscr{I}_c(k)$. When the controller is activated, the process emulator receives the measurements of the state from all the sensors $\underline{x}_{c,i}^{\mathrm{P}}(k)$, $i = 1, \ldots, N_c$;
- the *sensor triggering condition* is active only when the controller is active and compares the state $\underline{x}_{c,i}^{\mathrm{E}}(k)$ with the measured state $\underline{x}_{c,i}^{\mathrm{P}}(k)$; if such a difference is greater than a given threshold, the measured state is transmitted to the controller and to the other sensors;
- the *sensor activation/deactivation condition* is always active. When the controller is active, the sensor verifies if the measured state $\underline{x}_{c,i}^{\mathrm{P}}(k)$ is lower than a given threshold: in this case, the sensor sends a signal $\mu_{c,i}(k)$ to the controller and to the other sensors; if instead the controller is not active, the sensor sends a signal $\mu_{c,i}(k)$ if it verifies that the measured state $\underline{x}_{c,i}^{\mathrm{P}}(k)$ exceeds a suitable threshold and hence it requires the control to be activated;
- the *activation/deactivation rule* is the same present in the controller and is used so that each sensor knows if the controller is activated or not: in the former case, the measured state $\underline{x}_{c,i}^{\mathrm{P}}(k)$ is sent to the other sensors and to the controller.

It is worth noting that in each cluster c all the process emulators (in the sensors and in the controller) are synchronised since they use the same model, update the state simultaneously and consider the same control $\underline{u}_c(k)$. Analogously, all the sensors and the controller are served by the same activation/deactivation rule block, so that the decision on the activation/deactivation of the controller is simultaneously computed by all the system components.

Let us describe more in detail a cluster-based decentralised event-triggered control scheme for a freeway traffic system controlled via ramp metering, by referring to a generic freeway cluster c. Note that the control scheme of the generic cluster c is similar to the model-based event-triggered MPC scheme with sensor decentralised decision-making described in Sect. 9.4.2 with, in addition, the activation/deactivation rule in the controller.

The Event-Triggered Controller in a Freeway Cluster The *control law generator* is an MPC controller which computes the ramp metering control action, i.e. the flows entering the freeway from on-ramps, using the CTM in MLD form for the prediction in the FHOCP. Compared with the CTM in MLD form presented in Sect. 3.3.3, it is here necessary to use two indices, c to indicate the cluster and i to indicate the cell within the cluster. The optimal control sequence computed at time step k is denoted as $r^{\mathrm{C}}_{c,i}(h|k)$, referred to cell $i = 1, \ldots, N_c$ of cluster c and time steps $h = k, \ldots, k+K_p-1$. The predicted state variables computed at time step k are denoted as $\rho^{\mathrm{C}}_{c,i}(h|k)$ and $l^{\mathrm{C}}_{c,i}(h|k)$, for the traffic density and queue length, respectively, of cell i at time steps $h = k+1, \ldots, k+K_p$.

The *activation/deactivation rule* receives the signals $\mu_{c,i}(k)$ from the sensors $i \in \mathscr{M}_c(k)$ and, based on the current state of the controller (active or not), decides if activating/deactivating the controller itself. In particular, if the controller is not active, the following *activation* rule is verified at time step k:

$$|\mathscr{M}_c(k)| \geq 1 \tag{9.18}$$

which corresponds to verify if at least one sensor sends the signal $\mu_{c,i}(k)$ requiring the activation of the controller. If instead the controller of cluster c is active, the following *deactivation* condition is verified at time step k:

$$|\mathscr{M}_c(k)| = C \tag{9.19}$$

which means that all the sensors are requiring the deactivation.

The *process emulator* adopts a dynamic traffic model to represent the system dynamics. As previously observed, also in this case this model can be, for instance, a CTM with a sufficiently large number of cells, or any other type of model which results sufficiently accurate in reproducing the freeway traffic evolution in time. At time step k, the process emulator determines the state variables $\rho^{\mathrm{E}}_{c,i}(k)$ and $l^{\mathrm{E}}_{c,i}(k)$, referred to cell i. The emulator updates its state at time step k if some measurements $\rho^{\mathrm{P}}_{c,i}(k)$, $l^{\mathrm{P}}_{c,i}(k)$, $i \in \mathscr{I}_c(k)$, are sent to the controller. In particular, the updated state is the part of the state relative to the cells which have transmitted the measurements, i.e. $\rho^{\mathrm{E}}_{c,i}(k) = \rho^{\mathrm{P}}_{c,i}(k)$ and $l^{\mathrm{E}}_{c,i}(k) = l^{\mathrm{P}}_{c,i}(k)$, $i \in \mathscr{I}_c(k)$.

The *controller triggering condition* aims to ensure that the FHOCP is not solved at each time step but only when it is strictly necessary, that is, at time steps when the selected triggering rule is satisfied. As in Sect. 9.4.2, the triggering rule is verified when the controller triggering conditions are satisfied or when a new measurement is received from sensors. Specifically, considering that $\bar{k}_c$ is the previous triggering time step, the controller triggering conditions to be verified at a generic time step $k \geq \bar{k}_c$ can be written as

$$\exists i \in \{1, \ldots, N_c\} : \left|\rho^{\mathrm{E}}_{c,i}(k) - \rho^{\mathrm{C}}_{c,i}(k|\bar{k}_c)\right| > \varepsilon^{\rho}_{c,i} \tag{9.20}$$

$$\vee$$

$$\exists i \in \{1, \ldots, N_c\} : \left|l^{\mathrm{E}}_{c,i}(k) - l^{\mathrm{C}}_{c,i}(k|\bar{k}_c)\right| > \varepsilon^{l}_{c,i} \tag{9.21}$$

$$\vee$$

$$k \geq \bar{k}_c + K_p \tag{9.22}$$

where $\varepsilon^{\rho}_{c,i}$ and $\varepsilon^{l}_{c,i}$ are given threshold values.

The overall triggering rule, forcing the solution of a new FHOCP, is given by the controller triggering conditions (9.20)–(9.22) and the following conditions:

$$\exists i \in \{1, \dots, N_c\} : \rho^{\mathrm{P}}_{c,i}(k) \text{ is received from sensors} \tag{9.23}$$

$$\vee$$

$$\exists i \in \{1, \dots, N_c\} : l^{\mathrm{P}}_{c,i}(k) \text{ is received from sensors} \tag{9.24}$$

The *event-triggered MPC mechanism* can be defined as follows:

- at time step $k = 0$, the FHOCP is solved determining the optimal control sequence $r^{\mathrm{C}}_{c,i}(h|0)$, $i = 1, \dots, N_c$, $h = 0, \dots, K_p - 1$, and $r^{\mathrm{C}}_{c,i}(0|0)$ is applied;
- at a generic time step $k > 0$, the triggering rule is verified:
  - if it is not met, the already available control sequence $r^{\mathrm{C}}_{c,i}(k|\bar{k}_c)$ is applied, where $\bar{k}_c$ is the last triggering time step;
  - if it is met, the FHOCP is solved, the optimal control variables $r^{\mathrm{C}}_{c,i}(h|k)$, $i = 1, \dots, N_c$, $h = k, \dots, k + K_p - 1$ are derived and $r^{\mathrm{C}}_{c,i}(k|k)$ is applied. Note that if the triggering rule is met because triggering conditions (9.20)–(9.22) are verified, the initial state for the FHOCP is $\rho^{\mathrm{C}}_{c,i}(k|k) = \rho^{\mathrm{E}}_{c,i}(k)$ and $l^{\mathrm{C}}_{c,i}(k|k) = l^{\mathrm{E}}_{c,i}(k)$, $i = 1, \dots, N_c$. If instead conditions (9.23) and (9.24) are verified, the initial state for the FHOCP is $\rho^{\mathrm{C}}_{c,i}(k|k) = \rho^{\mathrm{P}}_{c,i}(k)$ and $l^{\mathrm{C}}_{c,i}(k|k) = l^{\mathrm{P}}_{c,i}(k)$ for cells $i \in \mathscr{I}_c(k)$ and $\rho^{\mathrm{C}}_{c,i}(k|k) = \rho^{\mathrm{E}}_{c,i}(k)$ and $l^{\mathrm{C}}_{c,i}(k|k) = l^{\mathrm{E}}_{c,i}(k)$ for cells $i \notin \mathscr{I}_c(k)$.

The Event-Triggered Sensors in a Freeway Cluster The *process emulator* in each sensor provides the same system state variables $\rho^{\mathrm{E}}_{c,i}(k)$ and $l^{\mathrm{E}}_{c,i}(k)$, generated by the process emulator in the controller. Analogously to the process emulator in the controller, the process emulators in the sensors update their state when a new transmission is received and are active only when the controller is active.

The *sensor triggering condition*, as the process emulator, works only if the controller is active. The meaning of the sensor triggering condition is to transmit the measured system state when this is necessary. Hence, in each sensor i, the measured system state is transmitted to the controller and to the other sensors when the system state determined by the process emulator and the measured system state differ significantly. More precisely, at a generic time step k, the sensor present in cell i verifies the following triggering condition:

$$\left|\rho^{\mathrm{P}}_{c,i}(k) - \rho^{\mathrm{E}}_{c,i}(k)\right| > \kappa^{\rho}_{c,i} \tag{9.25}$$

$$\vee$$

$$\left|l^{\mathrm{P}}_{c,i}(k) - l^{\mathrm{E}}_{c,i}(k)\right| > \kappa^{l}_{c,i} \tag{9.26}$$

with $\kappa^{\rho}_{c,i}$, $\kappa^{l}_{c,i}$ threshold values. If this condition is fulfilled, the sensor of cell i transmits the current measurement $\rho^{\mathrm{P}}_{c,i}(k)$ and $l^{\mathrm{P}}_{c,i}(k)$ to the controller and to all the other sensors.

The *sensor activation/deactivation condition* is always active. If the controller is active, each sensor i checks the following deactivation condition:

$$\rho^{\mathrm{P}}_{c,i}(k) \leq \zeta^{\rho}_{c,i}(k) \wedge l^{\mathrm{P}}_{c,i}(k) \leq \zeta^{l}_{c,i}(k) \tag{9.27}$$

where $\zeta^{\rho}_{c,i}(k)$ and $\zeta^{l}_{c,i}(k)$ are proper deactivation threshold values. If this deactivation condition is verified, sensor i sends the signal $\mu_{c,i}(k)$ to the controller and to the other sensors.

If instead the controller is not active, each sensor i checks the following activation condition:

$$\rho^{\mathrm{P}}_{c,i}(k) > \sigma^{\rho}_{c,i}(k) \vee l^{\mathrm{P}}_{c,i}(k) > \sigma^{l}_{c,i}(k) \tag{9.28}$$

where $\sigma^{\rho}_{c,i}(k)$ and $\sigma^{l}_{c,i}(k)$ are proper activation threshold values. If this activation condition is verified, sensor i sends the signal $\mu_{c,i}(k)$ to the controller and to the other sensors.

The *activation/deactivation rule* present in each sensor is the same present in the controller. It receives the signals $\mu_{c,i}(k)$ from the other sensors $i \in \mathcal{M}_c(k)$ and decides if activating/deactivating the controller itself, by verifying the activation rule (9.18) if the controller is not active, and the deactivation rule (9.19) if it is active. Note that when the activation rule is verified, all the sensors transmit the measured state $\rho^{\mathrm{P}}_{c,i}(k)$ and $l^{\mathrm{P}}_{c,i}(k)$ to the controller and to all the other sensors.

References

1. Bertsekas DP, Tsitsiklis JN (1997) Parallel and distributed computation. Athena Scientific, Belmont
2. Heemels WPMH, Johansson KH, Tabuada P (2012) An introduction to event-triggered and self-triggered control. In: Proceedings of the 51st IEEE conference on decision and control, pp 3270–3285
3. Hespanha JP, Naghshtabrizi P, Xu Y (2007) A survey of recent results in networked control system. Proc IEEE 95:138–162
4. Gupta RA, Chow M-Y (2010) Networked control system: overview and research trends. IEEE Trans Industr Electron 57:2527–2535
5. Scattolini R (2009) Architectures for distributed and hierarchical model predictive control: a review. J Process Control 19:723–731
6. Cui H, Jacobsen EW (2002) Performance limitations in decentralized control. J Process Control 12:485–494
7. Goldstein NB, Kumar KSP (1982) A decentralized control strategy for freeway regulation. Transp Res Part B 16:279–290
8. Maestre JM, Muñoz de la Peña D, Camacho EF, Alamo T (2011) Distributed model predictive control based on agent negotiation. J Process Control 21:685–697
9. Frejo JRD, Camacho EF (2011) Feasible cooperation based model predictive control for freeway traffic systems. In: Proceedings of the 50th IEEE conference on decision and control and European control conference, pp 5965–5970
10. Frejo JRD, Camacho EF (2012) Global versus local MPC algorithms in freeway traffic control with ramp metering and variable speed limits. IEEE Trans Intell Transp Syst 13:1556–1565

11. Majid H, Hajiahmadi M, De Schutter B, Abouaïssa H, Jolly D, (2014) Distributed model predictive control of freeway traffic networks: a serial partially cooperative approach. In: Proceedings of the 17th international IEEE conference on intelligent transportation systems, pp 1876–1881
12. Ferrara A, Sacone S, Siri S, (2014) Distributed model predictive control for MLD systems: application to freeway ramp metering. In: Proceedings of the American control conference, pp 5294–5299
13. Ferrara A, Nai Oleari A, Sacone S, Siri S (2015) Freeways as systems of systems: a distributed model predictive control scheme. IEEE Syst J 9:312–323
14. Reilly J, Bayen AM (2015) Distributed optimization for shared state systems: applications to decentralized freeway control via subnetwork splitting. IEEE Trans Intell Transp Syst 16:3465–3472
15. Kim B-Y, Ahn H-S (2016) Distributed coordination and control for a freeway traffic network using consensus algorithms. IEEE Syst J 10:162–168
16. Dabiri A, Kulcsár B (2017) Distributed ramp metering - a constrained discharge flow maximization approach. IEEE Trans Intell Transp Syst 18:2525–2538
17. Ghods AH, Fu L, Rahimi-Kian A (2010) An efficient optimization approach to real-time coordinated and integrated freeway traffic control. IEEE Trans Intell Transp Syst 11:873–884
18. Pisarski D, Canudas-de-Wit C (2016) Nash game-based distributed control design for balancing traffic density over freeway networks. IEEE Trans Control Netw Syst 3:149–161
19. Tabuada P (2007) Event-triggered real-time scheduling of stabilizing control tasks. IEEE Trans Autom Control 52:1680–1685
20. Wang X, Lemmon MD (2011) Event-triggering in distributed networked control systems. IEEE Trans Autom Control 56:586–601
21. Ferrara A, Nai Oleari A, Sacone S, Siri S, (2012) An event-triggered model predictive control scheme for freeway systems. In: Proceedings of the 51st IEEE conference on decision and control, pp 6975–6982
22. Ferrara A, Sacone S, Siri S (2015) Event-riggered model predictive schemes for freeway traffic control. Transp Res Part C 58:554–567
23. Ferrara A, Nai Oleari A, Sacone S, Siri S, (2013) Case-study based performance assessment of an event-triggered MPC scheme for freeway systems. In: Proceedings of the European control conference, pp 4027–4032
24. Ferrara A, Sacone S, Siri S, (2013) Supervisory model predictive control for freeway traffic systems. In: Proceedings of the 52nd IEEE conference on decision and control, pp 905–910
25. Groot N, De Schutter B, Zegeye S, Hellendoorn H, (2011) Model-based predictive traffic control: a piecewise-affine approach based on METANET. In: Proceedings of the 18th IFAC world congress, pp 10709–10714
26. Hajiahmadi M, De Schutter B, Hellendoorn H, (2012) Model predictive traffic control: a mixed-logical dynamic approach based on the link transmission model. In: Proceedings of the 13th IFAC symposium on control in transportation systems, pp 144–149
27. Maggi L, Maratea M, Sacone S, Siri S, (2013) Computational analysis of freeway traffic control based on a linearized prediction model. In: Proceedings of the 52nd IEEE conference on decision and control, pp 886–891
28. Maggi L, Sacone S, Siri S, (2015) Freeway traffic control considering capacity drop phenomena: comparison of different MPC schemes. In: Proceedings of the IEEE 18th international conference on intelligent transportation systems, pp 457–462
29. Hajiahmadi M, van de Weg GS, Tampère CMJ, Corthout R, Hegyi A, De Schutter B, Hellendoorn H (2016) Integrated predictive control of freeway networks using the extended link transmission model. IEEE Trans Intell Transp Syst 17:65–78
30. Garcia E, Antsaklis PJ, (2011) Model-based event-triggered control with time-varying network delays. In: Proceedings of the 50th IEEE conference on decision and control and European control conference, pp 1650–1655
31. Garcia E, Antsaklis PJ (2013) Model-based event-triggered control for systems with quantization and time-varying network delays. IEEE Trans Autom Control 58:422–434

32. Ferrara A, Sacone S, Siri S, (2014) Event-triggered strategies for the networked control of freeway traffic systems. In: Proceedings of the European control conference, pp 2594–2599
33. Ferrara A, Sacone S, Siri S (2016) Design of networked freeway traffic controllers based on event-triggered control concepts. Int J Robust Nonlinear Control 26:1162–1183
34. Ljesnjanin M, Quevedo DE, Nesic D (2014) Packetized MPC with dynamic scheduling constraints and bounded packet dropouts. Automatica 50:784–797
35. Bianchi D, Ferrara A, Di Benedetto MD, (2013) Networked model predictive traffic control with time varying optimization horizon: the Grenoble South ring case study. In: Proceedings of the European control conference, pp 4039–4044

Chapter 10
Control Strategies for Sustainable Mobility in Freeways

10.1 Sustainability Concepts for Freeway Traffic Control

The concepts of *sustainability* and *sustainable development* are worldwide recognised as of primary importance for the growth of individuals and organisations, in order to meet the needs of present and future generations. Various definitions of sustainability have been provided in the last decades, highlighting different specific aspects and considering different indicators and goals (see e.g. [1, 2] and the references therein for an overview on these concepts).

Several areas of development and various objectives can be defined for achieving sustainability. For instance, the Sustainable Development Goals Report [3], issued by the United Nations in 2017, has fixed seventeen general goals towards sustainability, regarding health, education, safety of people, as well as careful management of natural resources. These goals represent an ambitious challenge for the entire society in order to achieve an equitable and sustainable progress.

All the definitions of sustainable development, even with different peculiarities, agree on a common point, related to the necessity of strengthening actions now that do not neglect the possible negative consequences that will occur in the ecosystem in the long period. In other words, sustainability means satisfying the present needs of individuals and organisations without compromising the possibility of *future generations* to meet their own necessities.

Another common point in the various definitions of sustainability is related to three main areas of interest, which should be properly integrated and balanced to achieve a sustainable development. These three dimensions are as follows:

- *environment*: environmental protection and ecological integrity should be guaranteed, maintaining a balance among all the natural resources;
- *economy*: the economic sustainability must be preserved to allow that all the human communities have access to the resources they need;
- *society*: healthy, safe and secure systems should be realised to ensure the wellness of people worldwide.

A. Ferrara et al., *Freeway Traffic Modelling and Control*, Advances in Industrial Control, https://doi.org/10.1007/978-3-319-75961-6_10

In this very general and challenging framework, also *road transportation* has an important role. Indeed, it is undeniable that if, on the one hand, the increment in road transport systems allows the improvement of social and economic welfare, on the other hand, such increase produces several negative effects which have implications for the society in the form of social, environmental and, also, economic consequences.

Of course, when a driver is planning a travel or is moving on a road network, he takes his decisions considering his own and presently perceived costs, without estimating or forecasting more general costs related to the ecosystem and the impacts on the future generations. Nevertheless, a *traffic management and control tool*, which acts, instead, at a macroscopic level, should be devised to be *sustainable*, hence not neglecting these global factors and their future implications (see Fig. 10.1). A control tool designed for sustainable mobility should therefore regulate traffic in order to achieve system-wide objectives guaranteeing a high quality of life for citizens and ensuring environmental protection, but also taking into account the individual goals of travellers.

Despite this new sensibility for sustainability concepts, analysing the wide literature on freeway traffic control, it is worth noting that most of the research works are devoted to the sole reduction of congestion phenomena, i.e. to the minimisation of the total time spent by the drivers in the traffic network. However, in the last years, many other sustainability-devoted aspects have received attention, such as the reduction of pollutant emissions, as well as the increase of safety, and have been explicitly taken into consideration in the design of traffic control schemes for realising sustainable mobility systems.

Fig. 10.1 A road stretch in A4 freeway, close to Leiden, the Netherlands (courtesy of Rijkswaterstaat, Photo: Essencia Communication/Rob de Voogd)

In Sect. 10.2, the scientific literature on freeway traffic control explicitly addressing sustainable issues is revised. Then, Sects. 10.3 and 10.4 propose some possible freeway traffic control solutions which take into account, as control objectives, not only the reduction of congestion phenomena but also the mitigation of emissions in the environment, by distinguishing different typologies of vehicles. The control schemes described in Sects. 10.3 and 10.4 are of different types, from simple and easy-to-implement solutions to more sophisticated optimisation-based frameworks, and can constitute a basis for researchers to develop new traffic control strategies for sustainable mobility.

10.2 Overview of Traffic Control Schemes for Sustainable Freeways

Traffic control for sustainable mobility in freeway networks is a very recent research topic that is becoming more and more relevant within the scientific community of traffic control engineers. Sustainable issues can be taken into account in traffic control schemes in different ways. The most relevant directions followed so far to address sustainability-related factors are the following:

- considering *sustainable objectives* explicitly in the controller design;
- differentiating the traffic flow in different *vehicle categories*, so that it is possible to model them in a customised way (this is particularly relevant for instance for emission models) and to control them separately.

While some research works address these two aspects separately (see Sects. 10.2.1 and 10.2.2), some recent works consider them jointly, as discussed in Sect. 10.2.3 in general and addressed more in detail in Sects. 10.3 and 10.4.

10.2.1 Traffic Control with Sustainable Objectives

The idea of considering sustainable issues in the design of a traffic controller is rather recent and has been conceived, in most of the works, by including the reduction of traffic emissions among the objectives of the traffic control scheme. Another aspect that has been addressed explicitly is safety, generally expressed in terms of number of accidents expected to occur in the freeway.

The reduction of *traffic emissions* is explicitly considered as control objective in [4], where a receding-horizon parametrised traffic control strategy is proposed to jointly minimise travel times and emissions in the freeway through ramp metering and variable speed limits. In [5], a general framework is introduced to integrate the macroscopic METANET model with the microscopic emission and fuel consumption model called VT-micro, resulting in the so-called VT-macro model. The purpose of that modelling framework is to provide a prediction tool able to guarantee accurate estimates of the emissions and fuel consumptions in short computational times, as those required by freeway traffic controllers to be applied in real time. In [6], a model for the *dispersion* of traffic emissions along a freeway is proposed: such model should be adopted for control purposes in order to keep pollutant concentrations under legislation limits and, hence, it should require a low computational effort.

The VT-macro framework is adopted in [7], where a Model Predictive Control (MPC) scheme for a combined strategy of ramp metering and variable speed limits is proposed. In [7], in order to deal with an optimal control problem affordable from a computational point of view, the non-linear METANET model is approximated through a piecewise-affine formulation.

Besides ramp metering and variable speed limits, route guidance control has been investigated as well in order to reduce emissions in the freeways, corresponding to the so-called *eco-routing* strategies. For instance, in [8], the authors assess the environmental and energetic impacts produced by the route choice decisions using both a microscopic and a macroscopic tool, and they show that the faster routes preferably chosen by drivers are not always the best in terms of environmental issues and energy consumptions. In [9], a microscopic traffic assignment and simulation framework are proposed for setting eco-routing strategies for drivers.

In [10], game theory is applied for developing road pricing methods for routing drivers in urban and freeway traffic networks. Such methods should be used by traffic authorities to induce users to follow routes that are efficient from a system-optimum perspective, i.e. routes which minimise the total time spent by drivers in the network and reduce the total traffic emissions. In [11], an MPC scheme for real-time route guidance control is proposed, not only to improve traffic efficiency in terms of total time spent by drivers but also considering the reduction of emissions and fuel consumptions for all vehicles moving in the network.

Another very relevant issue addressed in freeway traffic towards sustainability is *road safety*. This aspect has been investigated in many papers and research reports, since it is undeniable that one of the major criticalities and consequences of congested roads is the high number of accidents, often serious or fatal, affecting many drivers every day. The causes of traffic accidents have been examined by researchers and are still under investigation. Many studies in this area rely on statistical analyses of real historical data of crashes, in order to correlate accidents with specific traffic states or conditions, as well as with other factors, such as road geometry, drivers' behaviours and environmental factors.

Among the works investigating the correlation between the safety level in a freeway and the present traffic conditions, it is possible to cite for instance [12], referring to the case of freeways in California, U.S. In that work, traffic data measured with loop detectors and detailed information about accidents, classified in different crash typologies, are used to highlight the relationships between traffic flow conditions and the likelihood of traffic accidents. Based on traffic and crash data from a Canadian case, the study developed in [13] aims at defining a relation between crashes and traffic data, such as flow and density, for both rural and urban freeway segments. A methodology to investigate the relation between traffic states and crash involvements in a freeway is discussed in [14].

Other researchers have focused their attention on analysing the impact on safety of the adoption of traffic control strategies in freeways. For instance, the benefits in terms of crash likelihood reduction due to the application of variable speed limits are analysed in [15, 16]. Analogously, the effects of the implementation of ramp metering strategies on safety are assessed in [17].

Very few research works are devoted to consider safety explicitly in the controller design. In [18], a freeway control algorithm adopting variable speed limits is defined to minimise the total crash risk in the system, while in [19] variable speed limits are applied specifically to reduce rear-end collision risks. In [20], a coordinated ramp metering strategy is proposed in order to both minimise the travel times for the drivers in the freeway and minimise the expected number of crashes in the system.

10.2.2 Multi-class Traffic Control

A relevant feature towards the definition of sustainable traffic control strategies includes the possibility of distinguishing different classes of vehicles, i.e. cars, trucks or other specific vehicles, since they generally present different dynamic behaviours and have different environmental impacts on the freeway system. Also, it is possible to distinguish vehicles in classes according to other aspects, e.g. one can distinguish among private vehicles, public means of transport, vehicles travelling for commercial uses and so on. A *multi-class traffic* framework consists not only in adopting multi-class traffic models but also in designing multi-class control strategies, so that specific control actions are defined for the different vehicle classes.

It is important to emphasise that the use of a *multi-class traffic model* allows to represent the traffic system behaviour more accurately than with a one-class model which assumes that the whole traffic is a homogeneous fluid (see Sects. 3.4 and 4.3 for a detailed discussion and some motivations for multi-class models). This is especially true for instance in case a high percentage of trucks is present in the freeway traffic system, since trucks have a strong impact on the overall traffic flow for many reasons, e.g. for their dimensions, low operating capabilities and so on.

The design of *multi-class traffic controllers* enables the adoption of specific policies for the different classes of vehicles, in order to assign them different priorities or different rules according to their characteristics. It is worth noting that multi-class control requires, from the *implementation* point of view, some specific features to be applied in the actuators. For instance, controlling separately different vehicle classes via ramp metering means that separate lanes and signals must be present at the on-ramps, while, for route guidance and variable speed limits, it means that specific indications must be given to the different vehicle typologies on Variable Message Signs (VMSs). Note that the increasing availability of on-board devices enables the communication of routing indications, as well as speed limits, directly to drivers, further motivating the development of multi-class control strategies.

The idea of proposing multi-class regulators is rather recent and has been developed in few research works. For instance, in [21], combined multi-class strategies relying on ramp metering and variable speed limits are investigated and an MPC control scheme is proposed. Multi-class ramp metering is also analysed in [22], also possibly considering different priorities for different vehicle classes.

10.2.3 Multi-class Sustainable Traffic Control

Some very recent works have combined the emission-related issues with the distinction of multiple vehicle types, leading to *multi-class sustainable control* frameworks. In [23], an MPC approach for multi-class coordinated ramp metering is developed, aiming at jointly reducing traffic emissions and travel times in freeway stretches. A two-class freeway traffic controller to reduce congestion and emissions is also presented in [24], while different multi-class traffic and emission models are compared in [25] for MPC schemes with end-point penalties in the objective function.

In [26], an optimal control scheme is proposed for reducing congestion and improving safety via multi-class coordinated ramp metering. The optimal control problem is solved with derivative-free solution algorithms.

Other multi-class sustainable control frameworks are for instance the local feedback control strategies, of ramp metering type, investigated in [27–29] and described in detail in Sect. 10.3.1. These control strategies are based on standard proportional–integral local controllers, extended to deal with a multi-class traffic flow and to reduce the emissions in the freeway.

These latter local ramp metering strategies were extended in [30, 31], leading to a supervisory coordinated ramp metering framework, in which local feedback controllers receive a communication from a supervisor about the control law to be applied. Specifically, a supervisor, acting at a higher level, receives measurements from the freeway network and periodically makes a prediction on the system evolution. At the lower level, local feedback controllers compute the control action on the basis of measurements in a given area close to the on-ramp and the parameters of the control law are communicated by the supervisor in real time, according to an event-triggered logic. This supervisory event-triggered control scheme for coordinated ramp metering is analysed in Sect. 10.3.2.

Optimal control techniques are adopted in [32, 33] for optimally reducing the total time spent by the drivers and the total emissions experienced by them in freeway systems, as discussed in Sect. 10.3.3. The optimal solution of this non-linear optimal control problem is obtained with gradient-based solution techniques and is used to verify if the reduction of traffic emissions and the reduction of congestion are conflicting objectives or not.

Finally, the combination of ramp metering and route guidance control strategies is exploited in [34, 35] to reduce the total time spent and the total emissions in a balanced way. Both the ramp metering and the route guidance controllers are of the multi-class type and are based on feedback predictive control laws, i.e. they compute the control actions not only on the basis of the measured system state but also on the basis of the prediction of the system evolution, in terms of traffic conditions and traffic emissions. This combined multi-class control framework is described in Sect. 10.4.

10.3 Multi-class Ramp Metering Strategies for Emission Reduction

This section describes three control schemes, having in common the multi-class nature, the adoption of ramp metering as control action, and the combined goal of reducing traffic emissions and mitigating congestion phenomena in a freeway stretch. The first control scheme is a simple local feedback control strategy (see Sect. 10.3.1). This feedback strategy is then extended to be included in a more sophisticated control framework, that is, the supervisory event-triggered control scheme described in Sect. 10.3.2. Finally, an optimal control approach is reported in Sect. 10.3.3, in which the solution found allows to reduce the traffic emissions and the total time spent by the drivers in the whole freeway.

10.3.1 Local Feedback Control

This section presents a *local feedback control strategy*, in which different classes of vehicles are considered, in order to better account for the fact that vehicles of different types present different dynamic behaviours and have different environmental impacts on the freeway system. Of course, the most practical and relevant example of multi-class traffic flow is the two-class case in which cars and trucks are distinguished. In the considered scheme, not only the macroscopic dynamic model is of the multi-class type but also the considered controllers are designed in order to define specific control actions for each vehicle category. More specifically, the adopted control strategy is ramp metering; hence, it is assumed that differently metered lanes are present at the on-ramps for each class of vehicles. It is straightforward that the implementation of *multi-class ramp metering strategies* is realistic with a small number of vehicle classes, surely for the two-class case of cars and trucks.

One of the main advantages of the present local ramp metering control scheme is that it is *simple* and *easily implementable* in real systems. The adopted regulator is a multi-class version of the well-known *PI-ALINEA* strategy, which has shown its effectiveness both theoretically and in practice [36], as discussed in Sect. 8.3.1. Generally speaking, PI-ALINEA is a feedback regulator of proportional–integral type, designed in order to track a set-point value of the density (or occupancy). If the goal of the controller is to reduce the total time spent by the drivers in the freeway system, the set-point is fixed equal to the critical density. Since we are dealing with multi-class control, PI-ALINEA is suitably extended to address the case in which different classes of vehicles are separately controlled.

Let us start by introducing the standard *one-class PI-ALINEA* in case the METANET model for a freeway stretch with on-ramps described in Sect. 4.2.2 is adopted. This is a discrete model, in which the freeway stretch is divided into N road sections (with index i indicating the generic road section of length L_i), while the time horizon is discretised into K time intervals (with k the index of the time step and T the sample time).

In that model, the generic ramp metering control variable is $r_i^{\mathrm{C}}(k) \in [r_i^{\min}, r_i^{\max}]$, representing the flow that should enter section i from the on-ramp during time interval $[kT, (k+1)T)$, $i = 1, \ldots, N$, $k = 0, \ldots, K$. The PI-ALINEA control law follows

$$r_i^{\mathrm{C}}(k) = r_i^{\mathrm{C}}(k-1) - K_P \left[\rho_i^{\mathrm{down}}(k) - \rho_i^{\mathrm{down}}(k-1)\right] + K_R \left[\rho_i^* - \rho_i^{\mathrm{down}}(k)\right] \quad (10.1)$$

where $\rho_i^{\mathrm{down}}(k)$ is the traffic density measured downstream the on-ramp, ρ_i^* is a set-point value for the downstream density, while K_P and K_R are regulator parameters.

Let us now describe the *multi-class PI-ALINEA* regulator, introduced for the first time in [37]. Let us rely on the multi-class METANET model for a freeway stretch described in Sect. 4.3.1, in which C classes of vehicles are explicitly modelled, with conversion factor η^c, $c = 1, \ldots, C$. Remind that $\rho_i^c(k)$ is the traffic density of class c in section i at time kT, while $l_i^c(k)$ is the queue length of vehicles of class c waiting in the on-ramp of section i at time kT. Moreover, the ramp metering control variable is referred to each class c. Specifically, the control variable is denoted as $r_i^{\mathrm{C},c}(k) \in [r_i^{\min,c}, r_i^{\max,c}]$, representing the flow of class c that should enter section i from the on-ramp during time interval $[kT, (k+1)T)$.

According to the multi-class PI-ALINEA regulator (see a generic scheme in Fig. 10.2), the on-ramp flow is computed by extending the control law (10.1) to the multi-class case and by taking into account that the set-point for the density ρ_i^* is still referred to the total density. Hence, the flow $r_i^{\mathrm{C},c}(k)$ at the on-ramp of section i, at time step k, for class c, is obtained as

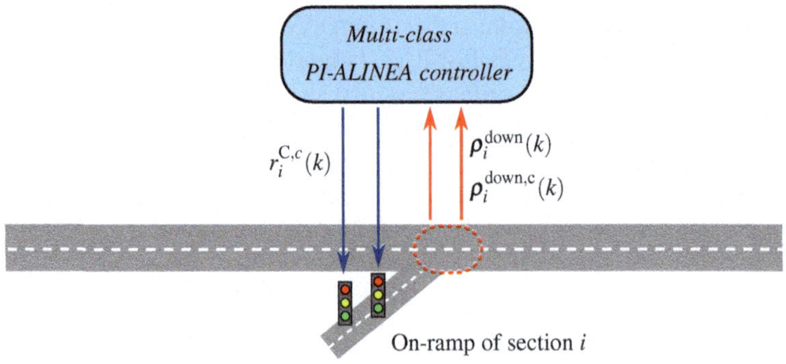

Fig. 10.2 Multi-class PI-ALINEA controller

$$r_i^{\mathrm{C},c}(k) = r_i^{\mathrm{C},c}(k-1) - K_P{}^c \left[\rho_i^{\mathrm{down},c}(k) - \rho_i^{\mathrm{down},c}(k-1)\right] \\ + K_R{}^c\, f_i^c(k) \left[\rho_i^* - \rho_i^{\mathrm{down}}(k)\right] \quad (10.2)$$

where $\rho_i^{c,\mathrm{down}}(k)$ is the traffic density of class c measured downstream the on-ramp, while $K_P{}^c$ and $K_R{}^c$ are parameters of the regulator depending on class c. Note that the multi-class PI-ALINEA control law (10.2) is based on density measurements, both referred to the total density and to the density of specific class c. In particular, the total density measurement $\rho_i^{\mathrm{down}}(k)$ is used to compute the difference with the set-point value ρ_i^*, since this latter is a reference value for the total density.

In (10.2), the term depending on the difference between the total density and the set-point value is split, among the different vehicle classes, by means of the *ratio* $f_i^c(k)$. Specifically, this ratio computes, for road section i and time step k, the quantity of vehicles of class c over the total vehicles, and is given by

$$f_i^c(k) = \frac{\eta^c \left[l_i^c(k) + \rho_i^c(k) L_i\right]}{\sum_{h=1}^{C} \eta^h \left[l_i^h(k) + \rho_i^h(k) L_i\right]} \quad (10.3)$$

The adoption of ramp metering control laws as the one expressed by (10.2) may cause the creation of long queues at the on-ramps, especially when the mainstream is congested. This phenomenon is often not feasible, because the on-ramps have physical limitations which impose a maximum queue limit. Nevertheless, also in the cases in which no restrictive physical limitations occur, too long queues are undesirable, for instance, because they can imply high concentrations of polluting emissions close to urban areas. Taking into account such motivations, the control law (10.2) can be extended in order to consider possible *maximum values of the queue lengths*.

Let us denote with $l_i^{\max,c}$ the maximum queue length for section i and class c. The limit on the queue length is activated in case the on-ramp flow computed by multi-class PI-ALINEA according to (10.2) creates a queue that is higher than $l_i^{\max,c}$. In this case, this on-ramp flow should be increased in order to reduce the queue to be not greater than its maximum value $l_i^{\max,c}$. The detailed algorithm for including the maximum queue lengths on the multi-class PI-ALINEA control law can be found in [27, 28].

It is worth noting that ALINEA and PI-ALINEA, in the standard one-class case, have been generally applied in order to reduce the Total Time Spent (TTS), i.e. to increase the system throughput, by fixing the set-point equal to the critical density, as deeply discussed in Sect. 8.3.1. In [27–29], different simulation analyses have been carried out to verify if these types of controllers, especially in the multi-class case, can be used also to reduce the Total Emissions (TE) of vehicles in the freeway. In particular, in those works, it has been analysed, first, which values of the downstream density set-point should be chosen in order to reduce the traffic emissions in the freeway and, second, if reducing the traffic emissions can also imply a reduction of the TTS by the drivers in the freeway or if, instead, the two control objectives are conflicting. The precise definitions and formulas for the TTS and TE can be found in Sects. 8.2.2 and 8.2.3, respectively.

In particular, this analysis has been carried out in [27, 28] by adopting the average-speed emission model COPERT for computing the emissions in the freeway, both in the mainstream and at the on-ramps (a detailed description of this model, as well as the mathematical formulation to adopt it within a freeway traffic model, is reported in Sect. 6.3). Note that the choice of this model is basically motivated by the fact that it is able to provide good estimations of the traffic emissions, while being simple to be applied. A similar analysis has been carried out in [29], where, instead, the VERSIT+ model is adopted to estimate the traffic emissions in the freeway system (see further details on VERSIT+ in Sect. 6.4).

All these tests have led to conclude that multi-class PI-ALINEA controllers represent an effective solution to reduce emissions and congestion in a freeway traffic system. In addition, the simulation analysis referred to many different traffic scenarios has shown that the reduction of emissions and the maximisation of the throughput are nonconflicting objectives, since both the total emissions and the congestion are reduced if this type of control actions is applied. Furthermore, the results reveal that the adoption of ramp metering control strategies may cause a high concentration of pollutants at the entering on-ramps that could be very critical, especially if the on-ramps are located in proximity of urban areas. As a consequence, the effect of these emissions should be expressly computed by means of models that calculate the emissions both in the mainstream and at the on-ramps, as done by both the COPERT and the VERSIT+ models, described in Sects. 6.3 and 6.4.

10.3.2 Supervisory Event-Triggered Control

The multi-class local feedback PI-ALINEA regulators described in Sect. 10.3.1 have shown to be effective in reducing congestion and emissions in freeways, since these two objectives have generally a nonconflicting nature. Nevertheless, as also discussed in Sect. 8.3.1, it is well known that the main weaknesses of ALINEA-like feedback regulators are due to their local nature, since they compute the control law only on the basis of measurements close to the on-ramp in which the control action is actuated. This aspect was addressed for instance in [36], where the authors analyse the application of ALINEA and PI-ALINEA in the presence of bottlenecks that are located far downstream the merge area.

The control framework described in this section goes further in the idea of considering distant bottlenecks, since it is based on *extended multi-class PI-ALINEA controllers* which compute the control law not only on the basis of the measurement downstream the on-ramp but also on the basis of measurements in a *neighbourhood* of the on-ramp. These further measurements refer to locations that are time-varying and are communicated to the local controllers by a *supervisor*, which acts according to an event-triggered nature, i.e. it changes the parameters of the control laws of the PI-ALINEA controllers only when suitable triggering conditions are met. The interested reader can find more details on event-triggered control, and its application to freeway traffic regulation, in Chap. 9.

The supervisory event-triggered control scheme based on extended PI-ALINEA controllers has been proposed for the first time in [30] for the one-class case, and then extended to the multi-class case in [31]. Note that the notation adopted here to describe the supervisory event-triggered control scheme is referred to a freeway stretch, to be more easily comparable with the other controllers described in this section, and then it is slightly different from the one used in [31], where a freeway network is instead considered.

The supervisory event-triggered control scheme is a hierarchical scheme composed of two levels (see a sketch in Fig. 10.3):

- at the *higher level*, the supervisor receives measurements from the network, periodically makes a prediction on the system evolution and, on the basis of this information, decides if the parameters of the control law for the lower level controllers should be updated or not, according to an event-triggered logic;
- at the *lower level*, local feedback controllers, specifically extended multi-class PI-ALINEA controllers, compute the control action on the basis of measurements in a neighbourhood of the on-ramp (the neighbourhood composition and the parameters of the control law are communicated by the supervisor).

A key point in this control scheme is the definition of the *neighbourhood* of a given on-ramp, from which measurements are taken. This neighbourhood is time-varying and is decided by the supervisor according to an event-triggered logic. It

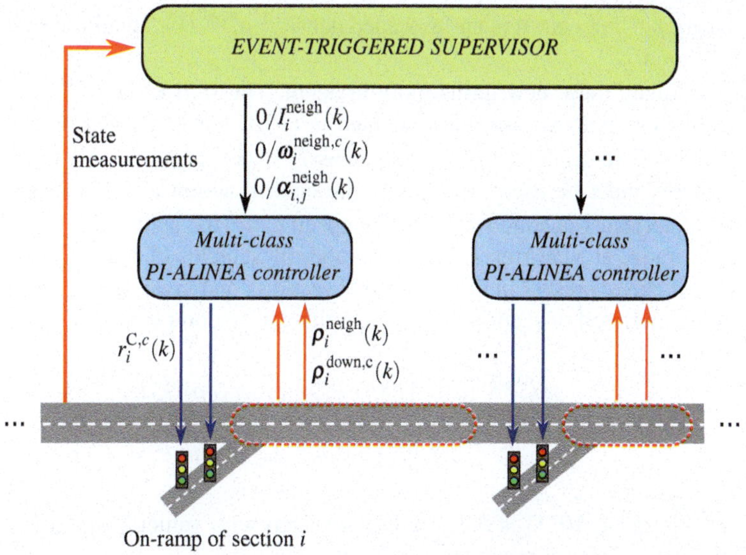

Fig. 10.3 Supervisory event-triggered control scheme based on extended multi-class PI-ALINEA controllers

is defined as a set of road sections downstream the on-ramp, i.e. more specifically, the neighbourhood associated with an on-ramp always starts from the section of the on-ramp and may last until a downstream section located before the next controlled on-ramp.

The two levels of the control scheme are described hereinafter.

Extended Multi-class PI-ALINEA Controllers As already done in Sect. 10.3.1, let us rely on the multi-class METANET model for a freeway stretch described in Sect. 4.3.1, in which the freeway stretch is divided in N road sections, the time horizon is discretised into K time intervals and C classes of vehicles are considered. According to this model, the ramp metering control variable is $r_i^{\mathrm{C},c}(k) \in [r_i^{\min,c}, r_i^{\max,c}]$, i.e. the flow of class c that should enter section i from the on-ramp during time interval $[kT, (k+1)T)$.

In the supervisory event-triggered scheme, the flow $r_i^{\mathrm{C},c}(k)$ at the on-ramp of section i, at time step k, for class c, is computed as

$$
\begin{aligned}
r_i^{\mathrm{C},c}(k) = r_i^{\mathrm{C},c}(k-1) - K_P{}^c\left[\rho_i^{\mathrm{down},c}(k) - \rho_i^{\mathrm{down},c}(k-1)\right] \\
+K_R{}^c\, f_i^{\mathrm{neigh},c}(k)\left[\rho_i^* - \rho_i^{\mathrm{neigh}}(k)\right]
\end{aligned}
\quad (10.4)
$$

where, as in (10.2), $\rho_i^{\mathrm{down},c}(k)$ is the traffic density of class c measured downstream the on-ramp, ρ_i^* is a set-point value for the downstream total density, $K_P{}^c$ and $K_R{}^c$ are parameters of the regulator depending on class c. Differently from (10.2), the split ratio $f_i^{\mathrm{neigh},c}(k)$ now depends on the traffic state in the neighbourhood of the on-ramp, and the density to be compared with the set-point value is no more the measured total density $\rho_i^{\mathrm{down}}(k)$ but it is an 'extended density' $\rho_i^{\mathrm{neigh}}(k)$, again referred to the neighbourhood.

Let us explain these new terms more in detail. The *split ratio* $f_i^{\mathrm{neigh},c}(k)$ has a meaning analogous to the split ratio defined in (10.3) but it is now referred to the *neighbourhood*, i.e. to the set of road sections $I_i^{\mathrm{neigh}}(k) \subseteq \{1, \ldots, N\}$ from which measurements must be used to compute the control action to be actuated at the on-ramp of section i at time step k. More specifically, the split ratio $f_i^{\mathrm{neigh},c}(k)$ is a weighted ratio, at time step k, of the number of vehicles of class c over all the vehicles, which are present in the on-ramp of section i and in the sections belonging to the neighbourhood of section i. Such quantity can be computed as

$$
f_i^{\mathrm{neigh},c}(k) = \frac{\omega_i^{\mathrm{neigh},c}(k)\eta^c\left[l_i^c(k) + \sum_{j\in I_i^{\mathrm{neigh}}(k)} \rho_j^c(k)L_j\right]}{\sum_{h=1}^{C} \omega_i^{\mathrm{neigh},h}(k)\eta^h\left[l_i^h(k) + \sum_{j\in I_i^{\mathrm{neigh}}(k)} \rho_j^h(k)L_j\right]} \quad (10.5)
$$

where $\omega_i^{\mathrm{neigh},c}(k) \in [0, 1]$ is the weight associated with vehicles of class c in the neighbourhood of the on-ramp of section i at time step k.

Moreover, in (10.4), the value of the *extended total density* to be compared with the set-point is a weighted sum of the total densities in the neighbourhood, i.e.

$$\rho_i^{\text{neigh}}(k) = \sum_{j \in I_i^{\text{neigh}}(k)} \alpha_{i,j}^{\text{neigh}}(k) \rho_j(k) \tag{10.6}$$

where $\alpha_{i,j}^{\text{neigh}}(k) \in [0, 1]$, $j \in I_i^{\text{neigh}}$, are parameters decided by the supervisor in order to properly weigh the measurements in the different sections belonging to the neighbourhood of the on-ramp of section i at time step k.

Event-Triggered Supervisor The supervisor receives measurements from the system state of the whole freeway stretch and makes periodic predictions. On the basis of the measured and predicted state variables and global performance indexes over the whole freeway stretch, the supervisor verifies specific triggering conditions and evaluates whether the parameters of the present control law of the extended multi-class PI-ALINEA controllers must be changed or not. The idea is that changes in the control laws are required if there are relevant variations in the system state and/or in the predicted system evolution, either locally or globally.

More specifically, at each time step $k = 0, \ldots, K - 1$ the supervisor receives *measurements* of the system state over the whole network, i.e. the traffic densities $\rho_i^c(k)$, the mean traffic speeds $v_i^c(k)$, and the queue lengths $l_i^c(k)$, $\forall c$, $\forall i$. Besides monitoring the single values of these state variables, at each time step k the supervisor also computes some *performance indexes* referred to the entire network and specifically defined for each vehicle class c. For instance, in [31], two global indicators have been defined, i.e. the instantaneous number of vehicles of class c in the network, denoted as $\eta^c(k)$, and the instantaneous emissions of vehicles of class c in the network, denoted as $\xi^c(k)$ (see [31] for the precise formula to compute these quantities).

The supervisor periodically makes a *prediction* of the system state evolution. In particular, the prediction of the system is computed at each time step $\bar{k} = nP$, where $n = 0, 1, 2, \ldots$ and P is an integer representing the number of time steps between one prediction and the next one. The prediction is realised over a given prediction horizon of K_p time steps. Note that different traffic and emission models can be used for the prediction; for instance, in [31], the multi-class METANET model for a freeway network described in Sect. 4.3.2 and the VERSIT+ emission model reported in Sect. 6.4 are used.

On the basis of the system state measured at time step $k = \bar{k}$ and using suitable traffic flow and emission models, the supervisor computes the *predicted state*, in terms of predicted traffic densities $\tilde{\rho}_i^c(k)$, predicted mean traffic speeds $\tilde{v}_i^c(k)$, and predicted queue lengths $\tilde{l}_i^c(k)$, $\forall c$, $\forall i$, $k = \bar{k} + 1, \ldots, \bar{k} + K_p$. With these predicted state variables, the supervisor also computes the *predicted* values of the considered *performance indexes*, e.g. the predicted instantaneous number of vehicles $\tilde{\eta}^c(k)$ and the predicted instantaneous emissions $\tilde{\xi}^c(k)$, $\forall c$, $k = \bar{k} + 1, \ldots, \bar{k} + K_p$.

If a change is required, the supervisor defines a new *neighbourhood* and the associated parameters, i.e. it properly communicates to the extended multi-class PI-ALINEA controller of the on-ramp of a generic section i the neighbourhood composition $I_i^{\text{neigh}}(k)$, the weights $\omega_i^{\text{neigh},c}(k)$, $c = 1, \ldots, C$, and $\alpha_{i,j}^{\text{neigh}}(k)$, $j \in I_i^{\text{neigh}}(k)$.

The event-triggered behaviour of the supervisor can be summarised as follows:

- at each time step $k \neq \bar{k}$, the supervisor verifies specific triggering conditions on the measured system state and on the global indicators;
- at each time step $k = \bar{k}$, the supervisor verifies specific triggering conditions on the measured system state, on the global indicators, as well as on the predicted system state and on the predicted global indicators;
- if at least one of the triggering conditions is met for the on-ramp of section i, the *neighbourhood* of section i is updated, i.e. the neighbourhood composition $I_i^{\text{neigh}}(k)$, the weights $\omega_i^{\text{neigh},c}(k)$, $c = 1, \ldots, C$, and $\alpha_{i,j}^{\text{neigh}}(k)$, $j \in I_i^{\text{neigh}}(k)$, are communicated to the extended multi-class PI-ALINEA controller in the on-ramp of section i, in order to compute (10.4), with (10.5) and (10.6);
- if none of the triggering conditions is met, the supervisor does not communicate any change to the extended multi-class PI-ALINEA controllers, which continue to apply the same control law (10.4) as before.

Different triggering conditions can be defined, and different logics to update the control law parameters can be formalised. The interested reader can find some examples for instance in [30, 31].

10.3.3 Coordinated Optimal Control

The combined reduction of traffic emissions and congestion in freeways is also the goal of the coordinated multi-class ramp metering control strategy described in this section, based on optimal control techniques. The control strategy is here sought by defining and solving an *optimal control problem* which turns out to be a finite-horizon non-linear optimal control problem with constrained control variables, that can be found also in [32, 33].

As introduced in Sect. 8.4.2, applying optimal control techniques for freeway traffic means that the control actions are computed by considering the dynamic evolution of the freeway traffic system over a given time horizon and by optimising its performance on the basis of specified control objectives. Hence, an optimal control problem is defined, being characterised by an objective function (the performance to be optimised), the state and control variables to be computed, and the constraints representing the dynamics of the system and bounds on the control variables. A sketch of the multi-class ramp metering optimal control strategy is reported in Fig. 10.4.

In the specific case considered here, the dynamic evolution of the system is expressed in terms of the multi-class METANET model for a freeway stretch described in Sect. 4.3.1 (in case the ramp metering control variables are the metering rates), which is a discrete-time non-linear model. The emission model COPERT described in Sect. 6.3 is used to compute the emissions in the freeway system.

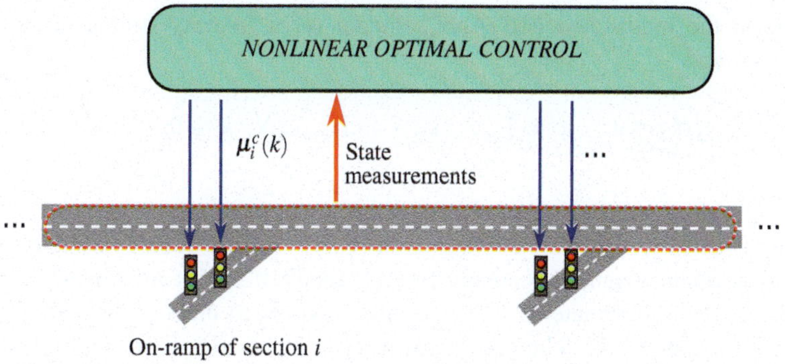

Fig. 10.4 Multi-class coordinated ramp metering optimal control strategy

The state variables are given by the traffic densities $\rho_i^c(k)$, the mean speeds $v_i^c(k)$, and the queue lengths $l_i^c(k)$ for each class $c = 1, \ldots, C$, for every section $i = 1, \ldots, N$, referred to time step $k = 0, \ldots, K$. The control variables are the ramp metering rates $\mu_i^c(k) \in [\mu_i^c(k), 1]$, $c = 1, \ldots, C$, $i = 1, \ldots, N$, $k = 0, \ldots, K$.

The objective function takes into account the minimisation of the TTS and the TE, according to the definition provided in Sects. 8.2.2 and 8.2.3, respectively.

Let us report the formulation of the optimal control problem for finding the optimal multi-class ramp metering control strategy which minimises traffic emissions and congestions in a freeway stretch over a finite horizon of K time steps.

Problem 10.1 Given the system initial conditions $\rho_i^c(0)$, $v_i^c(0)$, $l_i^c(0)$, $i = 1, \ldots, N$, $c = 1, \ldots, C$, given the estimated exogenous inputs $q_0^c(k)$, $v_0^c(k)$, $\rho_{N+1}^c(k)$, $s_i^c(k)$, $d_i^c(k)$, $c = 1, \ldots, C$, $i = 1, \ldots, N$, $k = 0, \ldots, K-1$, find the optimal control sequence $\mu_i^c(k)$, $c = 1, \ldots, C$, $i = 1, \ldots, N$, $k = 0, \ldots, K-1$, that minimises

$$J = \beta \Gamma TE + (1-\beta) TTS + J^{\mu} + J^{l} \tag{10.7}$$

with

$$TE = \sum_{k=0}^{K} \sum_{i=1}^{N} E_i^M(k) + \sum_{k=0}^{K} \sum_{i=1}^{N} E_i^R(k) \tag{10.8}$$

$$TTS = T \sum_{k=0}^{K} \sum_{i=1}^{N} \sum_{c=1}^{C} \eta^c \rho_i^c(k) L_i + T \sum_{k=0}^{K} \sum_{i=1}^{N} \sum_{c=1}^{C} \eta^c l_i^c(k) \tag{10.9}$$

$$J^{\mu} = \sum_{k=1}^{K-1} \sum_{i=1}^{N} \sum_{c=1}^{C} \omega_i^c \left[\mu_i^c(k) - \mu_i^c(k-1)\right]^2 \tag{10.10}$$

$$J^{l} = \sum_{k=0}^{K} \sum_{i=1}^{N} \sum_{c=1}^{C} \gamma_i^c \left[\max\left\{0, l_i^c(k) - l_i^{\max,c}\right\}\right]^2 \tag{10.11}$$

subject to the multi-class METANET model for a freeway stretch described in Sect. 4.3.1 and

$$\mu_i^{\min,c} \leq \mu_i^c(k) \leq 1 \quad c = 1, \ldots, C, \ i = 1, \ldots, N, \ k = 0, \ldots, K-1 \quad (10.12)$$

□

The first two terms in the cost function (10.11) are the Total Emissions and the Total Time Spent, given, respectively, by (10.8) and (10.9), that are properly weighted by means of the parameter $\beta \in [0, 1]$, and reported to the same order of magnitude with parameter Γ. The third term in (10.11), i.e. J^{μ}, is introduced in order to prevent oscillations of the control variables over consecutive time steps, and ω_i^c, $c = 1, \ldots, C$, $i = 1, \ldots, N$, are suitable weights. Finally, the last cost term J^l is included to penalise the cases in which the queue lengths at the on-ramps exceed their limits $l_i^{\max,c}$, with proper weights γ_i^c, $c = 1, \ldots, C, i = 1, \ldots, N$.

Problem 10.1 is a constrained non-linear optimal control problem. Problems of this kind often arise in the freeway traffic control domain, as deeply discussed in Sect. 8.4.2. Their numerical solution may be attempted by direct use of available non-linear programming codes, but this can present difficulties, especially in case of large-scale applications, since the problem dimensions become very high. An efficient numerical solution can be obtained by use of the *feasible direction algorithm*, which is a gradient-based algorithm adopted within the optimal freeway traffic control tool AMOC [38, 39]. A very efficient algorithm to solve this problem is the version of the feasible direction algorithm which applies the derivative back-propagation method *RPROP* (see [40] for further details on this algorithm and [41] for a recent application of this algorithm to another traffic control problem).

The feasible direction algorithm applying the derivative back-propagation method RPROP has been used to solve Problem 10.1 in [32, 33], where a detailed simulation analysis has been also carried out, for the specific case of two classes of vehicles, cars and trucks. In particular, different traffic scenarios have been considered, with and without limits on the maximum queue lengths, and by varying the parameter β assigning different importance levels to the minimisation of the TE and the TTS, respectively. As aforementioned in the previous sections, the results reported in [32, 33] show that the TE and the TTS are largely non-conflicting objectives, since both the average travel times and the emissions are reduced if the control actions manage to reduce or eliminate traffic congestion.

10.4 Multi-class Combined Strategies for Emission Reduction

A multi-class control strategy to reduce congestion and traffic emissions is reported in this section, based on the control scheme presented in [35], a preliminary version of which can be found in [34]. The main difference with the control strategies discussed

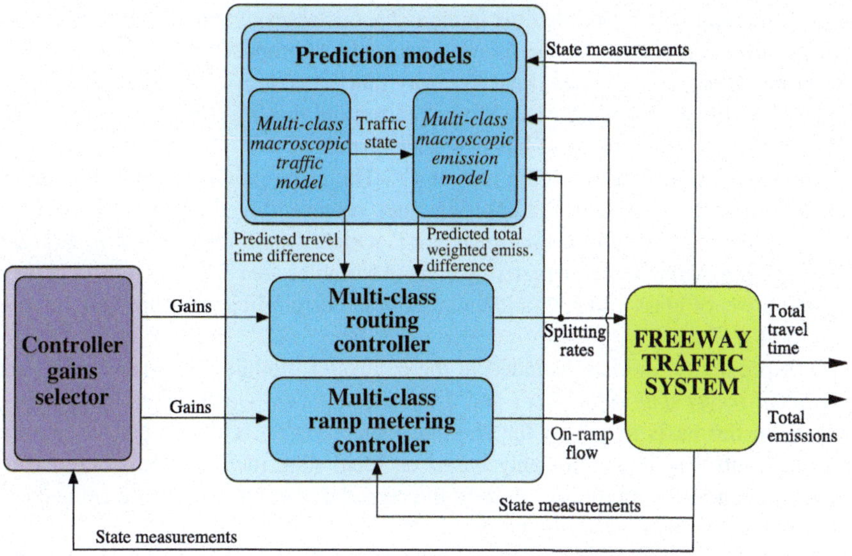

Fig. 10.5 Layout of the multi-class combined ramp metering and route guidance control framework

in Sect. 10.3, which rely on ramp metering for a freeway stretch, is that here a traffic network composed of interconnected stretches is considered and a combined control action given by ramp metering and route guidance is taken into account.

In particular, route guidance is actuated through VMSs (located near the freeway bifurcations) to inform the road users about alternative routes. These indications are assumed to be specifically differentiated for the different classes of vehicles. Moreover, ramp metering is applied in order to regulate the access of traffic to the mainstream through traffic signals installed at the on-ramps. Again, the ramp metering strategy is of the multi-class type, i.e. the different classes of vehicles have dedicated lanes and signals.

The layout of the proposed control framework is depicted in Fig. 10.5. The overall scheme consists of two main components:

- *multi-class controllers*: two types of controllers are adopted, a route guidance controller and a ramp metering controller;
- *gains selector*: this module computes the values of controller gains, according to a specified selection procedure.

The multi-class *routing* controller includes a traffic model and an emission model in order to make predictions of the system state evolution. In particular, the prediction models are run periodically and are initialised with the current system state. The multi-class METANET model for a freeway network described in Sect. 4.3.2 and the VERSIT+ model reported in Sect. 6.4 are here considered.

Let us briefly recall the main notation of the multi-class METANET model for a freeway network described in Sect. 4.3.2, in which the time horizon is divided into

K time intervals (with sample time interval T), C classes of vehicles are considered (with conversion factor η^c), the freeway network is composed of M freeway links, O origin links, and N nodes. Each freeway link $m = 1, \ldots, M$ is further divided into N_m sections with length L_m. Moreover, remind that $\rho^c_{m,i}(k)$ denotes the traffic density of class c in section i of link m at time instant kT and $l^c_o(k)$ is the queue length of class c at origin link o at time instant kT. The route guidance control variables are $\beta^{C,c}_{m,n,j}(k) \in [0, 1]$, i.e. the splitting rates representing the portion of flow of class c present in node n at time instant kT which should choose link m to reach destination j, while the ramp metering control variables are $r^{C,c}_o(k) \in [r^{\min,c}_o, r^{\max,c}_o]$, i.e. the flows of class c that should enter from the origin link o during time interval $[kT, (k+1)T)$.

The prediction models included in the controller module allow to compute the *predicted travel time differences* and the *predicted total weighted emission differences*, on the basis of which the routing control action is determined. Since the routing control action is not only based on local state measurements but also on traffic and emission prediction models, the considered controller can be defined as a *local feedback predictive* regulator.

The multi-class *ramp metering* controller computes the on-ramp flow on the basis of the local measurements obtained from the real system; hence, it is a *local feedback* regulator.

Both the controllers are characterised by some parameters, to be properly tuned, that are the *controller gains*. These gains are provided by the gains selector module (see Fig. 10.5), which includes a library of traffic scenarios (corresponding to specific traffic states and demand patterns), each of which has a set of associated controller gains. These gains are calibrated through a specific optimisation-based procedure which is applied offline. Moreover, the gains selector uses a classification algorithm which periodically chooses, on the basis of the present system state and the estimated demands, the most proper scenario and the corresponding controller gains.

The main components of the considered control scheme are described below, i.e. the multi-class routing controller, the multi-class ramp metering controller and the controller gains selector. The interested reader can find more details in [35].

The Multi-class Routing Controller The routing control strategy consists in informing the users about the preferred link to choose in a bifurcation, as deeply discussed in Sect. 8.3.3 on local route guidance strategies.

The routing controller relies on a prediction module, which runs the prediction periodically, and computes the routing control action with the same sample time. Specifically, let us denote with $\bar{k}$ the generic time step in which the prediction is run and the routing control law is updated. Such a prediction refers to alternative paths starting from bifurcation nodes, i.e. nodes having two outgoing freeway links. Therefore, it is carried out considering some *virtual test vehicles*, which leave the bifurcation node in order to reach their destination through alternative paths. The predicted behaviour of the virtual test vehicles is used to have information about such alternative paths.

Let us consider a generic bifurcation node n, from which it is possible to reach a generic destination j, and let us denote with m and m' the two exiting links (see also Fig. 8.9). Since the routing suggestion in node n is related to the choice of one of the two freeway links exiting the node, it is possible to gather the different paths connecting node n to destination j in two sets according to the freeway link exiting node n in each path. The two sets are denoted as the set of *primary* and *secondary* paths on the basis of the most common path choices made by the drivers: the primary paths having m as first freeway link and the secondary paths having m' as first freeway link.

For each pair of nodes (n, j), a number of virtual vehicles equal to the number of paths from n to j are introduced for each class of vehicles. The prediction at a generic time step $\bar{k}$ is realised assuming that the routing control actions are maintained constant for the whole prediction horizon, while the ramp metering control actions are computed with the multi-class PI-ALINEA control law.

The computations related to each virtual vehicle end when the vehicle itself reaches its destination. In particular, for each virtual vehicle, the following quantities are computed:

- the *predicted travel time* needed by the virtual vehicle to reach its destination;
- the *predicted total weighted emissions* experienced by the virtual vehicle to reach its destination.

The predicted travel time for the primary path and the one for the secondary path are then calculated as the minimum among all the predicted travel times of primary and secondary paths, respectively. Analogously, the predicted total weighted emissions are computed for the primary path and the secondary path. For each pair of nodes (n, j) and for each vehicle class c, it is possible to compute the *predicted travel time difference* at time step $\bar{k}$, denoted as $\Delta t^c_{n,j}(\bar{k})$, and the *predicted total weighted emission difference*, denoted as $\Delta e^c_{n,j}(\bar{k})$, being the difference computed between the secondary path and the primary path. These differences of travel times and total weighted emissions are used to calculate the control law, relying on equilibrium concepts.

Conditions of *Dynamic User Equilibrium* have been widely used in route guidance control schemes, by considering that traffic flows with the same origin and destination are distributed in the network so that the travel times on these routes are the same. At a generic time step $\bar{k}$ at which the routing control action at node n is computed, the conditions of Dynamic User Equilibrium relate the predicted travel time difference $\Delta t^c_{n,j}(\bar{k})$ with the splitting rates $\beta^{\mathrm{t,c}}_{m,n,j}(\bar{k})$, indicating the portion of flow of class c present in node n at time step $\bar{k}$ which should choose link m to reach destination j in order to reduce the travel times. Such conditions can be defined as

$$\Delta t^c_{n,j}(\bar{k}) > 0 \quad \Rightarrow \quad \beta^{\mathrm{t,c}}_{m,n,j}(\bar{k}) = 1 \tag{10.13}$$

$$\Delta t^c_{n,j}(\bar{k}) = 0 \quad \Rightarrow \quad 0 < \beta^{\mathrm{t,c}}_{m,n,j}(\bar{k}) < 1 \tag{10.14}$$

$$\Delta t_{n,j}^{c}(\bar{k}) < 0 \quad \Rightarrow \quad \beta_{m,n,j}^{\mathrm{t},c}(\bar{k}) = 0 \tag{10.15}$$

Analogously to the travel times, it is possible to consider a *Dynamic Emission Equilibrium*, aimed at balancing the weighted pollutant emissions along the suggested routes. The conditions of Dynamic Emission Equilibrium may be formulated as a relation between the predicted total weighted emission difference $\Delta e_{n,j}^{c}(\bar{k})$ and the splitting rates $\beta_{m,n,j}^{\mathrm{e},c}(\bar{k})$, indicating the portion of flow of class c present in node n at time step $\bar{k}$ which should choose link m to reach destination $j$in order to reduce the total weighted emissions, i.e.

$$\Delta e_{n,j}^{c}(\bar{k}) > 0 \quad \Rightarrow \quad \beta_{m,n,j}^{\mathrm{e},c}(\bar{k}) = 1 \tag{10.16}$$

$$\Delta e_{n,j}^{c}(\bar{k}) = 0 \quad \Rightarrow \quad 0 < \beta_{m,n,j}^{\mathrm{e},c}(\bar{k}) < 1 \tag{10.17}$$

$$\Delta e_{n,j}^{c}(\bar{k}) < 0 \quad \Rightarrow \quad \beta_{m,n,j}^{\mathrm{e},c}(\bar{k}) = 0 \tag{10.18}$$

The proposed feedback routing control strategy is based on PI-controllers, i.e. feedback controllers of the proportional–integral type. Let us consider the two PI-controllers control laws adopted at time step $\bar{k}$ to define the splitting rates $\beta_{m,n,j}^{\mathrm{t},c}(\bar{k})$ and $\beta_{m,n,j}^{\mathrm{e},c}(\bar{k})$, by taking into account the equilibrium conditions (10.13)–(10.15) and (10.16)–(10.18), i.e.

$$\beta_{m,n,j}^{\mathrm{t},c}(\bar{k}) = \beta_{m,n,j}^{\mathrm{t},c}(\bar{k}-1) + K_P^{\mathrm{t},c}[\Delta t_{n,j}^{c}(\bar{k}) - \Delta t_{n,j}^{c}(\bar{k}-1)] + K_I^{\mathrm{t},c}\Delta t_{n,j}^{c}(\bar{k}) \tag{10.19}$$

$$\beta_{m,n,j}^{\mathrm{e},c}(\bar{k}) = \beta_{m,n,j}^{\mathrm{e},c}(\bar{k}-1) + K_P^{\mathrm{e},c}[\Delta e_{n,j}^{c}(\bar{k}) - \Delta e_{n,j}^{c}(\bar{k}-1)] + K_I^{\mathrm{e},c}\Delta e_{n,j}^{c}(\bar{k}) \tag{10.20}$$

where $K_P^{\mathrm{t},c}$, $K_I^{\mathrm{t},c}$, $K_P^{\mathrm{e},c}$ and $K_I^{\mathrm{e},c}$, $c = 1, \ldots, C$, are controller gains. It is worth noting that the resulting splitting rates $\beta_{m,n,j}^{\mathrm{t},c}(\bar{k})$ and $\beta_{m,n,j}^{\mathrm{e},c}(\bar{k})$ should be truncated to the admissible interval [0, 1].

The route guidance control variables $\beta_{m,n,j}^{\mathrm{C},c}(\bar{k})$, i.e. the splitting rates representing the portion of flow of class c present in node n at time step $\bar{k}$ which should choose link m to reach destination j, are given by the following weighted sum:

$$\beta_{m,n,j}^{\mathrm{C},c}(\bar{k}) = \alpha^{c}\beta_{m,n,j}^{\mathrm{t},c}(\bar{k}) + (1-\alpha^{c})\beta_{m,n,j}^{\mathrm{e},c}(\bar{k}) \tag{10.21}$$

where α^{c} is a design parameter defined for class c, with $0 \le \alpha^{c} \le 1$. These parameters are fixed in order to apply specific control policies for each vehicle class, by properly balancing travel times and total weighted emissions.

The Multi-class Ramp Metering Controller The ramp metering control strategy is based on feedback controllers of the proportional–integral type, and in particular on the multi-class PI-ALINEA already described in Sect. 10.3.1 for a freeway stretch (the notation here is slightly different since it is referred to a freeway network instead of a freeway stretch). The control law is updated with a sample time T, which is equal to the model sample time, and allows to compute the ramp metering control variable $r_o^{\mathrm{C},c}(k)$, i.e. the flow of class c that should enter from the origin link o during time interval $[kT, (k+1)T)$.

In order to compute such flow, let us first of all introduce the variable $f_o^c(k)$ indicating the *ratio*, at time step k, of the number of vehicles of class c over the entire number of vehicles, which are present in origin link o and in the mainstream section immediately downstream link o (namely the first section of the downstream leaving link m). Such quantity can be computed as

$$f_o^c(k) = \frac{\eta^c[l_o^c(k) + \rho_{m,1}^c(k)L_m]}{\sum_{h=1}^{C} \eta^h[l_o^h(k) + \rho_{m,1}^h(k)L_m]} \tag{10.22}$$

Referring to a generic origin link o, the flow of class c that should enter at time step k is computed according to the following multi-class PI-ALINEA control law:

$$\begin{aligned} r_o^{\mathrm{C},c}(k) = r_o^{\mathrm{C},c}(k-1) - K_P{}^c\left[\rho_{m,1}^{\mathrm{down},c}(k) - \rho_{m,1}^{\mathrm{down},c}(k-1)\right] \\ +K_R{}^c\, f_o^c(k)\left[\rho_{m,1}^{*} - \rho_{m,1}^{\mathrm{down}}(k)\right] \end{aligned} \tag{10.23}$$

where $\rho_{m,1}^{\mathrm{down},c}(k)$ is the traffic density of class c measured downstream the origin link, i.e. in the first section of the downstream link m, $\rho_{m,1}^{\mathrm{down}}(k)$ is the total density measured in the same location, $\rho_{m,1}^{*}$ is the total density set-point of the first section of link m, K_P^c and K_R^c are gain parameters of the regulator.

The Controller Gains Selector The proposed controllers are characterised by some gains, which should be properly tuned. In particular, let us denote with $\mathbb{K} = \left\{K_P^{\mathrm{t},c}, K_I^{\mathrm{t},c}, K_P^{\mathrm{e},c}, K_I^{\mathrm{e},c}, K_P^c, K_R^c, c = 1, \ldots, C\right\}$ the set gathering these *controller gains*. In the considered control framework, these gains are selected according to a specific selection procedure.

In particular, a finite set Ξ of *traffic scenarios* is defined to take into account different traffic conditions. Each scenario $\sigma \in \Xi$ is characterised by a set of initial traffic conditions and a demand pattern and is associated with a set $\mathbb{K}^\sigma$ of controller gains. The controller gains to be used are chosen by the gains selector with a sampling time T^S [s], normally larger than T, i.e. every T^S seconds the selector identifies the most adequate scenario $\bar{\sigma}$ for representing the present traffic conditions using suitable classification techniques or clustering methods. On the basis of the chosen scenario $\bar{\sigma}$, the selector module feeds the controllers with the corresponding set $\mathbb{K}^{\bar{\sigma}}$ of gains.

The controller gains associated with each scenario are found by offline running an optimisation-based procedure. Specifically, the controller gains are found by solving an *optimisation problem* in which the minimisation of the TTS and the minimisation of the TE are explicitly taken into account in the objective function, while the system dynamics is included in the problem constraints, and the decision variables are represented by the gains. The statement of this optimisation problem and other detailed explanations can be found in [35].

References

1. Kates RW, Parris TM, Leiserowitz AA (2005) What is sustainable development? Goals, indicators, values, and practice. Environ Sci Policy Sustain Dev 47:8–21
2. Giovannoni E, Fabietti G (2014) What is sustainability? A review of the concept and its applications. In: Busco C et al (eds) Integrated reporting. Springer International Publishing, Switzerland
3. The sustainable development goals report (2017), United Nations, New York
4. Zegeye SK, De Schutter B, Hellendoorn J, Breunesse EA, Hegyi A (2012) A predictive traffic controller for sustainable mobility using parameterized control policies. IEEE Trans Intell Transp Syst 13:1420–1429
5. Zegeye SK, De Schutter B, Hellendoorn J, Breunesse EA, Hegyi A (2013) Integrated macroscopic traffic flow, emission, and fuel consumption model for control purposes. Transp Res Part C 31:158–171
6. Csikós A, Varga I, Hangos K-M (2015) Modeling of the dispersion of motorway traffic emission for control purposes. Transp Res Part C 58:598–616
7. Groot N, De Schutter B, Hellendoorn H (2013) Integrated model predictive traffic and emission control using a piecewise-affine approach. IEEE Trans Intell Transp Syst 14:587–598
8. Ahn K, Rakha H (2008) The effects of route choice decisions on vehicle energy consumption and emissions. Transp Res Part D 13:151–167
9. Rakha H, Ahn K, Moran K (2012) Integration framework for modeling eco-routing strategies: logic and preliminary results. Int J Transp Sci Technol 1:259–274
10. Groot N, De Schutter B, Hellendoorn H (2015) Toward system-optimal routing in traffic networks: a reverse Stackelberg game approach. IEEE Trans Intell Transp Syst 16:29–40
11. Luo L, Ge Y-E, Zhang F, Ban X (2016) Real-time route diversion control in a model predictive control framework with multiple objectives: traffic efficiency, emission reduction and fuel economy. Transp Res Part D 48:332–356
12. Golob TF, Recker WW, Alvarez VM (2004) Freeway safety as a function of traffic flow. Accid Anal Prev 36:933–946
13. Lord D, Manar A, Vizioli A (2005) Modeling crash-flow-density and crash-flow-V/C ratio relationships for rural and urban freeway segments. Accid Anal Prev 37:185–199
14. Yeo H, Jang K, Skabardonis A, Kang S (2013) Impact of traffic states on freeway crash involvement rates. Accid Anal Prev 50:713–723
15. Abdel-Aty M, Dilmore J, Dhindsa A (2006) Evaluation of variable speed limits for real-time freeway safety improvement. Accid Anal Prev 38:335–345
16. Lee C, Hellinga B, Saccomanno F (2006) Evaluation of variable speed limits to improve traffic safety. Transp Res Part C 14:213–228
17. Lee C, Hellinga B, Ozbay K (2006) Quantifying effects of ramp metering on freeway safety. Accid Anal Prev 38:279–288
18. Yu R, Abdel-Aty M (2014) An optimal variable speed limits system to ameliorate traffic safety risk. Transp Res Part C 46:235–246
19. Li Z, Liu P, Wang W, Xu C (2014) Development of a control strategy of variable speed limits to reduce rear-end collision risks near freeway recurrent bottlenecks. IEEE Trans Intell Transp Syst 15:866–877
20. Pasquale C, Sacone S, Siri S, Papageorgiou M (2017) A nonlinear optimal control approach to reduce travel times and to improve safety in freeway traffic systems. In: Proceedings of the 5th IEEE international conference on models and technologies for intelligent transportation systems, pp 634–639
21. Caligaris C, Sacone S, Siri S (2007) Optimal ramp metering and variable speed signs for multiclass freeway traffic. In: Proceedings of the European control conference, pp 1780–1785
22. Schreiter T, van Lint H, Hoogendoorn S (2011) Multi-class ramp metering: concepts and initial results. In: Proceedings of the 14th IEEE conference on intelligent transportation systems, pp 885–889

23. Liu S, De Schutter B, Hellendoorn H (2013) Multi-class traffic flow and emission control for freeway networks. In: Proceedings of 16th international ieee conference on intelligent transportation systems, pp 2223–2228
24. Liu S, De Schutter B, Hellendoorn H (2014) Integrated traffic flow and emission control based on FASTLANE and the multi-class VT-macro model. In: Proceedings of the European control conference, pp 2908–2913
25. Liu S, Hellendoorn H, De Schutter B (2017) Model predictive control for freeway networks based on multi-class traffic flow and emission models. IEEE Trans Intell Transp Syst 18:306–320
26. Pasquale C, Anghinolfi D, Sacone S, Siri S, Papageorgiou M (2016) A two-class traffic control scheme for reducing congestion and improving safety in freeway systems. In: Proceedings of the 19th IEEE intelligent transportation systems conference, pp 1767–1772
27. Pasquale C, Sacone S, Siri S (2014) Ramp metering control for two vehicle classes to reduce traffic emissions in freeway systems. In: Proceedings of the European control conference, pp 2588–2593
28. Pasquale C, Sacone S, Siri S (2014) Two-class emission traffic control for freeway systems. In: Proceedings of the 19th IFAC world congress, pp 936–941
29. Pasquale C, Liu S, Siri S, Sacone S, De Schutter B (2015) A new emission model including on-ramps for two-class freeway traffic control. In: Proceedings of the 18th IEEE international conference on intelligent transportation systems, pp 1143–1149
30. Ferrara A, Pasquale C, Sacone S, Siri S (2017) Congestion and emissions reduction in freeway traffic networks via supervisory event-triggered control. In: Proceedings of the 20th IFAC World congress, pp 4240–4245
31. Pasquale C, Sacone S, Siri S, Ferrara A (2017) Supervisory multi-class event-triggered control for congestion and emissions reduction in freeways. In: Proceedings of the 20th IEEE intelligent transportation systems conference, pp 1535–1540
32. Pasquale C, Papamichail I, Roncoli C, Sacone S, Siri S, Papageorgiou M (2015) A nonlinear optimal control approach for two-class freeway traffic regulation to reduce congestion and emissions. In: Proceedings of the European control conference, pp 2651–2656
33. Pasquale C, Papamichail I, Roncoli C, Sacone S, Siri S, Papageorgiou M (2015) Two-class freeway traffic regulation to reduce congestion and emissions via nonlinear optimal control. Transp Res Part C 55:85–99
34. Pasquale C, Sacone S, Siri S, De Schutter B (2016) A Multi-class ramp metering and routing control scheme to reduce congestion and traffic emissions in freeway networks. In: Proceedings of the 14th IFAC symposium on control in transportation systems, pp 329–334
35. Pasquale C, Sacone S, Siri S, De Schutter B (2017) A multi-class model-based control scheme for reducing congestion and emissions in freeway networks by combining ramp metering and route guidance. Transp Res Part C 80:384–408
36. Wang Y, Kosmatopoulos EB, Papageorgiou M, Papamichail I (2014) Local ramp metering in the presence of a distant downstream bottleneck: theoretical analysis and simulation study. IEEE Trans Intell Transp Syst 15:2024–2039
37. Pasquale C, Sacone S, Siri S (2013) Multi-class local ramp metering to reduce traffic emissions in freeway systems. In: Proceedings of the IFAC workshop on advances in control and automation theory for transportation applications, pp 43–48
38. Kotsialos A, Papageorgiou M, Middelham F (2001) Optimal coordinated ramp metering with AMOC. Transp Res Rec 1748:55–65
39. Papageorgiou M, Kotsialos A (2004) Nonlinear optimal control applied to coordinated ramp metering. IEEE Trans Intell Transp Syst 12:920–933
40. Papageorgiou M, Marinaki M, Typaldos P, Makantasis K (2016) A feasible direction algorithm for the numerical solution of optimal control problems - extended version (Internal report No 2016–26. Chania, Greece)
41. Jamshidnejad A, Papamichail I, Papageorgiou M, De Schutter B (2017) Sustainable model-predictive control in urban traffic networks: efficient solution based on general smoothening methods. IEEE Trans. Control. Syst. Technol. https://doi.org/10.1109/TCST.2017.2699160

Chapter 11
Emerging Freeway Traffic Control Strategies

11.1 Future Trends of Traffic-Oriented Technologies

New communication, control and information systems technologies, along with modern transportation infrastructures and equipment, are giving rise to a completely new scenario in traffic management and control. This is mainly due to the fact that the implementation of *Intelligent Transportation Systems* (ITS) or *Intelligent Vehicle Highway Systems* (IVHS) is becoming more and more feasible in principle [1].

ITS and IVHS are based on *smart vehicles* and *smart infrastructures* which cooperate, so that freeways can be seen as *Automated Highway Systems* (AHS). The actual implementation and penetration will also depend on non-technological aspects, such as costs, legal, ethical and insurance issues and, last but not least, the users' willingness to entrust highly automatic systems.

As a matter of fact, with the rapid advancement of technology and investments from car manufacturers, the penetration of *self-driving cars* is expected to grow rapidly. According to the reports of several research and investment institutes, semi-autonomous cars, which allow drivers to hand over full control to the car in certain situations, will likely become more and more popular through the 2020s. Optimistic analysts think that completely autonomous cars will become the norm over the years ranging from 2030 to 2040, with a penetration rate of about 75% around 2035. Also the estimates by Exane BNP Paribas, reported in [2] along with an interesting discussion of legal and insurance implications of the predicted scenario, concur on the fact that automated vehicles will reach a market share in the region of 60% or 70% since the early 30s of the millennium. This penetration growth will be accompanied by a formidable increase in reliability and efficiency of sensors and sensor networks, especially of wireless type, used in vehicular traffic monitoring systems [3–5].

The vehicles themselves can now be equipped with on-board limited-cost technology allowing them to efficiently communicate in a *Vehicle-to-Vehicle* (V2V) or *Vehicle-to-Infrastructure* (V2I) setting. As such, the possibility of using the vehicles themselves as moving sensors (i.e. regarding them as *probe vehicles*) capable of acquiring and transmitting data is now a reality [6]. Over the past few years,

A. Ferrara et al., *Freeway Traffic Modelling and Control*, Advances in Industrial Control, https://doi.org/10.1007/978-3-319-75961-6_11

researchers from different areas (e.g. telecommunications, computer science, electronics, vehicular technology) have made a commitment to investigate how to make the most of this scenario, which is totally different from the scenario experienced by the researchers who introduced the concept of AHS in the 90s [7–11]. Specifically, the huge potentialities and challenges of *Vehicular Ad hoc NETworks* (VANET) [12] are clear to people involved in traffic monitoring, control and forecast, and the main feature of such a wireless network made of vehicles is regarded as an element which is likely to have a pivotal role in a wide range of applications in freeway traffic systems.

11.2 Intelligent Vehicles and Autonomous Driving

Apart from being *connected*, today vehicles are *intelligent* in the sense that they are equipped with *Advanced Driver Assistance Systems* (ADAS). Such systems enable vehicles to perform the basic driving actions (braking, accelerating, cornering, changing lane, steering and so on) by assisting the driver or in a completely autonomous way. This is possible thanks to the last-generation sensors such as modular Global Positioning System (GPS), miniaturised multi-channel radar units, cameras for 360-degree vision, ultrasonic sensors, all associated with advanced data processing strategies.

Independently of the level of autonomy, *vehicle control* systems play a central role in guaranteeing safe and comfortable driving [13]. In addition, the quality of data acquired from the road or communicated by the infrastructure and received by the vehicular traffic controllers, as well as their reliability, are of primary importance. Data fusion procedures, real-time signal processing and data analysis tools are then the elements that complete the picture in order to obtain a *dependable automatic system*, i.e. a system which can justifiably be trusted [14]. Note that the notion of *dependability* is related to several aspects that the system needs to possess, in particular availability, reliability, safety, integrity and maintainability, all features of fundamental importance in a freeway traffic control system.

In order to provide a common platform for researchers and stakeholders working with autonomous driving, the Society of Automotive Engineers (SAE) Internationals On-Road Automated Vehicle Standards Committee published the SAE Information Report (J3016) on 'Taxonomy and Definitions for Terms Related to On-Road Motor Vehicle Automated Driving Systems' [15]. The SAE-levels subdivision is more descriptive than normative and describes the minimum features and capabilities for each level. According to it, there are six levels of driving automation.

1. *Level 0 (No automation)*: this is the level to which the vast majority of cars and trucks present on our roads nowadays belongs. The driver actively controls steering, throttle and braking, keeping into account the surroundings, as well as navigating and deciding when to use turn signals, change lanes and turn. There

can also be warning systems such as blind-spot and collision warning devices. All aspects of the dynamic driving task are under the human driver's responsibility.
2. *Level 1 (Driver assistance)*: vehicles in this level can assist the driver in steering, accelerating/decelerating, but not in all circumstances. This implies that the driver must be aware and ready to take over those functions if called upon by the vehicle or by the events that may occur.
3. *Level 2 (Partial assistance)*: the car autonomously handles steering, accelerating and braking, but immediately lets the driver take over if the driver identifies objects and events the car is not responding to. In these first three levels, the driver is responsible for monitoring the surroundings, including traffic, weather and road conditions.
4. *Level 3 (Conditional assistance)*: the car monitors the surroundings and performs steering, accelerating and braking in particular environments, such as freeways. The driver must be ready anyway to intervene if necessary.
5. *Level 4 (High automation)*: the car handles steering, accelerating/decelerating and monitors the surroundings in a wider range of environments, but not all, such as severe weather or severely congested traffic conditions. The driver switches on the automatic driving only when it is safe to do so.
6. *Level 5 (Full automation)*: the driver only sets the destination and starts the car. The car is fully autonomous in performing all the other tasks. The car can drive to any legal destination and make its own decisions on the way about routing and driving style. The automated driving system is completely in charge of the dynamic driving task. In case of driving conditions which the driver considers as risky, the driver can intervene and take over control.

Note that SAE J3016 is the standard adopted by the U.S. Department of Transportation and the National Highway Traffic Safety Administration.

Even if, considering the current level of technology maturity and legal and ethical implications, Level 4 automation seems at present the most likely to be accepted by users, fully autonomous vehicles are no longer a futuristic idea. Google has started its self-driving car project in 2009. In November 2017, Waymo, the autonomous car development company controlled by Google's parent company, Alphabet, announced to have started testing driverless cars without a safety driver at the driver position [16]. At the same time, Waymo invited residents in Phoenix, Arizona, to join the first public trial of self-driving cars named 'Waymo's Early Rider Program'.

In April 2016, Fiat-Chrysler began a partnership with Google and its Waymo project to create a fleet of 100 self-driving Pacifica minivans to be tested on the streets of Arizona, California and Michigan [17]. A Chrysler Pacifica hybrid outfitted with Waymo's suite of sensors and radar was shown at the North American International Auto Show in Detroit, in January 2017. Several other companies are making significant efforts to pursue autonomous driving technologies, with the goal of introducing fully self-driving cars in the near future. Tesla, in particular, claims on its website that 'all Tesla vehicles produced have the hardware equipment needed for autonomous driving with a substantially higher level of safety than that of a human driver'.

Fig. 11.1 Interior of MILA Future, technology carrier from the brand-independent automotive supplier Magna Steyr (courtesy of Magna Steyr)

What seems certain is that in the next decade we will have to get used to the idea of seeing more and more cars equipped with a touch screen instead of the steering wheel, as shown in Fig. 11.1.

11.3 Cooperative Vehicle Control for Traffic Improvement

The final goal of AHS is to improve vehicular traffic so as to ameliorate people's quality of life, to preserve environmental integrity and to reduce energy consumptions. Thanks to intelligent vehicles and smart infrastructures, it seems possible to change the perspective in traffic control. Instead of acting at a macroscopic level using road-based traffic control strategies, such as, for instance, ramp metering and variable speed limits (see Chap. 8), one can think of influencing the behaviour of individual vehicles in order to get a collective evolution of traffic which is in line with the control objectives.

Depending on the type and level of interaction between the roadside and the vehicles, it is possible to design different types of control strategies. Reference [18] provides an interesting classification of intelligent vehicles systems in AHS, just based on the nature of the interaction. That classification is hereafter reported for the reader's convenience:

- *autonomous vehicle systems*: vehicles are equipped with sensors and advanced processing capability but they operate as independent agents;

- *cooperative vehicle systems*: vehicles are equipped with sensors and wireless communication capabilities, so that they can operate in a coordinated way with neighbouring vehicles but with no interaction with the infrastructure;
- *infrastructure-supported systems*: vehicles communicate with each other, while the infrastructure provides rules and constraints to make decisions;
- *infrastructure-managed systems*: vehicles communicate to the infrastructure their future actions (e.g. lane changes, exits and entries) and the infrastructure replies with instructions for coordination among vehicles, so that the actions can be actually implemented;
- *infrastructure-controlled systems*: the infrastructure acts as a kind of supervisor in controlling the vehicle actions so as to optimise the traffic system performance.

Note that autonomous and cooperative systems typically operate in the milliseconds to seconds range and involve small groups of neighbouring vehicles, while the infrastructure-based systems typically work in a time range which goes from seconds to hours. Moreover, a much larger number of vehicles and a larger portion of the freeway, even arriving at an entire freeway network, is included in the entity named 'system' [18].

In the papers discussing the new scenarios related to the penetration of advanced traffic-oriented technologies, one of the most recurrent keyword is the term *coordination* [19]. As such, in the foregoing classification of intelligent vehicle systems in AHS, the last four typologies appear, at the moment, as the most promising frameworks to develop innovative macroscopic freeway traffic control systems acting at a microscopic level, i.e. at the vehicle level. This does not mean that one cannot think of controlling any single autonomous vehicle by implementing strategies, such as *reinforcement learning*, which allow the overall traffic system to reach the desired equilibria even without high-level coordination. The interested reader can find in [20–22], an overview of the theoretical aspects of reinforcement learning and applications of related algorithms to traffic engineering, especially in an ITS perspective.

11.3.1 Traffic Control via Coordinated Adaptive Cruise Control

Adaptive Cruise Control (ACC) is a commercially available radar-based or camera-based ADAS which can improve driving comfort and safety. The main feature is the capability of adjusting the vehicle speed taking into account the speed reference set by the driver, as well as the speed of the preceding vehicle. The concept was developed in the 90s [23], and further improved in subsequent years. The breakthrough in the ACC arrived, however, when communication among vehicles became easily available thanks to the development of specific V2V communication protocols.

V2V communication can be conveniently exploited to improve the performance of ACC systems [24]. An ACC system with V2V communication is called *Cooperative ACC* (CACC) system [25]. CACC systems were originally introduced in order to improve driving quality, comfort and safety. Yet, it was soon observed that CACC systems can significantly affect traffic throughput [26], by virtue of their intrinsic traffic regularisation properties. Their ability to influence global traffic dynamics is based on the knowledge of positions and velocities of nearby vehicles but also of aggregate traffic information, which enable the single CACC-equipped vehicle to have a perception of the overall evolution of the traffic system.

It seems reasonable to believe that the effect of CACC on macroscopic traffic control may be enhanced by using *robust control* strategies [27], since they can guarantee a strong invariance of the controlled vehicle performance even in presence of uncertainties in the vehicle dynamics model, in analogy with what observed in case of ACC systems (see, for instance [28, 29] and the reference therein), but also in case of communication delays, packet losses and jitters [13].

Optimisation will also play a key role in the presence of vehicles equipped with Vehicle Automation and Communication Systems (VACS). In [30], for instance the traffic control problem, under suitable assumptions, is faced via the formulation of a linearly constrained optimal control problem. The proposed solution allows to implement variable speed limit control per lane, lane changing control and ramp metering to obtain flow maximisation in an integrated way, and with an increase of efficacy with respect to solutions based on conventional actuators.

11.3.2 The Impact of Coordinated Adaptive Cruise Control on Traffic

The impact of CACC systems on traffic flow characteristics, such as traffic throughput or total time spent by vehicles in the system, is obviously tied to the penetration rate of CACC-equipped vehicles in the macroscopic traffic flow, as shown in [31–33], also considering different desired time-gap settings and different networks.

At the present stage of research, the design of innovative traffic control strategies based on the exploitation of autonomous vehicles must assume a sufficient penetration rate of vehicles equipped with VACS allowing V2I communication. As such, we may expect that several years have to pass before a sufficiently large number of vehicles are equipped with the necessary devices. Yet, the new traffic control concept is interesting and worthy of further insights since it does not require the creation of novel highway infrastructures fully dedicated to autonomous vehicles. In this, therefore, it seems adherent to what appears to be the most likely scenario in vehicle traffic systems of the next decade, i.e. a scenario in which there will be fully automated vehicles as well as vehicles featuring different levels of automation, up to the limit case of manually driven cars.

An interesting literature review on the impact of CACC systems on global traffic flow evolution can be found in [34]. Summarising the conclusions drawn in that paper, one has that:

- V2V communication is effective in providing the ACC system with more and better information about the preceding vehicle with respect to conventional radar or camera-based solutions; this implies that the vehicle controller can respond in an extremely fast way, thus allowing for shorter inter-vehicle distances;
- by reducing time gaps between vehicles, a better road capacity exploitation can be realised;
- traffic flow stability can be enhanced by improving the *string stability* of the platoon of coordinated vehicles [23].

CACC systems can be operated standalone or in an integrated way with conventional road-based traffic control strategies. This possibility was already investigated during the California PATH Program of the University of California, and the outcome was the so-called Integrated Roadway/Adaptive Cruise Control System [35]. Such system was meant to integrate ramp metering strategies and a speed control strategy, based on generalised versions of ALINEA (see Sect. 8.3.1), by taking into account highway-to-vehicle communication and ACC technologies on board of the vehicles. A more recent proposal of integration can be found in [36], where a control strategy is presented based on CACC and variable speed limits to reduce rear-end collision risks close to freeway bottlenecks. In that paper, only longitudinal car-following behaviours of CACC vehicles are considered, but the results encourage to think that, compared to the case of manually driven vehicles, the combination of CACC and variable speed limits can be beneficial to improve both safety and efficiency.

11.3.3 Ride Sharing as a Traffic Control Strategy

The reduction of the number of vehicles in the traffic network can also be achieved by increasing the number of passengers inside vehicles. One possibility is given by *ride sharing*, also called *demand reactive transportation* or *carpooling*, when it occurs at public level or at private level, respectively. As such, it can produce a positive impact on traffic, reducing congestions, crash risk, fuel and energy consumptions and environmental issues.

As a matter of fact, the design and establishment of autonomous vehicles for shared use (such as shared electric vans as collective taxis), namely *shared autonomous vehicles*, combining features of short-term on-demand usage with self-driving capabilities, can make ride sharing really attractive and convenient even in terms of traffic control and regularisation [37].

In principle, the supervisor of a ride sharing system can be designed so as to assign the transportation requests to the involved autonomous vehicles in order to minimise the total operational cost. However, the optimisation problem underlying the decision-making process can also be formulated in such a way to improve traffic conditions. For instance, taking into account real-time traffic information, runtime constraints could be added to the problem, so as to diverge the routes assigned to the shared vehicles from the sections of the road network which are congested or close to congestion. This can increase the total time spent by individual users in the traffic system, still maintaining a reasonable level of service, but can generate positive returns in terms of user comfort (the journey will be more regular and less marked by stop-and-go effects), as well as collective benefits such as a reduction of the global total time spent by drivers in the whole traffic network, with all the associated advantages (e.g. reduced pollution and energy consumption).

The most general formulation of the *ride sharing problem* involves a group of users, whose role (driver or passenger) should be determined, as well as the composition of the pools and the determination of the related routes. The problem solution method has to be compatible with real-time implementation constraints. It is worth noting that, in general, the problem is NP-hard [38] and is usually faced by means of heuristic methods. The problem can be formalised either as a many-source single-destination problem or as a single-source many-destination problem, to keep the computational costs limited with respect to the costs associated with the general many-to-many case.

An approach to the many-to-many ride sharing problem with automated passenger aggregation and reference to an urban scenario is presented in [39], where the proposed solution allows to optimally solve the related routing problem. Similar concepts could be soon extended to ride sharing over freeway networks, with the objective of reducing the number of vehicles hosting a number of passengers lower than the vehicle capacity. This approach would be extremely useful to improve the mobility conditions of the many commuters who daily travel by car along the freeways from the residential areas surrounding big cities, or from nearby towns, to the city centres.

Moreover, ride sharing as a traffic control strategy might complement the already existing *high-occupancy vehicle* lanes (also known as carpool or diamond lanes), present in many countries such as U.S., Canada, Indonesia, Australia and New Zealand. High-occupancy vehicle lanes are restricted traffic lanes reserved to vehicles with a driver and a prescribed number of passengers, including carpools.

Note that, besides the technical and technological difficulties related to optimal solutions of ride sharing problems, there are other aspects to consider. One critical aspect is associated with privacy-preserving information sharing. Privacy-preserving issues arise in ride sharing systems because of the necessity of transferring and handling trip data. This aspect will be addressed in more details in Sect. 11.5.1.

11.3.4 Coordination in Freight Transport

According to McKinsey's forecasts, it is likely that trucks will be the first vehicles to be equipped with full technology for autonomous driving on public roads, and this should happen in a couple of decades. With a predicted growth of transport activity of

Fig. 11.2 Spontaneous trucks platooning in A10 freeway, Italy (courtesy of Martina Bastianon)

57% by 2050, according to the European Commission [40], this creates the premises for several challenges also for control engineering methods applied to the transport sector.

With the increase in freight transport on the road, it is of primary importance to consider the increase in fuel consumption and greenhouse gas emissions, and devise efficient strategies to limit them, while guaranteeing the satisfaction of the demand for freight transport [41]. There are also other problems, such as road safety and the necessity to optimally use the infrastructure in order to avoid a detrimental impact on passenger transport and mobility, as well as to amortise the costs for their upgrading and modernisation.

Trucks platooning is a control challenge which has already attracted many research efforts [42–45]. The idea is to form a convoy of trucks or *Heavy-Duty Vehicles* (HDV), equipped with state-of-the-art driving support systems, driving close behind each other (see Fig. 11.2). One of the main benefits of platooning is associated with fuel consumption reduction for the follower vehicles, due to a natural reduction of aerodynamic drag [46, 47]. Apart from fuel consumption, trucks platooning determines a reduction of CO_2 emissions and positive impacts on the labour market, logistics and industry.

In Europe, for instance, a partnership among truck manufacturers, logistic service providers, research institutes and governments was established in 2016, giving birth to the *European Truck Platooning Challenge*. It has the objective of sharing knowledge and experience about truck platooning and realising truck platooning in the near future. The initiative was launched by the Dutch Ministry of Infrastructure and the Environment, the Directorate General Rijkswaterstaat, the Netherlands Vehicle Authority and the Conference of European Directors of Roads.

To form platoons of HDV, efficient and reliable coordination strategies need to be investigated [48]. At the same time, protocols have to be developed and standardisation/legislations issues have to be considered to reduce to the minimum the risks tied to miscommunications, faulty system operations or malicious attacks (see, Sect. 11.6).

An important aspect in this context is the necessity of suitably defining the paths followed by vehicles. The shortest path algorithm by Dijkstra [49] is the most commonly used algorithm to find in a graph, representing the transportation network, the path with the associated lowest total cost, according to the specified metrics. For instance, it allows to retrieve the shortest-distance path or the shortest-time path from the origin node to the destination node, which may be appropriate to make a decision about the most convenient and efficient route for heavy vehicles. Alternatively, path planning can be based on the most efficient path in terms of fuel consumptions, the so-called *eco-routing* [50]. In this case the edge cost is based on a vehicle model, on-road information and on a model to determine an estimate of the fuel consumption to travel from a node of the graph to another.

Clearly, HDV platoon formation and coordination have an impact on the overall vehicular traffic, in analogy to what discussed in Sect. 11.3.2, at the point that one could hypothesise to develop HDV platooning strategies that optimise freight transport, while simultaneously reducing freeway traffic congestion by creating intelligent moving bottlenecks [51–53]. Nevertheless, given a fleet of HDV, traffic conditions and their time evolutions are of paramount importance in making the correct decision about forming a new platoon, reaching and merging an already existent platoon or travel individually.

In other terms, such a decision has to consider the cost–benefit ratio associated with the different options. Note that this decision has to be made whenever a single vehicle begins its journey, but also periodically during the vehicle motion to account for possible variations in external conditions. The freeway traffic models discussed in Chaps. 3–5 can play an important role in the decision-making process [54]. Since such a process has to take place majorly in real time, there is the need to develop more and more scalable and extremely computational efficient versions of the traffic models, in order to make it possible to use them within the HDV fleet supervisors.

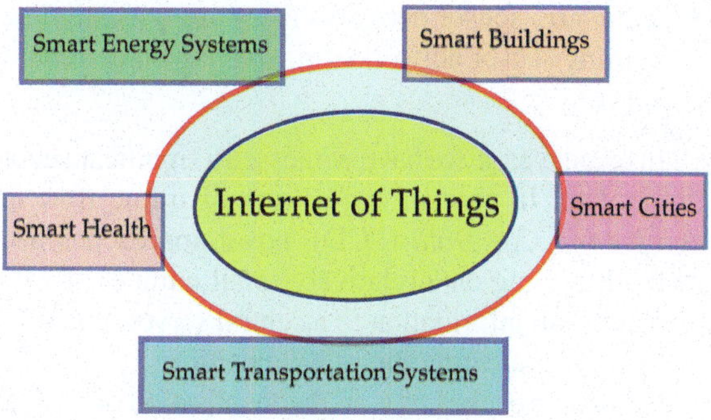

Fig. 11.3 Inter-networking of physical devices according to the IoT framework

11.4 Internet of Things Concepts in Traffic Control

Internet of Things (IoT) is the concept introduced with the Global Standards Initiative on Internet of Things in 2013 to define a trend which was already pervasive in the evolution of complex internet-based systems [55]. The concept encompasses the inter-networking of physical devices, including connected vehicles, smart sensors and actuators and other kinds of connected elements, such as data storage devices, as illustrated in Fig. 11.3. Smart transportation systems are one of the fields of application of the IoT concept. On the whole, such complex systems can be regarded as *cyber-physical systems* [56, 57] which can be remotely sensed and controlled [58, 59].

As a matter of fact, IoT provides a new technology platform for the realisation of intelligent transportation, also in view of environmental aspects [60]. Traffic control can greatly benefit from this new scenario, which is full of potentialities. In particular, decentralised and distributed traffic state estimation strategies and traffic control algorithms can be efficiently applied in an IoT setting.

The customisation of the IoT concept into the context of traffic systems has brought to define a novel paradigm, namely the *Internet of Vehicles* (IoV) [61]. Thanks to the new technologies and wireless short-range protocols, vehicles can be durably connected to the Internet, providing information for common services such as traffic management and road safety. Furthermore, the integration of social networking aspects into the IoV brings to the Social Internet of Vehicles (SIoV) paradigm, which can be regarded as an instance of the Social Internet of Things (SIoT) [62].

11.4.1 Sensor Networks and Vehicular Ad Hoc NETworks

The performance of the overall freeway traffic control system implemented in a IoT perspective is strictly related to the efficiency and reliability of the underlying sensor network. *Wireless sensor networks* for freeway traffic control [4] can be classified into two categories:

1. *stationary* sensor networks (they are typically integrated in the infrastructure);
2. *floating* sensor networks (in which the individual vehicles behave as mobile sensors).

In recent years, we have witnessed a significant evolution of methods for collecting traffic data. In case of acquisition of traffic flow data from vehicles, referred to as *Floating Car Data* (FCD), novel approaches need to be developed. There are methods relying on a relatively small number of vehicles which explicitly transmit their position information to a central server: this is the case, for instance, of taxis or buses communicating their GPS position.

Other approaches rely on the localisation of mobile phones. They need to involve the cellular network operators, requiring access to their real-time location databases. These approaches do not actually involve any sensing by the vehicle itself, but anyway make use of a wireless network (i.e. the existing cellular network) to estimate the current distribution of traffic flows.

Often, in the context of vehicular traffic control, especially in applications such as HDV platooning (see Sect. 11.3.4), the vehicular sensor network is called VANET, as previously mentioned in Sect. 11.1. In VANET, nodes can be mobile or static: mobile nodes are vehicles, while static nodes are Roadside Units (RSUs).

Generally speaking, monitoring a large area with stationary sensor networks, especially in transportation systems, requires a very large number of nodes which may imply a prohibitive cost. Then, it is reasonable to conceive distributed architectures where a set of mobile sensors collaborate with the stationary sensors in order to increase the overall sensor network *coverage* and enable the system to reliably detect and locate critical events, such as traffic jams or time-varying bottlenecks [63, 64].

Several technical challenges need to be faced at the level of sensor network management. In particular, there is the necessity of developing advanced skills to

- process potentially very large amounts of data;
- distinguish between useful and non-useful data;
- extrapolate the actual traffic flow data from the observation of only a subset of the vehicles present in the traffic system.

The FCD set can be complemented with information acquired thanks to the on-board electronics of the vehicles, obtaining *eXtended Floating Car Data* (XFCD). Modern road vehicles are equipped with sensors of various kinds such as temperature sensors, rain sensors and gyro sensors. They also possess advanced control systems which require to estimate the tire-road friction coefficient which depends on road conditions. Hence, the estimation algorithms used by the on-board vehicle controllers can be exploited to get indirect evaluations of the traffic system state, providing real-time information about road conditions and smoothness of traffic. XFCD can be made available to the public system and to the decision makers (including also traffic automatic control systems) so as to anticipate undesired traffic evolution and prevent congestions.

11.5 New Data Sources and Big Data

As discussed before, recent advances in technology have made available several new types of traffic measurements in freeway systems. Besides conventional traffic data coming from video cameras, radar-based detection systems and magnetometers, it is now possible to receive data from radars, mobile phones, Bluetooth-equipped vehicles, among other classes of less structured data, like those harvested through social networks. Freeway traffic control systems of the new generations will absolutely benefit from this large amount of heterogeneous data, provided that efficient and reliable methodologies to process them are conceived [65].

This new scenario in data collection requires to devise freeway traffic control systems which rely on specific architecture for *big data processing* in order to comply with time constraints typical of ITS [66]. Cloud-based data mining solutions, which allow to extract potentially useful information from raw data [67], are likely to have, in the near future, a strategic and pivotal role in enforcing implementability of traffic control strategies supplemented by big data usage.

11.5.1 Privacy-Preserving Data Sharing

In the scenario in which new data sources can be exploited, not only technical aspects but also privacy issues must be taken into consideration [68]. This means, for instance, that location of sensor data collected from private vehicles need to be processed so as to guarantee anonymity.

The privacy-preserving problem is particularly felt when multi-point traffic flow measurements are reconstructed via automatic traffic data collection based on *vehicular cyber-physical systems*, which provide information through V2I communications. Note that multi-point traffic statistics describe the number of vehicles travelling through multiple geographical locations during a measurement period. Vehicles are assumed to be able to communicate with RSUs in real time via dedicated short-range communications, that IEEE has standardised under IEEE 802.11p [69]. RSUs are connected to the central server through a wired or wireless communication line. They periodically report information collected from vehicular cyber-physical systems to the central server. The challenge of preserving the privacy of all participating vehicles is accompanied by the need to design a measurement scheme efficient and accurate enough to fit monitoring and control requirements of large-scale vehicular networks [70].

Another case in which privacy-preserving strategies need to be adopted is *ride sharing* or *carpooling*, also mentioned in Sect. 11.3.3. In ride sharing, some criticality arises from the fact that the users of the ride sharing system need to disclose their pick-up and drop-off locations and times, as well as to specify their complete routes. This is because the carpooling server typically uses a similarity measurement technique, which evaluates the similarity of the users' trip data and identify the users who can share rides. Note that the communicated information can include the users' residence address, the places of employment, the typical places where the users spend their leisure time, places of worship that reveal religious beliefs, and so on. Moreover, it is worth observing that hiding the users' identities is in general not enough because malicious intruders can identify the users from their pick-up and drop-off locations. In case of ride sharing, it is then very important to develop appropriate methods which rely on similarity measurement techniques over encrypted data [71, 72].

Well-known *anonymisation algorithms* may not meet the prescribed accuracy requirements and may fail to provide privacy guarantees for drivers especially in low-density areas. Indeed, when the user density is low, the space–temporal characteristics of the data can allow tracking and re-identification of anonymous vehicles. Then, it is necessary to design algorithms that hide location samples or modify the location traces, while maintaining the original data accuracy. Metrics describing how long an individual vehicle can be tracked in the data set need to be introduced, so that, relying on these metrics, reliable privacy algorithms capable of guaranteeing a specified maximum *time-to-oblivion* could be formulated.

Privacy issues are also associated with data harvested from *social networks*. As a matter of fact, powerful profiling algorithms are normally utilised to analyse users' navigations through the internet and enable browsers and social networks to propose them advertisements which are expected to be of interest. Data collected about travel habits and typical destinations, for instance via the localisation of smartphones and 'social' interaction with other drivers, could be improperly used for commercial or even malicious purposes [73].

Note that the concern about privacy also exists for traditional traffic monitoring and control schemes. In fact, whenever video cameras and image processing are used to derive information about traffic conditions, there is always the risk to violate the vehicle passengers' privacy.

11.6 Cybersecurity, Resilience and Dependability of Traffic Control

Within the context of road vehicles and traffic control systems, especially in case of traffic control strategies based on emerging technologies, such as autonomous and coordinated vehicles, *cybersecurity* is a fundamental aspect [57]. It consists in the protection of automotive electronic systems, communication networks, control algorithms, software, users and associated data from malicious attacks, damage, unauthorised access or manipulation.

In the traffic systems present nowadays in freeway networks, malicious attacks are possible mainly because the present trend in interconnecting supervisory control and data acquisition systems is characterised by the use of the standard Transmission Control Protocol (TCP) and the Internet Protocol (IP) (i.e. the TCP/IP suite of protocols). This is due to the benefits that this choice provides in terms of communication network design, maintenance and troubleshooting [74]. But even at the level of a single device there can be exploitable vulnerabilities.

For instance, recent studies have shown that many traffic lights used to implement ramp metering in practice may allow an attacker to tamper with the configuration of the signal. Also modifying variable speed limit prescriptions does not seem so difficult. This kind of vulnerability is hardly usable to cause accidents in a freeway traffic context, but it can be exploited to cause disastrous traffic congestions which can dramatically increase the total travel time in the transportation network [75].

Cybersecurity needs to be coped with by carefully monitoring, in the design phase, potential cyber vulnerabilities of all the subsystems which constitute the overall system. In case of vehicular traffic control systems, it implies that all the agents involved (e.g. vehicles, sensors, actuators, V2V and V2I communication devices) must be carefully checked and certified from this viewpoint. Then, it is necessary to assess the vulnerability of the entire transportation network and the traffic signal configuration [76, 77]. This can be done by suitably modelling the transportation network with suitable traffic models, but also by formulating a model of the attacker, to define the attacker's action space and objectives. It is also important to predict the control strategy which is likely to be employed by the attacker to create a desired disruption via complex congestion patterns [78].

Further, traffic control systems must be reliable under a large variety of circumstances, not only versus cyberattacks. *Faults* and *malfunctioning* of the devices can occur frequently in a possibly large-scale, heterogeneous, and highly interconnected system. This implies that emergent traffic control approaches have to encompass strategies to detect, isolate and identify all the critical events, including faults [79, 80], as it happens in other critical networks like power grids.

Finally, it is advisable to foresee strategies to reconfigure the network of sensors and actuators in order to maintain acceptable performance of the traffic control system even in the presence of faults or attacks. In other terms, the traffic system must be able to protect itself from the consequences of critical events by promptly changing its control paradigm and communication topology, so as to avoid cascading failures. In this way the traffic control system can be classified as a *resilient traffic system*, i.e. a system having the ability to resist and to recover from disturbances in traffic flow [81]. The fundamental step in designing resilient traffic networks and signal configurations is the assessment of the vulnerability of the entire transportation network mentioned before. Moreover, one has to identify critical signals, which have the highest impact on congestion, this in order to correctly foresee and activate defensive countermeasures and resources.

To enforce the prescribed properties in a traffic control system, it is also necessary to design control and communication strategies that give the traffic system the required *robustness* (i.e. the capability of maintaining the function for which it was originally designed in front of the vulnerabilities due to possible critical events) [82] and make it *tolerant* [83]. In this perspective, one can expect that robust and fault-tolerant control methodologies adopted in other fields, such as power and hydraulic networks, will be massively transferred and tailored to freeway traffic control systems in future years.

Reliability, robustness and resilience are major features of a *dependable control system* [84], a system that users can rely on and utilise getting the major benefits from it.

References

1. Papageorgiou M, Diakaki C, Nikolos I, Ntousakis I, Papamichail I, Roncoli C (2015) Freeway traffic management in presence of vehicle automation and communication systems (VACS). In: Meyer G, Beiker S (eds) Road vehicle automation, vol 2. Springer, pp 205–214
2. Schubert MN (2015) Autonomous cars - initial thoughts about reforming the liability regime. Insurance Issues
3. Daponte P, De Vito L, Picariello F, Rapuano S, Tudosa I (2012) Wireless sensor network for traffic safety. In: Proceedings of the IEEE workshop on environmental energy and structural monitoring systems, pp 42–49
4. Pascale A, Nicoli M, Deflorio F, Dalla Chiara B, Spagnolini U (2012) Wireless sensor networks for traffic management and road safety. IET Intell Transp Syst 6:67–77
5. Guevara J, Barrero F, Vargas E, Becerra J, Toral S (2012) Environmental wireless sensor network for road traffic applications. IET Intell Transp Syst 6:177–186
6. Knorr F, Baselt D, Schreckenberg M, Mauve M (2012) Reducing traffic jams via VANETs. IEEE Trans Veh Technol 61:3490–3498
7. Varaiya P (1993) Smart cars on smart roads: problems of control. IEEE Trans Autom Control 38:195–207
8. Hedrick JK, Tomizuka M, Varaiya P (1994) Control issues in automated highway systems. IEEE Control Syst 14:21–32
9. Broucke M, Varaiya P (1996) A theory of traffic flow in automated highway system. Transp Res Part C 4:181–210
10. Ioannou P (ed) (1997) Automated highway systems. Springer, US
11. Alvarez L, Horowitz R, Li P (1999) Traffic flow control in automated highways systems. Control Eng Pract 7:1071–1078
12. Hartenstein H, Laberteaux KP (2008) A Tutorial survey on vehicular ad hoc networks. IEEE Commun Mag 46:164–171
13. Watzening D, Horn M (eds) (2017) Automated driving: safer and more efficient future driving. Springer International, Berlin
14. Avizienis A, Laprie J-C, Randell B, Landwehr C (2004) Basic concepts and taxonomy of dependable and secure computing. Trans Dependable Sec Comput 1:11–33
15. SAE On-Road Automated Driving (ORAD) Committee (2014) *Taxonomy and definitions for terms related to on-road motor vehicle automated driving systems*, Standard: *J*3016_201401, SAE International
16. Hawkins AJ (2017) Waymo is first to put fully self-driving cars on US roads without a safety driver. The Verge, 7 Nov 2017
17. Durbin D-A (2017) FCA, Google begin offering rides in self-driving Pacifica hybrid minivan. Chicago Tribune, 25 April 2017
18. Baskar LD, De Schutter B, Hellendoorn J, Papp Z (2011) Traffic control and intelligent vehicle highway systems: a survey. IET Intell Transp Syst 5:38–52
19. Rios-Torres J, Malikopoulos AA (2017) A survey on the coordination of connected and automated vehicles at intersections and merging at highway on-ramps. IEEE Trans Intell Transp Syst 18:1066–1067
20. Sutton R, Barto A (1993) Reinforcement learning: an introduction, MIT Press, Cambridge, MA
21. Abdulhai B, Kattan L (2003) Reinforcement learning: introduction to theory and potential for transport applications. Can J Civil Eng 30:981–991

22. Yang Z, Wen K (2010) Multi-objective optimization of freeway traffic flow via a fuzzy reinforcement learning method. In: Proceedings of the 3rd international conference on advanced computer theory and engineering, vol V5, pp 530–534
23. Swaroop D, Hedrick JK (1996) String stability of interconnected systems. IEEE Trans Autom Control 41:349–357
24. Liu Y, Özgüner Ü, Acarman T (2006) Performance evaluation of inter-vehicle communication in highway systems and in urban areas. IEE Proc IntellTranspSyst 153:63–75
25. Shladover SE, Nowakowski C, Lu X-Y, Ferlis R (2015) Cooperative adaptive cruise control. Transp Res Rec 2489:145–152
26. Kim T, Jerath K (2016) Mitigation of self-organized traffic jams using cooperative adaptive cruise control. In: Proceedings of the international conference on connected vehicles and expo, pp 7–12
27. Ferrara A (ed.) (2017) Sliding mode control of vehicle dynamics, IET
28. Ferrara A, Vecchio C (2006) Cruise control with collision avoidance for cars via sliding modes. In: Proceedings of the IEEE international conference on control applications, pp 2808–2813
29. Ferrara A, Vecchio C (2008) Second-order sliding mode control of a platoon of vehicles. Int J Model Identif Control 3:277–285
30. Roncoli C, Papageorgiou M, Papamichail I (2015) Traffic flow optimisation in presence of vehicle automation and communication systems. Part II: optimal control for multi-lane motorways. Transp Res Part C 57:260–275
31. Ntousakis IA, Nikolos IK, Papageorgiou M (2015) On microscopic modelling of adaptive cruise control systems. Transp Res Procedia 6:111–127
32. Delis AI, Nikolos IK, Papageorgiou M (2015) Macroscopic traffic flow modeling with adaptive cruise control: development and numerical solution. Comput Math Appl 70:1921–1947
33. Delis AI, Nikolos IK, Papageorgiou M (2016) Simulation of the penetration rate effects of ACC and CACC on macroscopic traffic dynamics. In: Proceedings of the IEEE 19th international conference on intelligent transportation systems, pp 336–341
34. Van Arem B, Van Driel CJ, Visser R (2006) The impact of cooperative adaptive cruise control on traffic-flow characteristics. IEEE Trans Intell Transp Syst 7:429–436
35. Ioannou P, Wang Y, Chang H (2007) Integrated roadway/adaptive cruise control system: safety, performance, environmental and near term deployment considerations. California PATH Research Report UCB-ITS-PRR-2007-8
36. Li Y, Xu C, Xing L, Wang W (2017) Integrated cooperative adaptive cruise and variable speed limit controls for reducing rear-end collision risks near freeway bottlenecks based on microsimulations. IEEE Trans Intell Transp Syst 18:3157–3167
37. Fagnant DJ (2015) Shared autonomous vehicles: model formulation, sub-problem definitions, implementation details, and anticipated impacts. In: Proceeding of the 2015 American control conference, pp 2593–2593
38. Hartman IB, Keren D, Dbai AA, Cohen E, Knapen L, Yasar AU, Janssens D (2014) Theory and practice in large carpooling problems. Procedia Comput Sci 32:339–347
39. Celsi LR, Di Giorgio A, Gambuti R, Tortorelli A, Delli Priscoli F (2017) On the many-to-many carpooling problem in the context of multi-modal trip planning. In: Proceedings of the 25th Mediterranean conference on control and automation, pp 303–309
40. European Commission (2013) Trends to 2050: Reference Scenario 2013, ISBN: 978-92-79-33728-4
41. European Automobile Manufacturers Association (ACEA) Position Paper (2016) Reducing CO2 emissions from heavy-duty vehicles
42. Alam A, Besselink B, Turri V, Martensson J, Johansson KH (2015) Heavy-duty vehicle platooning for sustainable freight transportation: a cooperative method to enhance safety and efficiency. IEEE Control Syst 35:34–56
43. Besselink B, Turri V, van de Hoef S, Liang KY, Alam A, Maartensson J, Johansson KH (2016) Cyber-physical control of road freight transport. Proc IEEE 104:1128–1141
44. Liang KY, van de Hoef S, Terelius H, Turri V, Besselink B, Mrtensson J, Johansson KH (2016) Networked control challenges in collaborative road freight transport. Eur J Control 30:2–14

45. Liang KY (2016) Fuel-efficient heavy-duty vehicle platoon formation, Ph.D. Thesis, Kungliga Tekniska Hgskolan KTH, Stockholm, Sweden
46. Liang KY, Mrtensson J, Johansson KH (2016) Heavy-duty vehicle platoon formation for fuel efficiency. IEEE Trans Intell Transp Syst 17:1051–1061
47. Turri V, Besselink B, Johansson KH (2017) Cooperative look-ahead control for fuel-efficient and safe heavy-duty vehicle platooning. IEEE Trans Control Syst Technol 25:12–28
48. Larson J, Liang KY, Johansson KH (2015) A distributed framework for coordinated heavy-duty vehicle platooning. IEEE Trans Intell Transp Syst 16:419–429
49. Dijikstra EW (1959) A note on two problems in connexion with graphs. Numerische Mathematik 1:269–271
50. Dhaou IB (2011) Fuel estimation model for eco-driving and eco-routing. In: Proceedings of the IEEE Intelligent Vehicles Symposium, pp 37–42
51. Muñoz JC, Daganzo CF (2002) Moving bottlenecks: a theory grounded on experimental observation. In: Taylor MAP (ed) Transportation and traffic theory in the 21st century, pp 441–461
52. Delle Monache ML, Goatin P (2016) A numerical scheme for moving bottlenecks in traffic flow. Bull Brazilian Math Soc New Ser 47:605–617
53. Piacentini G, Goatin P, Ferrara A (2018) Traffic control via moving bottleneck of coordinated vehicles. In: Accepted to 15th IFAC symposium on control in transportation systems
54. Johansson I, Jin J, Ma X, Pettersson H (2017) Look-ahead speed planning for heavy-duty vehicle platoons using traffic information. Transp Res Procedia 22:561–569
55. Atzori L, Iera A, Morabito G (2010) The Internet of things: a survey. Comput Netw 54:2787–2805
56. M. Jamshidi (ed.) (2008) Systems of systems engineering: principles and applications. CRC Press, Boca Raton
57. Engell S, Paulen R, Reniers MA, Sonntag C, Thompson H (2015) Core research and innovation areas in cyber-physical systems of systems, cyber physical systems. Design, Modeling, and Evaluation. In: Mousavi M, Berger C (eds) Lecture Notes in Computer Science, vol 9361. Springer, Cham
58. Gubbi J, Buyya R, Marusic S, Palaniswami M (2013) Internet of things (IoT): a vision, architectural elements, and future directions. Future Gener Comput Syst 29:1645–1660
59. Al-Fuqaha A, Guizani M, Mohammadi M, Aledhari M, Ayyash M (2015) Internet of things: a survey on enabling technologies, protocols, and applications. IEEE Commun Surv Tutor 17:2347–2376
60. Guerrero-ibanez JA, Zeadally S, Contreras-Castillo J (2015) Integration challenges of intelligent transportation systems with connected vehicle, cloud computing, and internet of things technologies. IEEE Wirel Commun 22:122–128
61. Yang F, Wang S, Li J, Liu Z, Sun Q (2014) An overview of internet of vehicles. China Commun 11:1–15
62. Atzori L, Iera A, Morabito G, Nitti M (2012) The social internet of things (SIoT) - when social networks meet the internet of things: concept, architecture and network characterization. Comput Netw 56:3594–3608
63. Li W, Cassandras CG (2005) Distributed cooperative coverage control of sensor networks. In: Proceedings of the 44th IEEE conference on decision and control and the European control conference, pp 2542–2547
64. Lambrou TP, Panayiotou CG (2009) Collaborative area monitoring using wireless sensor networks with stationary and mobile nodes. EURASIP J Adv Signal Process 2009:750657
65. Lovisari E, Canudas de Wit C, Kibangou AY (2016) Density/flow reconstruction via heterogeneous sources and optimal sensor placement in road networks. Transp Res Part C 69:451–476
66. Guerreiro G, Figueiras P, Silva R, Costa R, Jardim-Goncalves R (2016) An architecture for big data processing on intelligent transportation systems. An application scenario on highway traffic flows. In: Proceedings of the IEEE 8th international conference on intelligent systems, pp 65–72
67. Grossman RL, Kamath C, Kegelmeyer P, Kumar V, Namburu R (eds) (2001) Data mining for scientific and engineering applications. Springer Science & Business Media, Berlin

68. Kaufman C, Perlman R, Speciner M (2002) Network security, private communication in a public world, 2nd edn. Prentice Hall, Upper Saddle River
69. Morgan YL (2010) Notes on DSRC & WAVE standards suite. IEEE Commun Surv Tutor 12:504–518
70. Zhou Y, Chen S, Zhou Y, Chen M, Xiao Q (2015) Privacy-preserving multi-point traffic volume measurement through vehicle-to-infrastructure communications. IEEE Trans Veh Technol 64:5619–5630
71. Xia Z, Wang X, Sun X, Wang Q (2016) A secure and dynamic multi-keyword ranked search scheme over encrypted cloud data. Trans Parallel Distrib Syst 27:340–352
72. Sherif ABT, Rabieh K, Mahmoud MMEA, Liang X (2017) Privacy-preserving ride sharing scheme for autonomous vehicles in big data era. IEEE Internet of Things J 4:611–618
73. Jeske T (2013) Floating car data from smartphones: what Google and Waze know about you and how hackers can control traffic. In: Proceedings of the BlackHat Europe
74. Tanenbaum AS (2003) Computer networks. Prentice Hall, Upper Saddle River
75. Laszka A, Potteiger B, Vorobeychik Y, Amin S, Koutsoukos X (2016) Vulnerability of transportation networks to traffic-signal tampering. In: Proceedings of the ACM/IEEE 7th international conference on cyber-physical systems
76. Sullivan J, Novak D, Aultman-Hall L, Scott DM (2010) Identifying critical road segments and measuring system-wide robustness in transportation networks with isolating links: a link-based capacity-reduction approach. Transp Res Part A 44:323–336
77. Jenelius E, Mattsson L-G (2012) Road network vulnerability analysis of area-covering disruptions: a grid-based approach with case study. Transp Res Part A 46:746–760
78. Reilly J, Martin S, Payer M, Bayen AM (2016) Creating complex congestion patterns via multi-objective optimal freeway traffic control with application to cyber-security. Transp Res Part B 91:366–382
79. Kwon J, Chen C, Varaiya P (1870) Statistical methods for detecting spatial configuration errors in traffic surveillance. Transp Res Rec 2005:124–132
80. Phegley B, Horowitz R, Gomes G (2016) Model-based fault detection among freeway loop sensors. In: Proceedings of the ASME dynamic systems and control conference, vol 2
81. Calver SC, Snelder M (2018) A methodology for road traffic resilience analysis and review of related concepts. Transp A: Transp Sci 14:130–154
82. Snelder M, Van Zuylen HJ, Immers LH (2012) A framework for robustness analysis of road networks for short term variations in supply. Transp Res Part A 46:828–842
83. Albert R, Jeong H, Barabsi A-L (2000) Error and attack tolerance of complex networks. Nature 406:378–382
84. Tran T, Ha QP (2015) Dependable control systems with internet of things. ISA Trans 59:303–313

Index

A. Ferrara et al., *Freeway Traffic Modelling and Control*, Advances in Industrial Control, https://doi.org/10.1007/978-3-319-75961-6

The manufacturer's authorised representative in the EU is Springer Nature Customer Service Centre GmbH, Europaplatz 3, 69115 Heidelberg, Germany. If you have any concerns regarding our products, please contact ProductSafety@springernature.com

Printed and bound by CPI Group (UK) Ltd, Croydon, CR0 4YY
01/10/2026
02225860-0004